Innovations in Environmental Legislation and Justice:

Environmental and Water-Energy-Food Nexus Laws

Nima Norouzi
Islamic Azad University, UAE & Law and Political Science Department, University of Tehran, Tehran, Iran

Hussein Movahedian
Islamic Azad University, UAE & Department of Private Law, Islamic Studies and Law Faculty, Imam Sadiq University, Tehran, Iran

IGI Global
PUBLISHER of TIMELY KNOWLEDGE

A volume in the Practice, Progress, and Proficiency in Sustainability (PPPS) Book Series

Published in the United States of America by
IGI Global
Information Science Reference (an imprint of IGI Global)
701 E. Chocolate Avenue
Hershey PA, USA 17033
Tel: 717-533-8845
Fax: 717-533-8661
E-mail: cust@igi-global.com
Web site: http://www.igi-global.com

Library of Congress Cataloging-in-Publication Data

Names: Norouzi, Nima, 1994- author. | Movahedian, Hussein, 1991- editor.
Title: Innovations in environmental legislation and justice : environmental and water-energy-food Nexus laws / Nima Norouzi, and Hussein Movahedian, editor.
Description: Hershey, PA : Information Science Reference, 2022. | Includes bibliographical references and index. | Summary: "The purpose of this text is to examine all of the issues discussed in this broad definition of "environmental law" in a logical and systematic way so that the reader can build up his or her knowledge in a logical way"-- Provided by publisher.
Identifiers: LCCN 2022032957 (print) | LCCN 2022032958 (ebook) | ISBN 9781668471883 (hardcover) | ISBN 9781668471890 (paperback) | ISBN 9781668471906 (ebook)
Subjects: LCSH: Environmental law, International. | Environmental law. | Environmental justice.
Classification: LCC K3585 .N669 2022 (print) | LCC K3585 (ebook) | DDC 344.04/6--dc23/eng/20220924
LC record available at https://lccn.loc.gov/2022032957
LC ebook record available at https://lccn.loc.gov/2022032958

This book is published in the IGI Global book series Practice, Progress, and Proficiency in Sustainability (PPPS) (ISSN: 2330-3271; eISSN: 2330-328X)

British Cataloguing in Publication Data
A Cataloguing in Publication record for this book is available from the British Library.

Practice, Progress, and Proficiency in Sustainability (PPPS) Book Series

Ayman Batisha
International Sustainability Institute, Egypt

ISSN:2330-3271
EISSN:2330-328X

Mission

In a world where traditional business practices are reconsidered and economic activity is performed in a global context, new areas of economic developments are recognized as the key enablers of wealth and income production. This knowledge of information technologies provides infrastructures, systems, and services towards sustainable development.

The **Practices, Progress, and Proficiency in Sustainability (PPPS) Book Series** focuses on the local and global challenges, business opportunities, and societal needs surrounding international collaboration and sustainable development of technology. This series brings together academics, researchers, entrepreneurs, policy makers and government officers aiming to contribute to the progress and proficiency in sustainability.

Coverage

- Socio-Economic
- Global Business
- Strategic Management of IT
- ICT and knowledge for development
- Knowledge clusters
- E-Development
- Green Technology
- Innovation Networks
- Technological learning
- Global Content and Knowledge Repositories

IGI Global is currently accepting manuscripts for publication within this series. To submit a proposal for a volume in this series, please contact our Acquisition Editors at Acquisitions@igi-global.com or visit: http://www.igi-global.com/publish/.

The Practice, Progress, and Proficiency in Sustainability (PPPS) Book Series (ISSN 2330-3271) is published by IGI Global, 701 E. Chocolate Avenue, Hershey, PA 17033-1240, USA, www.igi-global.com. This series is composed of titles available for purchase individually; each title is edited to be contextually exclusive from any other title within the series. For pricing and ordering information please visit http://www.igi-global.com/book-series/practice-progress-proficiency-sustainability/73810. Postmaster: Send all address changes to above address.

Titles in this Series

For a list of additional titles in this series, please visit:
http://www.igi-global.com/book-series/practice-progress-proficiency-sustainability/73810

Sustainable Approaches and Strategies for E-Waste Management and Utilization
A. M. Rawani (National Institute of Technology, Raipur, India) Mithilesh Kumar Sahu (O.P. Jindal University, India) Siddharth S. Chakarabarti (O.P. Jindal University, India) and Ajit Kumar Singh (GLA Universiy, India)
Engineering Science Reference • ©2023 • 300pp • H/C (ISBN: 9781668475737) • US $240.00

Handbook of Research on Solving Societal Challenges Through Sustainability-Oriented Innovation
Luísa Cagica Carvalho (Instituto Politécnico de Setúbal, Portugal & CEFAGE, University of Évora, Portugal) Paulo Bogas (Instituto Politécnico de Setúbal, Portugal) Jordana Kneipp (Federal University of Santa Maria, Brazil) Lucas Avila (Federal University of Santa Maria, Brazil) and Elis Ossmane (Universidade Aberta, Portugal)
Engineering Science Reference • ©2023 • 419pp • H/C (ISBN: 9781668461235) • US $295.00

Handbook of Research on Socio-Economic Sustainability in the Post-Pandemic Era
Jozef Oleński (High School Technology and Economics, Jaroslaw, Poland) Jeffrey Sachs (Columbia University, USA) Masayuki Susai (Nagasaki University, Japan) Yannis Tsekouras (University of Macedonia, Greece) and Arjan Gjonça (London School of Economics and Political Science, UK)
Information Science Reference • ©2023 • 400pp • H/C (ISBN: 9781799897606) • US $270.00

Handbook of Research on Promoting Sustainable Public Transportation Strategies in Urban Environments
Zafer Yilmaz (TED University, Turkey) Silvia Golem (University of Split, Croatia) and Dorinela Costescu (Polytechnic University of Bucharest, Romania)
Engineering Science Reference • ©2023 • 401pp • H/C (ISBN: 9781668459966) • US $295.00

Energy Transition in the African Economy Post 2050
Olayinka Ohunakin (Covenant University, Nigeria)
Engineering Science Reference • ©2023 • 300pp • H/C (ISBN: 9781799886389) • US $215.00

Implications of Industry 5.0 on Environmental Sustainability
Muhammad Jawad Sajid (Xuzhou University of Technology, China) Syed Abdul Rehman Khan (Xuzhou University of Technology, China) and Zhang Yu (ILMA University, Pakistan)
Business Science Reference • ©2023 • 328pp • H/C (ISBN: 9781668461136) • US $250.00

IGI Global
PUBLISHER of TIMELY KNOWLEDGE

701 East Chocolate Avenue, Hershey, PA 17033, USA
Tel: 717-533-8845 x100 • Fax: 717-533-8661
E-Mail: cust@igi-global.com • www.igi-global.com

Table of Contents

Preface

Students enrolling in environmental law courses at the undergraduate and graduate levels usually do not have a background in legal studies, and they seem more motivated by their passion for protecting the natural environment than their love for law. As a consequence, this book addresses the needs of those students by enabling them to access the many legal issues that arise in relation to the natural environment without compromising comprehensiveness. Despite its ease of reading, the book contains comprehensive information.

Getting a full understanding of substantive environmental law requires familiarity with the basics of law, which is one of the most challenging components of learning environmental law. To that end, sufficient introductory material has been included, along with plenty of advanced content. Environmental literacy is developed in an engaging and challenging manner.

An ever-evolving field, environmental law is constantly changing. The issues are constantly changing. There are a number of questions this book can help answer, such as whether environmental degradation should be permitted at all, who should pay for the restoration, and if it can be done. But how? As a result, the reader has a clear picture of the legal environment in place, as well as the potential possibilities that can be created from its foundation. Additionally, laws themselves change with time. The decisions made by the court have a significant impact on public expectations, industry reactions, and legislative agendas. Policy and executive changes occur as a result. The legislative process continues.

The contents address basic foundations of international law that are essential to understanding our legal system as it pertains to the natural environment. Though the book's primary focus is on international law, it addresses other jurisdictions as well. The book follows the development and the application of environmental laws to the present day. International laws and regulations dealing with global air, water, and land pollution, respectively, are undoubtedly among the most important sources of environmental law in the world. Environmental law, in its broadest sense, is not limited to the enforcement of statutory or regulatory mandates for pollution control. Additionally, it goes beyond merely conceptualizing natural resources as items that can be managed, protected, or conserved - an approach that major global treaties have attempted to do. Although these things are important, environmental law also includes the concepts of property rights and property ownership. This is in addition to how those fundamental understandings of human relationships to the land affect our understanding and belief about what we are allowed to do with it.

The purpose of this text is to examine all of the issues discussed in this broad definition of "environmental law" in a logical and systematic way so that the reader can build up his or her knowledge in a logical way. In my opinion, it is best to study the chapters in the order in which they are presented. In other words, each chapter builds upon the concepts presented in the chapter preceding it. In the first few chapters of part 1 of this book entitled "Pragmatic Legal Implications and Policy Implications of Environmental Lawmaking", students who are new to studying law will be introduced to basic law concepts while developing a comprehensive understanding of environmental law as they go. It is because, through the use of environmental law examples, each chapter illustrates the concepts covered in the chapter by using real-life examples. In the second part of our book, which is called *Innovations in Environmental Legislation and Justice: Environmental and Water-Energy-Food Nexus Laws*, we are going to discuss the most innovative technologies and issues in environmental law.

Chapter 1 is aimed to discuss that the international community is concerned about food security as a global issue. Economic, social, cultural, and environmental factors contribute to food security. The linkages between food security and environmental sustainability are inextricably linked. Sustainable food security requires a balanced environment.

In Chapter 2, we will look at the guidelines International Human Rights and Environment provides for the State with respect to the appropriate structure of its Water Laws. Moreover, it discusses the relevant aspects of the human right to water and sanitation, as well as the conceptual connection between human rights and the environment. In addition, the report emphasizes the importance of the principles of international law for ensuring the right management of water resources as a guarantee of the human right to water and sanitation.

Human life and survival on Earth depend on the exploitation of diverse resources, including water. Improper use of environmental resources will lead to pollution and destruction. As one of the most sensitive areas of the environment to which human life depends, water is exposed to a variety of environmental pollutants. The protection of the health of water resources has created the need for intervention and the use of legal and criminal solutions in organizing their use. Domestic penal policy in the field of legislation, inspired by the provisions of Sharia law and local and national considerations for the protection of water resources, has directly and indirectly affected the requirements of accession to international instruments and has enacted regulations on the protection of small water resources. Chapter 3 is aimed to study this field.

The purpose of Chapter 4 is to examine some of the opportunities and threats of globalization to the environment and the possible positive or negative effects of economic growth resulting from the globalized economy. Also, solutions to reduce the negative effects of globalization on the environment include paying special attention to sustainable development, supporting environmentally friendly technologies, reforming energy consumption patterns, encouraging renewable and clean energy, imposing carbon taxes, and environmental labels suggested.

In Chapter 5, considering the World Trade Organization's recent practice, criteria will be presented to balance governments' environmental and commercial commitments supporting renewable energy.

Chapter 6 will examine the principles and rules of international law in this field, considering the importance of developing renewable energy as a global issue.

The general objective of Chapter 7 is to establish, based on a theoretical and legislative study, the need to adhere to the international law on Climate Change with a view to mitigation and adaptation to its effects on an international scale.

Chapter 8 examines some of the most important jurisprudential and legal considerations arising from waste management with an analytical-descriptive method. Finally, it concludes that recycling should be done with the principle of precaution and prevention. But if it leads to damages, the fault-based system cannot compensate for the damages due to its difficulties in the proof process. Therefore, using a pure/absolute liability system and promoting specialized insurance in this regard are recommended.

Chapter 9 examines the new paradigms, military, and hardware variables, and political, economic, socio-cultural, and environmental factors that can threaten international security. In the age of globalization, climate change is one of the biggest and most complex international challenges. These small changes in global warming could exacerbate climate change. Climate change directly impacts our lives today.

Chapter 10 examines the procedure of the regional courts of human rights and the domestic courts in some countries regarding the possibility of public litigation.

Chapter 11 is aimed to analyze how gender is incorporated into international environmental protection and to make adequate conclusions about its current state, the most relevant juridical instruments have been considered.

Chapter 12 states nongovernmental actors play a crucial role both internationally and nationally. Since these new actors are geographically and functionally diverse, it is impossible for government actors to ignore them. This chapter is aimed to study this impact.

In Chapter 13, an attempt has been made to analyze the content of the constitutions of Latin America in the context of environmental issues and the conservation of natural resources with a comparative approach and as an example. The basic hypothesis of this article is based on the fact that today, paying attention to the issue of the environment at the level of constitutions means following sub-norms and the legal system from the norm of reference and the effects that this right can have in the legal system. This is well reflected in the constitutions of Latin America.

In Chapter 14, an attempt has been made to analyze the content of the constitutions of Latin America in the context of environmental issues and the conservation of natural resources with a comparative approach and as an example. The basic hypothesis of this article is based on the fact that today, paying attention to the issue of the environment at the level of constitutions means following sub-norms and the legal system from the norm of reference and the effects that this right can have in the legal system. This is well reflected in the constitutions of Latin America.

While implementing existing strategies to combat environmental terrorism, chapter 15 seeks to provide an effective strategy to combat "green terrorism" or "ecosystem" to provide an effective strategy in the light of new criminal law approaches. Provide emerging threats. The leading research is strengthening practical strategies to combat green terrorism and implementing effective strategies such as environmental litigation, both nationally and internationally, considering legal means. In this regard, green criminalization as a green strategy can be considered an action to combat environmental terrorism.

Nima Norouzi
Islamic Azad University, UAE & Law and Political Science Department, University of Tehran, Tehran, Iran

Hussein Movahedian
Islamic Azad University, UAE & Department of Private Law, Islamic Studies and Law Faculty, Imam Sadiq University, Tehran, Iran

Introduction

Human treatment of the nonhuman world as regulated by local, national, and international laws, policies, and regulations. Regulatory standards for emissions from coal-fired power plants in Germany, initiatives in China to create a "Green Great Wall" (a shelter belt of trees) to protect Beijing from sandstorms, and international agreements for the protection of biological diversity and the ozone layer are among the many topics covered in this vast field. Late in the 20th century, environmental law developed from a modest adjunct to public health regulations into an almost universally recognized independent field protecting both human and nonhuman health.

Human health has been protected from environmental contamination by national laws throughout history. In AD 80, the Senate of Rome passed legislation protecting the city's clean water supply. England prohibited the burning of coal in London in the 14th century, as well as disposing of waste into waterways. William Penn, the Quaker leader of the English colony of Pennsylvania, ordered in 1681 that one acre of forest be preserved for every five acres cleared for settlement, and Benjamin Franklin led campaigns to curb waste dumps in the following century. The British government passed regulations to reduce the negative effects of coal burning and chemical manufacturing on public health and the environment during the Industrial Revolution in the 19th century.

Despite international declarations, conventions, and treaties, environmental law exists at different levels. Statutory - meaning enacted by legislative bodies - and regulatory - meaning created by government agencies tasked with protecting the environment - make up the bulk of environmental law.

Many national constitutions also include some right to environmental quality. Environmental protection has been enshrined in the German Grundgesetz ("Basic Law") since 1994, which stipulates that the government must preserve "the natural foundations of life." Similarly, the Chinese constitution declares that the state "ensures the rational use of natural resources and protects rare animals and plants"; the South African constitution recognizes a right to "an environment that is not harmful to health or well-being; and to have the environment protected, for the benefit of present and future generations"; the Bulgarian constitution provides for a "right to a healthy and favourable environment, consistent with stipulated standards and regulations"; and the Chilean constitution contains a "right to live in an environment free from contamination."

International, national, and local courts also make significant contributions to environmental law. The Trail Smelter arbitration of 1941, for example, enjoined the operation of a smelter in British Columbia, Canada, near the international border with Washington and held that "no State will use or permit the use of its territory in such a way as to cause injury by fumes to the territory of another or to the properties or people therein." Some environmental law also appears in the decisions of national courts. In Scenic Hudson Preservation Conference v. Federal Power Commission (1965), a U.S. federal appeals court overturned the Federal Power Commission's license for building an environmentally damaging pumped-storage hydroelectric plant (a plant that pumps water from a lower to an upper reservoir) in an area of stunning natural beauty, proving that judicial challenges could succeed. The California Supreme Court severely limited Los Angeles' ability to divert water from Mono Lake in California's eastern desert in National Audubon Society v. Superior Court (1976).

Even though numerous international environmental treaties have been concluded, effective agreements remain difficult to achieve for a variety of reasons. Due to the fact that environmental problems disregard political boundaries, they can only be adequately addressed by a wide range of governments, some of which may have serious disagreements on important points of policy. Many countries, particularly those in the developing world, have been reluctant to enter into environmental treaties because the measures necessary to address environmental problems typically result in social and economic hardships. A growing number of environmental treaties have incorporated provisions aimed at encouraging developing countries to adopt them. Financial cooperation, technology transfer, and different schedules and obligations for implementation are examples of such measures.

Environmental treaties are most effective when they are complied with. Treaties may attempt to enforce compliance by imposing sanctions, but these measures are usually of limited use, in part because countries complying with a treaty may be unwilling or unable to impose the sanctions required by the treaty. Most countries are less concerned about sanctions than the possibility that by violating their international obligations they may lose their good standing in the international community. In the past, other enforcement mechanisms have been difficult to establish, usually due to the fact that they would require countries to cede significant aspects of their national sovereignty to foreign governments or international organizations. Enforcement is generally treated as a domestic issue in most agreements, allowing countries to define compliance in whatever way they deem most appropriate. As international environmental problems become more acute, international environmental treaties and agreements will continue to be important.

There are still many areas of international environmental law that are underdeveloped. Some types of environmentally harmful activity are subject to more or less consistent laws and regulations across countries due to international agreements, while others differ dramatically. Due to the fact that environmental damage is rarely contained within national boundaries, the lack of consistency in law has led to situations in which activities that are legal in one country result in illegal or otherwise unacceptable levels of environmental damage in neighbouring countries.

In the early 1990s, free trade agreements exacerbated this problem. Under the North American Free Trade Agreement (NAFTA), for example, large numbers of maquiladoras were opened along the U.S.-Mexican border in a 60-mile-wide (100-km) wide free trade zone. Mexico's government lacked the resources and political will to enforce environmental laws, allowing the maquiladoras to pollute surrounding areas with relative impunity, often dumping hazardous wastes directly into waterways, where they were carried into the United States. After NAFTA's adoption in 1992, negotiators added a "side agreement" to the treaty, which pledged environmental cooperation between the countries. In Europe, opposition to the Maastricht Treaty, which created the EU and expanded its authority, was fueled by concerns about free trade agreements and environmental degradation.

The purpose of this text is to examine all of the issues discussed in this broad definition of "environmental law" in a logical and systematic way so that the reader can build up his or her knowledge in a logical way. In my opinion, it is best to study the chapters in the order in which they are presented. In other words, each chapter builds upon the concepts presented in the chapter preceding it. In the first few chapters of part 1 of this book entitled “Pragmatic Legal Implications and Policy Implications of Environmental Lawmaking”, students who are new to studying law will be introduced to basic law concepts while developing a comprehensive understanding of environmental law as they go. It is because, through the use of environmental law examples, each chapter illustrates the concepts covered in the chapter by using real-life examples. In the second part of our book, which is called “Innovations in Environmental Legislation and Justice Environmental and Water-Energy-Food Nexus Laws” we are going to discuss the most innovative technologies and issues in environmental law.

Chapter 1
Environmental Law and Food Security

ABSTRACT

The international community is concerned about food security as a global issue. Economic, social, cultural, and environmental factors contribute to food security. The linkages between food security and environmental sustainability are inextricably linked. Sustainable food security requires a balanced environment. In contrast, efforts to achieve food security need to be environmentally friendly. Food security is regulated by several international environmental laws. In order to achieve sustainable food security, international environmental law needs to be implemented.

INTRODUCTION

At the start of the 21st century, there were many parts of the world that suffered from poverty and were affected by the degradation of the environment as well as the depletion of natural resources. By the end of the decade, the world will have fallen into a far worse state than it is today. If something becomes less valuable or damaged as a result of degradation, it is said to be degraded. There are many ways in which the environment can be degraded. It is well known that habitat destruction, biodiversity loss, and depletion of natural resources are damaging to the environment. Degradation of the environment occurs both naturally and as a result of human activity. As of right now, the greatest problems are the loss of rain forests, air pollution, smog, ozone depletion, and damage to the marine environment. There is a widespread problem of pollution across the globe that has affected marine as well. Even remote areas have been affected by the degradation of marine life. The presence of hazardous waste in some of these locations has been reported. It has also been reported that oil spills caused environmental damage in other locations(Bullock et al., 2017).

Degradation of the environment will lead to aggravation of human beings and insecurity. changed over the past few decades. In the past, external threats have always been the main source of concern. The state security policy has therefore always been focused on protecting the nation. In the last decades of the 20th century, the concept of human security was transformed into one of "internal threats" by

DOI: 10.4018/978-1-6684-7188-3.ch001

globalization. Security has increasingly become a global concern, and it has become more important to concentrate on individuals rather than states. Humanity is faced with a number of threats, such as the degradation of the environment, poverty, hunger, and endemic diseases such as the HIV/AIDS epidemic. such as the HIV/AIDS epidemic. Human security and quality of life can only be achieved when the basic needs of humans are met while enhancing the quality of the environment as a whole. The degradation of the environment is the result of a dynamic interest play between social, economic, institutional, and technological activities. As a result, the environment can be affected by a number of factors, including economic growth, population growth, urbanization, intensification of agriculture, and increased energy use. Poverty is linked to several environmental problems(Johnson & Walters, 2014).

The international community is still struggling with finding an effective solution to the problem of global food security. Preventing future food crises is paramount. A global solution is almost universally acknowledged. There are disagreements, however, about the specifics of these international solutions, and tensions continue to make it difficult for effective action to take place. Tensions between governments, international institutions, and scholars are evident in the competing policy prescriptions. In terms of the measures necessary to promote food security, states themselves act inconsistently and disagree. A number of overlapping branches of international law further exacerbate this dissonance by imperfectly promoting food security throughout the world. Norms relating to human security, refugee law, natural disaster law, international environmental law, and human rights all have something to say about the global food system. Agrarian reform, property tenure, agricultural patents, resource management, freedom of trade and investment, and population control are just some of the rules that are governed by these regimes. Nevertheless, nothing is a perfect fit (Nevitt, 2020).

Throughout this chapter, we will analyze the current legal and institutional governance structure of the international food system. A convincing account is provided of the disparate norms and institutional frameworks that affect food security from the perspectives of environmental, human rights, agriculture, development, trade, finance, disaster, and security. Considerating global food governance holistically is a necessary first step. In addition, it is crucial to develop an understanding of how all of these functionally differentiated regimes interact with each other, the conflicts between priorities and values, and the ways in which states and other actors use these regimes to best suit their interests (to protect food security or otherwise). A clear understanding of the overlapping legal regimes governing food production, distribution, and famine dynamics must be developed before any intelligent attempt at normative integration of global food governance can succeed.

Global problems can be attributed to the rapid growth of world population. Population growth creates environmental, social, and economic problems worldwide. As technologies and industries increase, these problems are aggravated as well. Environment-related problems are affecting nearly every country nowadays. This refers to a global trend of environmental problems occurring on a transnational scale. Environment has become a major topic of international discussion. It shows that people around the world are concerned about these issues, and more importantly, they realize that we will face some threats in the near future. The truth is, threats are not threats any longer, because they are taking place right now. As states consider each other's problems, they will become responsible for each other(Subramaniam et al., 2020).

As countries struggle to find fertile land to grow crops and raise animals, environmental degradation and population growth could lead to a global food crisis in the next half century. Over the next 50 years, more food will need to be produced than over the last 10,000 years combined in order to keep up with population growth. But in many countries, a combination of poor farming practices and deforestation aggravated by climate change will result in a degradation of soil fertility and render vast areas unfit for crops or grazing. Humans need food to survive. People cannot live without food just like they can't live without air or water, so it makes food a basic right for all humans. Malnutrition is typically a result of a food crisis, which is defined as a widespread shortage of food. In many countries, food crises are no longer a threat, but a reality, including Djibouti, Somalia, Kenya, and Ethiopia. Famine will affect more than 11 million people in those countries, according to a January 2006 FAO news release. Losses in biological diversity could also contribute to food shortages. Global warming has resulted in the loss of biological diversity. Only food security can prevent food crises(Lawrence, 2013).

854 million people do not have access to enough food for a healthy and active lifestyle, according to the UN's Food and Agriculture Program. Many of them are hungry and malnourished. Eighty-five percent live in economically developing countries, 76% in rural areas. There has been an increase in the number of hungry and malnourished people in the last decade, according to all available data. Despite the fact that enough food is being produced globally to feed the world's population, this problem persists. The environment has nothing to do with hunger and malnutrition today. It will also address shortage, or scarcity: hunger relates to access to food, to an adequate income, or to productive resources that enable poor people to produce or acquire enough food. Food, land, and other resources are inequitably distributed, causing hunger and malnutrition. The purpose of this chapter is to discuss food security under international environmental law and how it relates to international environmental law in terms of food security (Gonzalez, 2011).

FOOD SECURITY AND ENVIRONMENT

One strategy for increasing yields through intensification and/or extending the land base used for agricultural production can help meet a growing global demand for food and fodder. The most efficient use of land may be to intensify and concentrate food production in the most productive regions. A weakened supply chain and pollution may, however, increase food security risks. Pollinator decline, loss of crop diversity, and increased disease vulnerability of monocultures are additional stress factors. Alternatively, regional or local self-sufficiency and the reliance on extensive farming systems would necessitate more cultivated land at the expense of natural habitats.

Increases in total food production are not enough. Quality standards must be met as well as local availability, affordability, and affordability. Trade patterns and distribution channels are crucial in this regard. We may not see immediate risks to European food security, regardless of how much support we give to European agriculture, as long as we can afford to import food from other parts of the world. But the choices we make will affect trade and global food security, as well as the availability of local food products with implications for chain control, food safety, and other quality concerns. The EU is currently self-sufficient in cereals, butter, and beef, but is a net importer of fodder for livestock.

From a consumption perspective, food security can also be addressed by looking at efficiency gains from changing diets. Meat diets are associated with higher land use and nutrients losses due to livestock production, which is six times as inefficient as crop production(Ringler et al., 2010).

In addition to reducing waste in households and distribution chains, efficiency gains can be achieved by reducing energy consumption. According to Eurostat and national data, around 89 million tonnes of food waste was generated in the EU27 in 2006, of which 42-43 percent was generated by households, 39 percent was generated by manufacturing, and the rest was generated by retailers, wholesalers, and food service providers (excluding agricultural waste). 137 kg/person or 25 percent of food purchased by households in the United Kingdom is wasted, according to a recent study.

Agriculture has a direct impact on land cover and ecosystems, as well as on global and regional cycles of carbon, nutrients, and water. Globally, agriculture contributes to climate change by emitting greenhouse gases and reducing carbon storage in vegetation and soil. Land conversion, eutrophication, pesticides, irrigation, and drainage are examples of how agriculture reduces biodiversity and affects natural habitats locally. The use of excessive pesticides can result in soil erosion and loss of pollinators (due to unsustainable agricultural practices) (McBeath et al., 2010).

A major cause of biodiversity loss and ecosystem dysfunction is nutrient loading (mainly by phosphorus and nitrogen). Nitrogen is the most studied element. The amount of reactive nitrogen in the environment has doubled globally since the pre-industrial era, and has tripled in Europe. The cause is primarily the combustion of fossil fuels and the application of fertilizers produced by industrial processes. Air pollution and eutrophication of terrestrial, aquatic, and coastal ecosystems are caused by an excess of reactive nitrogen. It is estimated that agriculture contributes 50–80 percent of the nitrogen load transported into Europe's freshwater ecosystems and, ultimately, seas and coasts.

Significant reductions in nitrogen pollution have been achieved from key polluting sectors and sources over the last two decades. However, critical nitrogen loads are still exceeded throughout much of Europe. In 2010, atmospheric nitrogen deposition exceeded critical loads in more than 40% of sensitive terrestrial and freshwater ecosystems.

A critical load is defined as a quantitative estimate of an exposure to one or more pollutants that, according to present knowledge, does not produce significant harmful effects on specified sensitive elements of the environment.

As a result of agricultural pressure on the environment, natural capital is being lost. It is very worrying that agricultural habitats protected under the Habitats Directive are in a less desirable condition than average. Favorable conservation status was reported in only 7% of assessments, compared to 17% across all habitat types. About half of the agricultural habitats are considered to be in poor condition. Ecosystems in lakes and rivers fare slightly better, but their conservation status is also worse than average. The North Sea and the Baltic Sea are both considered to have poor or inadequate conditions as far as the marine environment is concerned(Vidar, 2022).

Agriculture may also contribute to maintaining species-rich semi-natural habitats. It is an explicit goal of the EU biodiversity and agriculture policy to conserve this 'high nature value farmland', which is mostly found in peripheral regions of northern and south-eastern Europe. Most of the species listed under the Bird and Habitat Directives are found on farmland (some almost exclusively), and many of the habitats that are targeted are semi-natural and rely on continued (extensive) agricultural management. The conservation of these extensive farming systems is increasingly difficult due to socio-economic factors (lifestyle changes, demographic trends, economic marginalization)(Kattelus et al., 2014).

EMERGENCE OF THE PRECAUTIONARY PRINCIPE

Environmental policy is today one of the most important social challenges for public authorities and economic agents. In addition, it is a very sensitive issue in public opinion since it directly affects well-being and health(Cameron & Abouchar, 1991).

The precautionary principle arises as a consequence of seeking the protection of the environment and human health against certain activities characterized by scientific uncertainty about their possible consequences.

Some historical antecedents that we can highlight in relation to environmental concern began after the Second World War, with some conventional instruments for the protection of freshwater and seawater, for example, the protocol signed by France, Belgium and Luxembourg for the protection of frontier waters of April 8, 1950, and the conventions to combat pollution of the Moselle River of October 27, 1956, Lake Leman of November 16, 1962 and the Rhine River of April 29, 1963. At the end of the 1960s, faced with scientific alarm, the most intense reactions of public opinion achieved a more generalized awareness of the dangers that lurked and still hang over the environment. This current of opinion was undoubtedly a phenomenon without precedent in history, becoming a philosophical current on the conception of the world that implied new individual and social values in reaction to the deterioration of the biosphere. Such criticism arose in Germany because certain chemical pollutants, in weak concentration, could have negative consequences for human health and there was uncertainty regarding the effect of such substances. In this way, the Vorsorgeprinzip or precautionary principle was gradually developed in German politics(Boutillon, 2001).

The precautionary principle arises, as can be seen, from the search for analytical tools that can be transformed into more effective political, legal and planning instruments. Although there is a great variety of formulations of the principle, its meaning consists in the idea that in the face of the threat of damage to the environment or human health, it is not necessary to wait for complete scientific certainty to take the appropriate protective measures.

The precautionary principle was initially introduced in sectoral regulations, for example, in the German law on chemical products (Chemikaliengesetz of 1980) or the law on the use of atomic energy (Atomgesetz of 1985). Subsequently, it achieved greater prominence, becoming one of the guiding principles of German environmental policy(Whiteside, 2006).

In German doctrine, the Vorsorgeprinzip is related to two other principles of its environmental policy. The first is Verursacherprinzip, which is generally translated as "the polluter pays", but which means "principle of causality" or "principle of responsibility". The second, Kooperationprinzip which means cooperation or consensus. In addition, two other principles helped in the development of German policy, namely the Wirtschaftliche Vertretbarkeit or "principle of economic viability", which holds that costs and benefits must be proportional and can be applied to all economic activities, and the Gemeinlastprinzip or "common burden principle", which allows the state to overcome the unintended consequences of inequality.

On the basis of these principles, the precautionary principle made it possible to justify the use of "best available technologies" (BAT) in German policy, which were subsequently introduced into European Union law. The BAT system seeks to reduce air and water pollution to the lowest technically accessible level, without putting the economic activity involved at risk(Kriebel et al., 2001).

Beginning with German law, the precautionary principle was extended to various international regulations. The Council of Europe, for example, adopted two texts in 1968, the first declared by an international organization in the field of the environment. We refer to the Declaration on the fight against air pollution and the European Water Charter, since neither air nor water knows borders. That same year, the Council of Europe approved the European Agreement on the limitation of the use of some detergents in washing and cleaning products(Read & O'Riordan, 2017).

FOOD SECURITY IN LAW

Food Utilization, Food Availability, and Food Access are all elements of food security and the food system. In Food Utilization, value is attached to nutrients, social value is attached to food, and food safety is attached to food. Hence, food is used both biologically and economically. When food is available and accessible, a household will have to make decisions about what food to consume and how to allocate the food within the household. Providing young children and mothers with a balanced diet that is rich in micronutrients is crucial to their nutritional status(Gottlieb & Fisher, 1996). The process of utilization requires not only an adequate diet, but also a healthy physical environment, including safe drinking water and sanitary facilities (to avoid disease) as well as a thorough understanding of health care, food preparation, and storage. Furthermore, health-care capacity, practices, and behaviors also play an important role. Access to food refers to the distribution, exchange, and production of food. Food availability refers to the availability of food, either from one's own production or from the market. Domestic food stocks, commercial food imports, food aid, and domestic food production all contribute to national food availability. Food Access is concerned with affordability, allocation, and preference.In order to ensure access to food, all household members and all individuals within those households must have sufficient resources to obtain appropriate food for a nutritious diet(Santilli, 2012).

Food security does not have a universal definition. In addition to conventions, treaties, protocols, and agreements, many international instruments use it. Food security can be defined as enough food for everyone to eat, produced from a sustainable food system[. Food security is defined by the UN's Food and Agriculture Organization (FAO) as "a state in which all people have access to sufficient, safe and nutritious food at all times(Carolan, 2013)."

It should be a safe and nutritious food to meet people's needs and ensure their health. How we achieve food security is also important. Sustainable food systems are crucial to achieving food security. Several factors contribute to sustainability. Legal debates over social and economic development, on one side, and environmental protection, on the other, have been dominated by the term sustainable development. Finally, in the Report of the World Commission of Environment and Development (WCED) in 1987, known as "Our Common Future" or the Brundtland Report, sustainable development was defined as "development that meets the requirements of today without endangering the ability of future generations to meet their own needs(Bardgett & Gibson, 2017)."

FOOD SECURITY IN ENVIRONMENTAL LAW

Environment has an impact on food security. Environment degradation has an effect on the quantity and quality of food. Agriculture is a way to achieve food security. But some agricultural activities may cause environmental damage(Venancio et al., 2018). A variety of agricultural practices pose threats to the environment, including soil degradation, water supply and quality problems, air pollution, misuse of chemicals, and threats to biological diversity. Excessive use of chemical fertilizers can degrade soil, water, and air quality. Additionally, agricultural pesticides cause negative impacts because they kill not only unwanted pests, but also important organisms, resulting in a reduction in biodiversity. The need to protect the environment led to the creation of international environmental law. Public consciousness became increasingly aware of the dangers threatening the earth following the expression of scientific alarm at the end of the 1960s. Protecting the environment must be an integral part of development, and cannot be treated as a separate issue(Smith et al., 2021). Additionally, peace, development, and environmental protection are interconnected and indivisible. Traditional international law is challenged by many of the fundamental concepts of international environmental law. It puts new limits on the sovereignty of States, it encroaches upon the territorial integrity and domestic jurisdiction of States, it creates greater responsibilities for States, and it involves many non-State actors as well. Because environmental issues are global in scope, national action alone is unlikely to be sufficient, and international cooperation is essential. Environmental awareness and concern have increased at both the national and international levels. The law, which structures and regulates behavior, can help put this concern into practice. There are many environmental treaties and declarations in international law, as well as some State practices and mechanisms for compliance(Krasnova, 2014). Food security and environmental protection are closely linked. There are the following international instruments:

Nagoya Protocol 2010

Nagoya Protocol on Access to Genetic Resources and the Fair and Equitable Sharing of Benefits Emerging from their Utilization to the Convention on Biological Diversity, Nagoya 2010. It is important to note that on 29 October 2011, representatives from over 100 countries signed the Nagoya Protocol on Access to Genetic Resources and the Fair and Equitable Sharing of Benefits that Flow from Their Utilization. There is no doubt that the Nagoya Protocol outlines how countries are to allow access to genetic resources, share in the benefits from their use, and cooperate in the event of allegations of misuse. In order to be able to enter into force, the agreement must be ratified by at least fifty parties. There have been a number of issues related to the Protocol between the North and the South for quite some time. With the adoption of the Protocol, these wounds will be healed. Thus, as a result, there will be increased transparency and trust between countries, and trust is absolutely necessary for countries to cooperate and use genetic resources in a way that will promote food security and economic development. Among the most contentious issues in the last round of negotiations included the measures "user countries" would take to monitor and enforce compliance with agreements granting access to genetic resources from other countries; whether the scope of the Protocol would extend beyond genetic resources to biological resources more generally; and how traditional knowledge holders would be included in procedures of access to genetic resources. According to the Protocol, the International Treaty on Plant Genetic Resources for Food and Agriculture establishes precedents for access and benefit-sharing. The Nagoya Protocol also explicitly allows for the development of future access and benefit-sharing regimes consistent with the Convention on Biological Diversity and the Protocol. There is good news here because it seems likely that in the future the international community will need to agree to multilateral access and benefit-sharing norms for other genetic resources used in agriculture that are not covered by the International Treaty, for example, agricultural microbial genetic resources and farm animal genetic resources(Smith, 2018).

Cartagena Protocol 1996

The Protocol of Cartagena on Biosafety to the Convention on Biological Diversity, Cartagena 1996. According to the CBD, 1992, biotechnology is defined as any application of technology using biological systems, living organisms, or derivatives thereof, to create or modify products or processes in order to fulfill a desired purpose. It would appear that one of the most controversial topics in this science is the introduction and handling of genetically modified organisms (GMOs) into the environment. To ensure biosafety, two important issues need to be addressed: first, the handling of GMOs in the laboratory, so as to protect work and prevent the risk of accidental release into the environment ("contained use"); and, second, the use of a regulatory system to govern the deliberate release of GMOs into the environment, whether for laboratory testing purposes or on a commercial scale. The use of genetically modified organisms in agricultural research is not uncommon. It is believed that research will be conducted in order to introduce herbicide resistance into virtually all major crops, so that weeds will be easier to control in the future. In addition, because long-term use of pesticides has many adverse effects on humans and animals, genetic engineering of microorganisms has become an alternative strategy to improve pest control. It is estimated that more than 100 fungi and bacterial species exist that are capable of destroying insects(Gupta, 2000).

Rio de Janeiro Protocol UN Conventions on Biological Diversity 1992

The purpose of this convention is to protect the genetic pool of all species of plants, animals, and microorganisms. In order to conserve and sustain biological diversity, an integrated approach rather than a sectoral approach should be applied. It is in this context that the convention focuses on the preservation of biological resources, assuring their sustainable use, ensuring access to genetic resources, sharing the benefits of using genetic material as well as ensuring access to technology, including biotechnology. This convention is the first time that humanity is addressing the issue of biological diversity as a whole. In addition to identifying and monitoring processes and activities that may have significant adverse effects on biological diversity, states should identify important aspects of biological diversity and prioritize those that are likely to require special conservation measures. In addition, they must develop national strategies and plans that integrate conservation of biological diversity into relevant sectoral plans and programs, as well as into national decision-making. A major focus of the convention is the promotion of sustainable use. It is part of the agreement that sustainable fishing and harvesting practices will be developed for the harvesting of biological resources. There are several general principles of international environmental law contained in the convention, including responsibility for transborder environmental damage, information, cooperation, repair, and prevention(Jacquemont & Caparrós, 2002).

Kyoto Protocol UN Convention 1997

In the Kyoto Protocol, agriculture is mentioned in several articles, such as in Article 2.1, which states that every party included in Annex I, in achieving its quantified emission limitation and reduction commitments, must do certain actions in order to promote sustainable development, which includes in sub article (iii) promoting sustainable agriculture in light of climate change. Article 10 of the Protocol provides that, on the basis of common but differentiated responsibilities, all parties should have specific national and regional development priorities, objectives, and circumstances that will enable them to mitigate climate change and facilitate adequate adaptation to climate change, and this includes agriculture. Among the categories of sources listed in Annex A of the Protocol, agriculture includes enteric fermentation, manure management, changes in precipitation patterns, and increased rice cultivation, agricultural soils, prescribed burning of savannas, and field burning of agricultural residues(de Chazournes, 1998).

Rio de Janeiro Protocol UN Convention on Climate Change 1992

There is a close relationship between climate change and agriculture. According to the U.S. Department of Agriculture, 13.5 percent of annual greenhouse gas (GHG) emissions are attributed to agriculture (with forestry contributing an additional 19 percent), compared with 13.1 percent attributed to transportation. As a secondary option to carbon sequestration, land and land use management, and biomass production, agriculture is also a viable option for reducing GHG emissions. Agricultural production is threatened by climate change through increasing temperatures, more extreme weather events such as droughts and floods, which are all caused by climate change. At the 15th Conference of Parties (COP) to the Convention, in Copenhagen in December 2009, a number of sectors, among them the agriculture sector, was discussed. There will be an AdHoc Working Group of Long-term Co-operative Action (AWC-LCA) composed of some countries attending that conference. Workshop papers, presentations, and discussions provide information on agriculture's role in mitigation under the LCA of the AWG(Dasgupta, 2012).

Ramsar Convention 1971

Despite problems such as soil degradation, erosion, flooding, and desertification, international soil protection is relatively recent. Nearly 100 parties have ratified the agreement since it took effect in 1975. In it, all countries are required to designate at least one wetland area for protection, and it recognizes the important role wetlands play in maintaining ecological equilibrium. The Convention's mission is "the conservation and wise use of all wetlands through local, national, and international action and cooperation as a contribution to the achievement of sustainable development on a global scale." The Convention uses a broad definition of the types of wetlands covered in its mission, including lakes and rivers, swamps and marshes, wet grasslands and peatlands, oases, estuaries, deltas and tidal flats, near-shore marine areas, mangroves and coral reefs, and human-made sites such as fish ponds, rice paddies, reservoirs, and salt pans. Then, farming and agricultural practices must be related to this convention(Mombo et al., 2011).

FOA Treaty 2001

The PGRFA Treaty places a high priority on ensuring food security, agricultural productivity, and agricultural resources for food and agriculture. There is a great deal of significance to this agreement, as it allows the free flow of the PGRFA, one of the most important factors contributing to food security worldwide, and heavily dependent on countries. A comprehensive framework of protection and sustainable use is also provided by the Treaty to all PGRFAs as part of its provisions. By the Treaty, a Multilateral System of Access and Benefit Sharing will be established for PGRFAs of crops that are essential for food security and to the interdependence of countries(Argumedo et al., 2011).

BINDING EFFECTIVENESS OF ENVIRONMENTAL REGULATIONS

The emergence of international environmental law is a recent legal phenomenon. Its implementation has required the progressive overcoming of the postulates of maximum permissiveness derived from what have been called the two great principles of laissez-faire in ecological matters: state sovereignty and freedom of the high seas and, in general, of common spaces. of the planet(McInerney, 2007). However, in recent years, a diversified and complex international environmental regulation has progressively been developed. This regulation presents some particular characteristics that must be highlighted, since they give the whole a particular legal physiognomy, namely, functionality and predominance of soft law. But it should be underlined, as Juste Ruiz warns, that the flexibility and ductile nature of international environmental law have not prevented the occasional emergence of norms that have the rigorous profiles of hard law. When trying to establish whether the precautionary principle is has become a rule of customary international law, it is first necessary to examine what soft law and hard law are. This is so, because no other branch of international law is influenced by such a multitude of declarations, resolutions and other instruments that lead us to soft law(Abate et al., 1990).

For international environmental law, the so-called soft law represents an instrument or resource that allows projecting environmental legal principles and criteria that, without being binding or mandatory yet, set the standard for international regulations. In reality, soft law is not a traditional source of law. However, it is an unequivocal meaning or concept, since aspects such as the authority from which these instruments emanate, the limits with respect to other international instruments, perhaps more assimilable to the doctrine as a source, such as, for example, the Bruntland Report or Agenda 21. The term soft law was introduced into International Law by Lord Mcnair, who used it to distinguish between propositions de lege lata and de lege ferenda, and not to distinguish a complex phenomenon that supposes the existence of variations regulations ranging from the non-binding to the binding(Matlock et al., 2002).

Initially, the expression soft law tried to describe statements formulated as abstract principles, present in all legal systems, which became operational through their judicial application. Subsequently, what the term soft law sought was to describe the existence of legal phenomena characterized by a lack of binding legal force. This means admitting the existence of a very "relative" regulation in the international system, in the sense that it lacks binding effectiveness, but aspires to achieve it(Koutalakis et al., 2010).

The different formulations of the precautionary principle are found in a wide variety of international instruments, binding agreements and non-binding declarations, that is, in all kinds of instruments of global and regional application.

The soft law character manifests itself in three ways in the international legal world. The first, through international instruments that are distinguished by not having, per se, binding legal force, that is, instruments with normative aspirations but that are not binding. The second, through norms or provisions that are established in instruments not considered soft law, that is, regardless of the legal nature of the instrument, mandatory or not, there are rules that have this soft nature due to their political, programmatic, declarative or goodwill content, rather than legal-binding. And the third manifestation, through regulations that are in the process of gestation, not yet consolidated, without having entered into force.73 This type of soft instruments can be bilateral, regional or multilateral(Kostka, 2016).

The appearance of soft law norms in non-binding legal instruments on environmental matters, such as resolutions, declarations, international conferences, etc., have had a predominant character. Juste Ruiz explains the presence of soft law in international environmental law based on sociological, political and legal reasons, namely: the impact of the regulatory methods used by international organizations, the divergences of interests between developed countries and developing countries. in developing countries, and the evolution of science and technology that advise adopting flexible standards, capable of adapting to changes.

CONCLUSION

Managing environmental challenges requires a legal system that is effective in all three levels, i.e. bilaterally, regionally, and internationally. It is in this sense that International Environmental Law plays an important role. International Environmental Law is weak because of the number of soft law instruments equal to or even more than the number of hard law instruments. A characteristic of soft law instruments is that they do not bind states to implement them, but instead act more as suggestions. When this occurs, the word "should" may be used instead of the word "must". The other thing is that this kind of instrument has moral sanction rather than legal sanction, so it is more of a morally binding instrument rather than a legally binding one.

Environmental protection and food security are interdependent. At the same time, food security efforts should not harm the environment, yet environmental conservation is crucial to support food security. Food security can be achieved through sustainable agriculture without harming the environment. A sustainable farm is one that produces food for extended periods without harming the environment. Agriculture that is environmentally sustainable is also physically sustainable.

REFERENCES

Abate, C., Patel, L., Rauscher, F. J. III, & Curran, T. (1990). Redox regulation of fos and jun DNA-binding activity in vitro. *Science, 249*(4973), 1157–1161. doi:10.1126cience.2118682 PMID:2118682

Argumedo, A., Swiderska, K., Pimbert, M., Song, Y., & Pant, R. (2011). *Implementing Farmers Rights under the FAO International Treaty on PGRFA: The Need for a Broad Approach Based on Biocultural Heritage. International Institute for Environment and Development.* IIED.

Bardgett, R. D., & Gibson, D. J. (2017). Plant ecological solutions to global food security. *Journal of Ecology, 105*(4), 859–864. doi:10.1111/1365-2745.12812

Boutillon, S. (2001). The precautionary principle: Development of an international standard. *Mich. J. Int'l L., 23*, 429.

Bullock, J. M., Dhanjal-Adams, K. L., Milne, A., Oliver, T. H., Todman, L. C., Whitmore, A. P., & Pywell, R. F. (2017). Resilience and food security: Rethinking an ecological concept. *Journal of Ecology, 105*(4), 880–884. doi:10.1111/1365-2745.12791

Cameron, J., & Abouchar, J. (1991). The precautionary principle: A fundamental principle of law and policy for the protection of the global environment. *BC Int'l & Comp. L. Rev., 14*, 1.

Carolan, M. S. (2013). *Reclaiming food security*. Routledge. doi:10.4324/9780203387931

Dasgupta, C. (2012). Present at the creation: the making of the UN Framework Convention on Climate Change. In *Handbook of Climate Change and India* (pp. 113–122). Routledge.

de Chazournes, L. B. (1998). *Kyoto protocol to the united nations framework convention on climate change.* UN's Audiovisual Library of International Law. http://untreaty. un.org/cod/avl/ha/kpccc/kpccc.html

Gonzalez, C. G. (2011). Climate change, food security, and agrobiodiversity: Toward a just, resilient, and sustainable food system. *Fordham Environmental Law Review*, 493-522.

Gottlieb, R., & Fisher, A. (1996). Community food security and environmental justice: Searching for a common discourse. *Agriculture and Human Values*, *13*(3), 23–32. doi:10.1007/BF01538224

Gupta, A. (2000). Governing trade in genetically modified organisms: The Cartagena Protocol on Biosafety. *Environment*, *42*(4), 22–33. doi:10.1080/00139150009604881

Jacquemont, F., & Caparrós, A. (2002). The convention on biological diversity and the climate change convention 10 years after Rio: Towards a synergy of the two regimes. *Rev. Eur. Comp. & Int'l Envtl. L.*, *11*(2), 169–180. doi:10.1111/1467-9388.00315

Johnson, H., & Walters, R. (2014). Food security. In *The Handbook of Security* (pp. 404–426). Palgrave Macmillan. doi:10.1007/978-1-349-67284-4_19

Kattelus, M., Rahaman, M. M., & Varis, O. (2014, May). M yanmar under reform: Emerging pressures on water, energy and food security. *Natural Resources Forum*, *38*(2), 85–98. doi:10.1111/1477-8947.12032

Kostka, G. (2016). Command without control: The case of C hina's environmental target system. *Regulation & Governance*, *10*(1), 58–74. doi:10.1111/rego.12082

Koutalakis, C., Buzogany, A., & Börzel, T. A. (2010). When soft regulation is not enough: The integrated pollution prevention and control directive of the European Union. *Regulation & Governance*, *4*(3), 329–344. doi:10.1111/j.1748-5991.2010.01084.x

Krasnova, I. O. (2014). Environmental security as a legal category. *Lex Russica*, (5), 543–555.

Kriebel, D., Tickner, J., Epstein, P., Lemons, J., Levins, R., Loechler, E. L., Quinn, M., Rudel, R., Schettler, T., & Stoto, M. (2001). The precautionary principle in environmental science. *Environmental Health Perspectives*, *109*(9), 871–876. doi:10.1289/ehp.01109871 PMID:11673114

Lawrence, G. (2013). *Food security, nutrition and sustainability*. Earthscan. doi:10.4324/9781849774499

Matlock, M. M., Henke, K. R., & Atwood, D. A. (2002). Effectiveness of commercial reagents for heavy metal removal from water with new insights for future chelate designs. *Journal of Hazardous Materials*, *92*(2), 129–142. doi:10.1016/S0304-3894(01)00389-2 PMID:11992699

McBeath, J. H., McBeath, J., & McBeath, J. (2010). *Environmental change and food security in China*. Springer. doi:10.1007/978-1-4020-9180-3

McInerney, T. (2007). Putting regulation before responsibility: Towards binding norms of corporate social responsibility. *Cornell Int'l LJ*, *40*, 171.

Mombo, F., Speelman, S., Huylenbroeck, G. V., Hella, J., & Pantaleo, M. (2011). *Ratification of the Ramsar convention and sustainable wetlands management: Situation analysis of the Kilombero Valley wetlands in Tanzania*. Academic Press.

Nevitt, M. P. (2020). On Environmental Law, Climate Change, & National Security Law. *Harv. Envtl. L. Rev.*, *44*, 321.

Read, R., & O'Riordan, T. (2017). The precautionary principle under fire. *Environment, 59*(5), 4–15. doi:10.1080/00139157.2017.1350005

Ringler, C., Biswas, A. K., & Cline, S. A. (Eds.). (2010). *Global change: Impacts on water and food security*. Springer. doi:10.1007/978-3-642-04615-5

Santilli, J. (2012). *Agrobiodiversity and the Law: Regulating genetic resources, food security and cultural diversity*. Routledge. doi:10.4324/9780203155257

Smith, D., Hinz, H., Mulema, J., Weyl, P., & Ryan, M. J. (2018). Biological control and the Nagoya Protocol on access and benefit sharing–a case of effective due diligence. *Biocontrol Science and Technology, 28*(10), 914–926. doi:10.1080/09583157.2018.1460317

Smith, S., Nickson, T. E., & Challender, M. (2021). Germplasm exchange is critical to conservation of biodiversity and global food security. *Agronomy Journal, 113*(4), 2969–2979. doi:10.1002/agj2.20761

Subramaniam, Y., Masron, T. A., & Azman, N. H. N. (2020). Biofuels, environmental sustainability, and food security: A review of 51 countries. *Energy Research & Social Science, 68*, 101549. doi:10.1016/j.erss.2020.101549

Venancio, M. D., Pope, K., & Sieber, S. (2018). Brazil's new government threatens food security and biodiversity. *Nature, 564*(7734), 39–40. doi:10.1038/d41586-018-07611-7 PMID:30518897

Vidar, M. (2022). Soil and agriculture governance and food security. *Soil Security, 6*, 100027. doi:10.1016/j.soisec.2021.100027

Whiteside, K. H. (2006). *Precautionary politics: principle and practice in confronting environmental risk*. Mit Press.

ADDITIONAL READING

McInerney, T. (2007). Putting regulation before responsibility: Towards binding norms of corporate social responsibility. *Cornell Int'l LJ, 40*, 171.

Nevitt, M. P. (2020). On Environmental Law, Climate Change, & National Security Law. *Harv. Envtl. L. Rev., 44*, 321.

Read, R., & O'Riordan, T. (2017). The precautionary principle under fire. *Environment, 59*(5), 4–15. doi:10.1080/00139157.2017.1350005

Ringler, C., Biswas, A. K., & Cline, S. A. (Eds.). (2010). *Global change: impacts on water and food security*. Springer. doi:10.1007/978-3-642-04615-5

Santilli, J. (2012). *Agrobiodiversity and the Law: regulating genetic resources, food security and cultural diversity*. Routledge. doi:10.4324/9780203155257

Smith, D., Hinz, H., Mulema, J., Weyl, P., & Ryan, M. J. (2018). Biological control and the Nagoya Protocol on access and benefit sharing–a case of effective due diligence. *Biocontrol Science and Technology*, *28*(10), 914–926. doi:10.1080/09583157.2018.1460317

Smith, S., Nickson, T. E., & Challender, M. (2021). Germplasm exchange is critical to conservation of biodiversity and global food security. *Agronomy Journal*, *113*(4), 2969–2979. doi:10.1002/agj2.20761

Subramaniam, Y., Masron, T. A., & Azman, N. H. N. (2020). Biofuels, environmental sustainability, and food security: A review of 51 countries. *Energy Research & Social Science*, *68*, 101549. doi:10.1016/j.erss.2020.101549

Venancio, M. D., Pope, K., & Sieber, S. (2018). Brazil's new government threatens food security and biodiversity. *Nature*, *564*(7734), 39–40. doi:10.1038/d41586-018-07611-7 PMID:30518897

Vidar, M. (2022). Soil and agriculture governance and food security. *Soil Security*, *6*, 100027. doi:10.1016/j.soisec.2021.100027

Whiteside, K. H. (2006). *Precautionary politics: principle and practice in confronting environmental risk*. MIT Press.

KEY TERMS AND DEFINITIONS

Equity: Defined by UNEP to include intergenerational equity - "the right of future generations to enjoy a fair level of the common patrimony" - and intragenerational equity - "the right of all people within the current generation to fair access to the current generation's entitlement to the Earth's natural resources" - environmental equity considers the present generation under an obligation to account for long-term impacts of activities and to act to sustain the global environment and resource base for future generations. Pollution control and resource management laws may be assessed against this principle.

Polluter Pays Principle: The polluter pays principle stands for the idea that "the environmental costs of economic activities, including the cost of preventing potential harm, should be internalized rather than imposed upon society at large." All issues related to responsibility for environmental remediation costs and compliance with pollution control regulations involve this principle.

Precautionary Principle: One of the most commonly encountered and controversial principles of environmental law, the Rio Declaration formulated the precautionary principle: To protect the environment, the precautionary approach shall be widely applied by States according to their capabilities. Where there are threats of serious or irreversible damage, lack of complete scientific certainty shall not be used as a reason for postponing cost-effective measures to prevent environmental degradation. The principle may play a role in any debate over the need for environmental regulation.

Prevention: The concept of prevention can perhaps better be considered an overarching aim that gives rise to a multitude of legal mechanisms, including prior assessment of environmental harm, licensing or authorization that set out the conditions for operation and the consequences for violation of the conditions, as well as the adoption of strategies and policies. Emission limits and other product or process standards, the use of best available techniques, and similar techniques can all be seen as applications of the concept of prevention.

Public Participation and Transparency: Identified as necessary conditions for "accountable governments...industrial concerns," and organizations generally, public participation and transparency are presented by UNEP as requiring "effective protection of the human right to hold and express opinions and to seek, receive and impart ideas...a right of access to appropriate, comprehensible and timely information held by governments and industrial concerns on economic and social policies regarding the sustainable use of natural resources and the protection of the environment, without imposing undue financial burdens upon the applicants and with adequate protection of privacy and business confidentiality," and "effective judicial and administrative proceedings." These principles are present in environmental impact assessment, laws requiring publication and access to relevant environmental data, and administrative procedures.

Transboundary Responsibility: Defined in the international law context as an obligation to protect one's environment and prevent damage to neighboring environments, UNEP considers transboundary responsibility at the international level as a potential limitation on the sovereign state's rights. Laws that limit externalities imposed upon human health and the environment may be assessed against this principle.

Chapter 2
Environmental Law and Water Legislation

ABSTRACT

In this chapter, the authors look at the guidelines International Human Rights and Environment provides for the state with respect to the appropriate structure of its water laws. Moreover, it discusses the relevant aspects of the human right to water and sanitation, as well as the conceptual connection between human rights and the environment. In addition, the report emphasizes the importance of the principles of international law for ensuring the right management of water resources as a guarantee of the human right to water and sanitation.

INTRODUCTION

Through their respective legislation, the States regulate with certain freedom the various aspects related to water resources. Thus, they determine the public ownership of the waters, regulate the access of individuals to this resource, define hydrological planning, create organisms and assign functions to them. In this regard, International Law, especially in the field of human rights and the environment, establishes multiple obligations that directly or indirectly condition national water regulation, setting the State a margin of discretionary powers, and whose non-compliance can do so. incur international responsibility.

Thus, the present work focuses on water as an element of the environment and as a human right, with the purpose of determining the standard that International Law sets to the State for a configuration of its water legislation that allows it to guarantee compliance with human rights. and environmental protection.

Thus, in the first part of the study, reference is made to the conceptual link between the legal right to the environment and other human rights, as well as to the existence and legal status of the right to an adequate environment. The foregoing will make it possible to deliver the general obligations of international regulations and the reasons why they should be incorporated into state water legislation(Gleick, 1998).

In turn, the second part examines the most important issues of the recognition of the human right to water and sanitation, its legal status and its content, the obligations derived from said recognition and its environmental dimension. Likewise, the essential principles of International Law regarding the

DOI: 10.4018/978-1-6684-7188-3.ch002

sustainable management of water resources are analyzed, which will finally deliver the international standard in terms of human rights and the environment. This first part aims to determine those human rights obligations that, because they are related to the enjoyment of an adequate environment, constitute a general requirement for domestic Water Law. To do this, the various perspectives in which the legal good of the environment is related to other human rights are examined, as well as the existence and legal status of the adequate environment as an autonomous human right (Alan, 2017).

ENVIRONMENT AS A SINE QUA NON-CONDITION FOR HUMAN RIGHTS

There is unanimous recognition on the part of the doctrine that environmental deterioration can mean the violation of a series of fundamental rights such as life, health, the right to water and sanitation, to food, to adequate housing, to private and family life, self-determination and even property. In this way, it is understood that environmental damage hinders States in their task of guaranteeing the full and effective enjoyment of human rights. Therefore, a first approach alludes to the environment as a sine qua non requirement for the enjoyment of other human rights cataloged as substantive, which has been expressed in numerous international instruments and by different bodies for the promotion and protection of human rights, which are analyzed below (Orakhelashvili, 2018).

International Instruments Referring to the Adjective Nature of the Environment

In multiple international conferences, the international community has referred to the environment as an indispensable prerequisite for the fulfillment of other rights. However, States have not always focused exclusively on human rights as the basis for environmental protection, but rather the link between both legal rights has been evolving hand in hand with other concepts, such as sustainable development, the right to development and the right of future generations.

It was in the Declaration of the United Nations Conference on the Human Environment, adopted in Stockholm in 1972, that for the first time the transcendence of the link between the protection of the environment and the full and effective enjoyment of human rights is alluded to, also expressing the duty of States to guarantee the rights of future generations.

Two decades later, the Declaration of the United Nations Conference on Environment and Development, adopted in Rio de Janeiro, establishes important principles on the matter and affirms that the human person constitutes the central axis of decisions aimed at sustainable development, explicitly recognizing "...the right to a healthy and productive life in harmony with nature" (principle 1).

The following year, with the Declaration of the World Conference on Human Rights, adopted in Vienna, the States maintain that the right to development must be met in an equitable manner with the needs of the environment, both of present and future generations. Likewise, they admit that the dumping and disposal of toxic substances can generate a serious risk to the right to life and health (paragraph 11).

Finally, mention should be made of the Final Report of the Special Rapporteur Fatma Zohra Ksentini on Human Rights and the Environment, presented to the then United Nations Commission on Human Rights in 1994. In this document, the idea that the environmental damage directly harms the enjoyment of a large number of human rights (paragraph 248) and that the violation of any of these rights affects the environment (paragraph 250).

In short, various international instruments highlight the obligation of States to guarantee adequate protection of the environment, since its physical composition -material environment- constitutes the basic assumption for the full enjoyment and exercise of a series of human rights (Boyle & Redgwell, 2021).

Decisions of International Jurisdictional and Quasi-Jurisdictional Bodies

In the sphere of the International Court of Justice, the opinion of the then vice-president, Mr. Weeramantry, stands out. He refers to environmental protection as a crucial piece in the discourse of human rights, since, in his opinion, the enjoy multiple rights, such as the right to life and health. However, the various universal and regional bodies for the promotion and protection of human rights have ruled on this link on numerous occasions.

In the American system, although the American Convention on Human Rights (hereinafter, ACHR) does not refer to the legal good of the environment, the Inter-American Commission on Human Rights (hereinafter, IACHR) has pointed out that deficient legislation or non-compliance with adequate regulations can cause environmental deterioration that means the violation of other human rights protected by the ACHR. Meanwhile, the Inter-American Court of Human Rights (hereinafter, Inter-American Court) has expressly recognized the existence of an undeniable link between the protection of the environment and the enjoyment of other human rights(Koskenniemi, 2007).

For its part, the European Convention for the Protection of Human Rights and Fundamental Freedoms of 1950 (hereinafter, ECHR) does not expressly establish the right to an adequate environment. However, the European Court of Human Rights (hereinafter, ECHR) in numerous judgments has developed an evolutionary interpretation of the ECHR to recognize environmental protection, through the right to life (article 2) and, particularly, the right to private and family life (article 8.1).

Thus, and due to the adjective role of the legal right of the environment in relation to other human rights, each State is obliged to configure its legal system -including its water legislation- based on all the necessary considerations to allow optimal environmental conservation.

Consequently, the Water Law must be aimed at guaranteeing that the use of water resources does not constitute an impact on the environment, which means or may mean a violation of other fundamental rights, such as the human right to water and sanitation. Otherwise, the respective State would eventually be failing to comply with its obligations and, consequently, subject to international responsibility. Likewise, civil society can take advantage of all the international mechanisms established for the defense of human rights that could be affected by environmental degradation. In the same way, citizens can make use of human rights whose exercise allows the protection of other rights, and whose link is analyzed below(Evans, 2014).

PROCEDURAL HUMAN RIGHTS FOR PROTECTING THE ENVIRONMENT

A second perspective refers to the enormous importance that the exercise of certain rights, such as freedom of expression and association, education and, particularly, access to information, public participation in decision-making, and access to Justice and effective remedies matter for the legal protection of the environment.nIt is important to keep in mind that when the exercise of procedural rights is maximized, the authority is in a better position to conduct environmental policies and, likewise, there are more possibilities for civil society to demand greater environmental protection. In accordance with the principles of indivisibility, interrelationship and interdependence, it is understood that these procedural rights constitute a guarantee mechanism for the legal right of the environment, being the cornerstone of an environmental democracy, achieved only by the consent of citizens in the adoption of environmental decisions.

At the universal level, Convention No. 169 of the International Labor Organization (hereinafter, ILO) enshrines participation as the complement of the right of indigenous peoples to land, territory and natural resources, establishing that these peoples have right to participate in the use, administration and conservation of natural resources (article 15).

In the inter-American system, both the IACHR and the Inter-American Court have alluded to the transcendental importance of the rights of access to information, participation, and access to justice in environmental matters. Thus, in cases in which the respective State has granted concessions for the development of economic activities that may cause an environmental impact, in general, and water, in particular, the IACHR has made clear the obligation of the States to "consult in prior, full, free and informed manner" to civil society (Boyd, 2012).

Likewise, the Inter-American Court has held that, in order to ensure effective participation in development and investment plans, the State must provide the pertinent information, which enables the community to become aware of possible risks to the environment and the health of the community. population. Likewise, the State is required to actively consult indigenous communities, according to their own customs and traditions, from the first stages of the projects. In addition, the State is obliged to make simple, urgent, informal, accessible resources available to the population and processed by independent bodies, which allow the protection of said rights in an adequate and effective manner.

Even the IACHR has ruled on other rights, such as freedom of expression and freedom of association, which, because they are violated due to repression, persecution or arbitrary arrests, would constitute, in turn, a direct impediment to the protection of environment.

In relation to the European system, the broad exercise of the human rights to information, to participation in decision-making processes and to access to justice, constitutes an essential factor for environmental protection. In this sense, the jurisprudence of the ECHR has considered procedural protection, which includes the recognition of a firm obligation of governments to provide access to information, citizen participation and access to justice. Thus, for example, in the cases of López Ostra with Spain and Guerra with Italy, the absence of delivery of information by the respective governments on environmental risks constituted for the ECHR a violation of Article 8 of the ECHR.

In this way, although in Europe the States have a considerable margin to determine the measures they deem necessary to ensure the rights related to the protection of the environment, the anomalies in the exercise of the procedural rights, which end up affecting the rights recognized by the ECHR, give rise to international responsibility for the offending State. Similarly, in the case of Claude Reyes and others with Chile, the Inter-American Court ruled that the lack of access to information constitutes a violation of Article 13 of the ACHR.

Consequently, International Law requires States to respect and ensure these human rights of a procedural nature, so that the consecration and strengthening of them in national law is key when guaranteeing adequate cooperation from all sectors of society. in the main issues related to the management and sustainable planning of water resources(Rourke, 2008).

A LOOK FROM THE PRINCIPLE OF EQUALITY AND NON-DISCRIMINATION

A third approach examines the negative effects that environmental damage means for the fulfillment of the substantive rights of certain vulnerable or subordinate groups, and, therefore, the transgression of the principle of equality and non-discrimination.

Just as there are essentially individual rights, which can be protected only because of a person, there are other rights that, in turn, can be protected collectively, especially when a group is affected. In this way, there are groups that are more suitable for collective protection, and for this reason International Law contemplates protection mechanisms for certain vulnerable groups such as women, children, people with disabilities, the elderly, indigenous peoples and workers under certain conditions.

Thus, for example, the close link between migration and the environment has been highlighted, proposing to leave behind the perception of adaptive frustration that migration would have due to climate change and adopt a position that perceives this phenomenon as an opportunity to correctly address the need of environmental refugees to adapt to the current living conditions that the environment demands. In this way, the enjoyment of rights by subordinate groups is an issue that can be hampered by changing environmental conditions. Similarly, it is argued that often, whether due to socioeconomic issues, nationality or racial origin, environmental services, on the one hand, and risks and damage to the environment, on the other, are not equitably distributed among the population., which would be a matter of environmental justice.

For the IACHR "... the violation of the right to equality occurs by granting unequal environmental protection based on race [or other social condition], without any reasonable justification, creating a form of environmental racism that serves no objective legitimate and establishing inadequate and ineffective means to obtain environmental protection, knowing that people of color [or other vulnerable groups] bear the disproportionately substantial burden of such inadequate and ineffective measures."

In this way, and due to various international instruments, the protection of vulnerable groups would not be given only in the field of human rights, nor in that of International Environmental Law, but in both legal fields jointly (Malanczuk, 2002).

By virtue of the foregoing, the States are obliged to respect and guarantee the protection of the environment and human rights, but also to ensure that all the mechanisms provided in this regard are implemented on the principle of equality and non-discrimination. This implies that national legislation must develop all actions aimed at accelerating the de facto equality of those communities that have been affected in the enjoyment and exercise of their human rights as a result of inadequate environmental protection, thereby exacerbating a situation of discrimination. prolonged in time.

Thus, the challenge for state water legislation is to make a system more flexible that allows it to adapt to communities that historically, or recently, have been affected by access to water, either due to overexploitation or quality of resources. hydric. This will make it possible to protect groups that require preferential attention, such as people who live in desert areas and rural areas, people dedicated to small-scale agriculture, as well as members of indigenous peoples, since all these groups are very susceptible to scarcity and poor water quality. All of the above with special emphasis on women, children, people with disabilities, the elderly, migrants, among others (Weissbrodt & Kruger, 2017).

ENVIRONMENT AS A LIMIT TO OTHER HUMAN RIGHTS

It is due to the very use and enjoyment of the elements that make up the environment that a link is generated with other human rights, where the environmental legal asset establishes or may establish a restriction on different fundamental rights and freedoms, such as the right to property, freedom of movement, freedom of entrepreneurship or development of economic activities.

For the IACHR, it is essential that the development of the various economic activities be subject to certain measures to ensure that said activities are not carried out at the expense of the enjoyment of the human rights of people and their environment, on which their physical well-being depends., cultural and even spiritual. In any case, for the Inter-American Court, the declaration of a territory as a protected wilderness area under private domain cannot disregard the right of the members of indigenous communities over their traditional lands.

In the European system, through Article 8 of the ECHR, the positive obligation arises on the part of the State to adopt reasonable and appropriate measures that guarantee the rights of the plaintiffs. Thus, the ECHR must weigh the collective advantages against the particular interests of individuals. In both cases, the State has a certain margin of appreciation when deciding what measures must be adopted to guarantee respect for the ECHR, and which may eventually restrict the exercise of certain rights by other individuals, which are especially related to the exploitation of natural resources and the development of industrial or commercial activities (Goldfarb, 2020).

Environmental protection requires respect both from individuals—when developing various economic activities—and from the State in its role of ensuring that such activities are carried out in compliance with current regulations. In this sense, it is affirmed that the complaints for violation of human rights in environmental matters seek to rightly hold the State responsible for not having adequately regulated an issue, or else not having demanded compliance with the law, since the generality of the damages to the environment comes from private activities.

In this way, environmental protection constitutes a counterweight to the unrestricted exercise of those rights that allow the development of multiple economic and industrial activities and, therefore, the exploitation of natural resources. The foregoing does not mean that the right to an adequate environment cannot, in turn, be restricted by the exercise of these rights. On the contrary, since there is no hierarchy of rights in the international human rights system, there are no absolute rights, so the restriction must be subject to the standards of proportionality (Benjamin et al., 2004).

Consequently, the Water Law must in no case be configured on the basis of the free exploitation of water resources by individuals, but must also incorporate the necessary mechanisms to guarantee the sustainable use and management of water. that allows, likewise, to ensure the protection of the environment, and with it the respect and guarantee of the human rights that depend on it. The foregoing requires the incorporation of figures that in a certain way limit the development of economic activities, such as the norm of effective and beneficial use; the rule that establishes an order of preference based on the different uses of water; the provisions that recognize and develop the principle of hydrographic basin unit, and those that establish ecological flows.

AUTONOMOUS RIGHT TO AN ADEQUATE ENVIRONMENT

At the universal level, the existence of an autonomous right to an adequate environment has not yet been recognized by any binding instrument, which has conditioned the legal status of this right and has generated a doctrinal discrepancy about the content of it.

In accordance with the already mentioned Report of the Special Rapporteur Fatma Zohra, since the Stockholm Declaration in 1972, a series of instruments have been elaborated at the universal, regional and national levels that have developed the legal foundation of the right to an adequate environment. The foregoing, based mainly on the connection between the protection of the environment, development and the enjoyment of human rights (paragraphs 236-238), with which the universal and regional recognition of the human right to an adequate environment would be affirmed (paragraph 240).

Thus, for example, the Additional Protocol to the ACHR on Economic, Social, and Cultural Rights (hereinafter, Protocol of San Salvador) expressly recognizes that "[e]very person has the right to live in a healthy environment and to have basic public services," and that "[t]he States Parties shall promote the protection, preservation, and improvement of the environment" (Article 11). On the other hand, from 1972 to 2008, 119 States had enshrined in their respective constitutions some type of guarantee on this right, which would account for an emerging rule of customary International Law (Klein et al., 2009).

However, and without prejudice to that recognition, it is argued that the multiple constitutional provisions use terms that are very dissimilar to each other, and on few occasions are rights independent of other legal assets already protected by existing human rights, which would prevent a precise definition of the scope and content of this right, thereby hindering the uniformity of state practice. Likewise, it is affirmed that the mere incorporation of these provisions in the national constitutions is not enough to demonstrate an effective opinio juris. Being able to conclude that both international regulations and constitutional recognition reaffirm the existence of the principles of human rights, rather than establishing a new right.

Now, although the autonomous right to an adequate environment has not yet been recognized at the universal level, there is no doubt that it has already entered the legal discourse of human rights, at least with the status of soft law, finding itself waiting to be crystallized by conventional or customary International Law. In national sphere, the 1980 Constitution explicitly recognizes the right to live in a pollution-free environment, which has been defined, both by the Supreme Court and by the Constitutional Court, as a human right of constitutional hierarchy, with a dual character of subjective and collective public law at the same time. Consequently, and by virtue of the ACHR, the Inter-American Court would be obliged to decide environmental cases brought against Chile, since it is not allowed "... to limit the enjoyment and exercise of any right or freedom that may be recognized in accordance with the laws of any of the States Parties or in accordance with another convention to which one of said States is a party" (article 29) (Godden, 2005).

HUMAN RIGHT TO WATER AND SANITATION

The genesis of the human right to water and sanitation is found in a series of international conferences held in the context of strategies for the environment and development. Thus, in the Declaration of the United Nations Conference on Water, signed in Mar del Plata in 1977, the power "...to have drinking water in sufficient quantity and quality for their basic needs is recognized as a right of all peoples."

Likewise, the Declaration of the International Conference on Water and Environment, adopted in Dublin in 1992, confirms that "... it is essential to recognize above all the fundamental right of every human being to have access to clean water and sanitation for a affordable price" (principle 4). Likewise, in the Declaration of the International Conference on Freshwater, adopted in Bonn in 2001, it is understood that in order to achieve sustainable development, the integration of the multiple uses of the resource with social, environmental and economic aspects is required (Ball & Bell, 1994).

For its part, the United Nations Conference on Environment and Development, held in Rio de Janeiro in 1992, derived a series of international instruments on environmental matters, including Agenda 21, which recognizes that stress water and the progressive contamination of water require planning and integrated management of water resources (paragraph 18.3).

In this way, these international instruments place a particular emphasis on the link between development, water conservation and the environment, where environmental protection is considered as the essential budget to guarantee the availability of the resource for human consumption (McIntyre, 2010).

Environmental Dimension of the Human Right to Water and Sanitation

The protection of the environment is essential for the enjoyment of the human right to water and sanitation, since at least the availability and quality of the resource depends on it. In this way, to guarantee the exercise of this right, the adoption of political and legal measures is required, integrated with a sustainable management of water resources. In addition, it must be taken into account that the violation of the human right to water and sanitation hinders sustainable development and environmental justice.

Similarly, the ESCR Committee understands that the exercise of the human right to water "... must be sustainable so that this right can be exercised by current and future generations." With this same objective, the obligation of States to adopt strategies and programs to avoid overexploitation and contamination of water resources is established, considering factors such as climate change, desertification and deforestation, as well as the loss of biodiversity (Wilkinson, 2005).

However, the human right to water and sanitation, being essentially anthropocentric and conceptually limited to a minimum -minimum amount for personal and domestic use-, its protection can only be projected to the water resources necessary to satisfy the basic needs of the population. population. Consequently, the management and preservation of water must necessarily be assumed by the (International) Environmental Law and the (International) Water Law. In this way, and considering that the main difficulties of access to water are related to its status and conservation, it is that the guarantee of this right is closely linked to the protection -or degradation- of the environment and economic growth, rather than with aspects of human rights. This has led to the proposal of building a single right that includes water and sanitation as constituent elements together with environmental protection.

Consequently, and due to the interdependence, interrelationship and indivisibility of human rights, and the importance of water as a component of the physical phase of these rights, it is that International Law requires that national Laws implement international regulations in a manner comprehensive and not partial. For this, the role of Water Law as a sectoral branch is key, in the sense of merging the regulation of the content of the right to water and sanitation with the protection and conservation of the environment (Adler, 2010).

INTERNATIONAL LAW AND SUSTAINABLE WATER RESOURCES MANAGEMENT

The conventional obligations assumed by the States in the International Law of Human Rights (hereinafter, IHRL) coincide with those commitments adopted in the international treaties on the environment, since in both cases the final objective -or at least one of its objectives ends - is constituted by the need to guarantee the well-being of the human person64.

In this way, it is pertinent to study both the fundamental principles of International Environmental Law (hereinafter, DIMA) and the particular principles of International Water Law (hereinafter, DIA). The foregoing with the purpose of incorporating in the national legislation the necessary mechanisms for a sustainable management of water resources, which allow an adequate protection of the environment and a guarantee of the other related human rights, with strict compliance with the principle of equality and discrimination (Eckstein, 2009).

Prevention and Precaution Principles

It is essential for the States to harmonize the development of economic activities with the conservation of the environment as a guarantee for the survival and well-being of all people, allowing the authorities to adopt decisions that make it possible to reach a point of balance between both interests.

From DIMA's perspective, the Rio Declaration enshrines the principles of prevention and precaution that seek to avoid environmental damage. According to the principle of prevention, States have the obligation to ensure that activities carried out within their jurisdiction or under their control do not cause damage to the environment of other States or of areas beyond their jurisdiction (principle 2) . In turn, said international instrument establishes that, with the purpose of adequate protection of the environment, States must abide by the precautionary principle, by virtue of which the absence of scientific certainty cannot be used as an excuse for not adopting the measures necessary to avoid environmental deterioration (principle 15). The 1992 United Nations Framework Convention on Climate Change provides that States parties "...should take precautionary measures to anticipate, prevent or minimize the causes of climate change and mitigate its adverse effects. When there is a threat of serious or irreversible damage, the lack of full scientific certainty should not be used as a reason for postponing such measures" (article 3.3).

Thus, the regulation of uncertain risks is the cornerstone of the precautionary principle, where the causal relationship between a given activity and the risk it produces is ignored, unlike the duty of prevention, where the risks caused by an activity are previously known. Therefore, public authorities must manage both certain threats -prevention- and uncertain ones -precaution-, considering not only the activities that will surely cause damage, but also those that could eventually cause it, that is, where there is scientific uncertainty. In this sense, the ECHR has expressly stated that the precautionary principle recommends that States not delay in adopting preventive measures, simply due to the existence of scientific uncertainty, and must manage the risks to the environment and the health of the population. The foregoing implies a review of the influence exerted by the scientific manifesto on governments, and which ends up depriving science of its character as an unequivocal guarantor of the truth of public decisions (Eckstein, 2009).

In this way, the precautionary principle conforms the legal materialization of the judgment on the risk, in such a way that the protective decisions of the environment are limited by law, as well as all those economic activities that may threaten the preservation of the environment. Therefore, the need arises to configure the various legal instruments aimed at effectively preventing the emergence of new risks for society and the environment. In this sense, access to information, by civil society, regarding the analysis and responsibility of social risk, is crucial when legitimizing the regulation strategies that have been adopted.

For its part, the United Nations Convention on the Law of the Uses of International Watercourses for purposes other than navigation, adopted in 1997 (hereinafter, the Convention on International Watercourses), enshrines the obligation of States of use international watercourses, adopting all appropriate measures to prevent significant damage from being caused to said watercourse in the territory of other States (article 7.1), making reference to the protection and preservation of ecosystems and the prevention, reduction and pollution control (articles 20 and 21). In this way, this obligation must be interpreted from the point of view of the precautionary principle, in such a way that it allows adequate management of water resources in the face of scientific uncertainty regarding whether or not a certain activity may cause significant damage to water.

Thus, in compliance with the principles of prevention and precaution, the domestic Water Law must provide the legal mechanisms that allow the Administration to act in a preventive and precautionary manner, and not only when there is a risk of imminent damage to water resources. In this way, the public body will be able to count on the necessary tools to plan and manage water in a sustainable manner in accordance with environmental objectives. Likewise, the water legislation must empower the administration to condition the rights to use water to an effective and beneficial use of these resources, giving it the possibility of making said rights expire in the event of non-compliance (Eckstein, 2009).

Cooperation Principle

At the universal level, the principle of cooperation or collaboration in environmental matters is widely recognized in principles 7, 9, 10 and 14 of the Rio Declaration. Thus, principles 7 and 14 refer to international cooperation between states, while principle 9 refers to the need for state cooperation within the state itself. For its part, principle 10 addresses the collaboration of civil society, stating that "[t]he best way to deal with environmental issues is with the participation of all interested citizens, at the appropriate level."

In turn, the United Nations Convention to Combat Desertification establishes that cooperation is a principle for the protection and conservation of land and water resources, which must be inserted at all levels (government, civil society and users) in order to better understand the characteristics and value of land and water, as well as the sustainable promotion of said resources (articles 3 letter c and 4.2 letter c) (Eckstein, 2009).

Likewise, the Convention on International Watercourses provides that States have the right to participate in an equitable and reasonable manner in the use of an international watercourse and, in turn, the obligation to cooperate in its use and protection with the purpose of to achieve optimal and sustainable use and adequate protection of the respective water course (articles 5 and 8). In this way, the obligation of the states to deliver mutually, and on a regular basis, the available information on the state and quality of the watercourse, as well as the corresponding preventive measures, is established (articles 9 and 31). Furthermore, this convention provides that to make fair and reasonable use and participation effective, and when necessary, States must consult in a spirit of cooperation (article 6). Likewise, it establishes the duty of the States to guarantee, without discrimination, access to justice for natural or legal persons, to request compensation or reparation for significant transboundary damage as a consequence of activities related to the respective international watercourse (article 32).

In Europe, for its part, the Aarhus Convention on Access to Information, Public Participation in Decision-Making and Access to Justice in Environmental Matters of 1998 is of great importance, which recognizes the rights described above with the purpose of protecting the right of every person, and of present and future generations, to live in an environment that guarantees the enjoyment and exercise of their right to health and well-being (article 1).

Thus, starting from the basis that decisions regarding the use of natural resources are related to a social risk, it is reasonable to estimate that it is society itself as a whole that determines the risks it is willing to assume. Therefore, each State has the international obligation to implement and ensure all the mechanisms that are necessary for an adequate collaboration of civil society in water matters, for which the water legislation must integrate and strengthen the rights of access to information., participation and access to justice (van Rijswick, 2010).

Consequently, the national water legislation must establish the mechanisms that ensure citizens have access to the main data related to the availability and quality of water resources, with the granting of concessions or water rights and the various uses that they can have. private companies give them, as well as all the information essential for the design of public policies on the matter. In this way, the design of the planning and management of water resources must have the participation of all social actors (administration, civil society and users), which enables effective cooperation in the protection of water resources and in safeguarding of the human right to water.

Regarding access to justice, domestic law, together with establishing a judicial action, at the constitutional or legal level, that establishes a quick, flexible and simple action to guarantee the full enjoyment and exercise of the human right to water and sanitation, it must strengthen the procedural mechanisms that ensure the rights of access to information and participation in decision-making (Blumm, 1998).

Responsibility Principle

According to principle 16 of the Rio Declaration, "[t]he national authorities should seek to promote the internalization of environmental costs and the use of economic instruments, taking into account the criterion that the polluter should, in principle, bear with pollution costs, with due regard to the public interest and without distorting trade or international accounts". Likewise, principle 13 establishes that "[t]he States shall develop national legislation regarding responsibility and compensation for the victims of pollution and other environmental damage."

In this way, this principle is broader than the simple fact that 'who pollutes, pays', since it includes not only the costs derived from pollution itself, but also those caused by the degradation of the environment, including those from (over)exploitation of natural resources (Sherk, 1990).

Thus, according to the Declaration of the United Nations Conference on the Human Environment and "[i]n accordance with the Charter of the United Nations and the principles of international law, States have the sovereign right to exploit their own resources in application of its own environmental policy, and the obligation to ensure that the activities carried out within its jurisdiction or under its control do not harm the environment of other States or of areas located beyond any national jurisdiction" (principle 21). Consequently, "[t]he States must cooperate to further develop international law with regard to liability and compensation to victims of pollution and other environmental damage that activities carried out within the jurisdiction or under the control of such States cause to areas outside their jurisdiction" (Principle 22).

For its part, under the Convention on International Watercourses, when a State, by using a watercourse, causes significant damage to another, it must, in consultation with the affected State, adopt all necessary measures"... to eliminate or mitigate such damage and, where appropriate, examine the issue of compensation" (article 7.2).

This is how the principle of responsibility arises, by virtue of which the coordination of accepted and rejected risks allows society to define the responsibilities that must correspond to each of its actors for the negative effects that could originate due to of the exercise of certain economic activities. In this way, the private sphere ceases to have an unconditional freedom, where natural resources are exploited without assuming any cost for environmental deterioration, to give way to a responsible freedom where individuals must internalize the costs for such damage, committing themselves in this way in the achievement of environmental purposes. Through this principle, not only the costs derived from repairing the damage must be internalized, but also those necessary to prevent it. In addition, diligent action by public authorities requires responsibility from the State itself, so that if, while complying with all the standards and parameters allowed for the exercise of economic activities, harm is also caused to the environment, the State must respond (Tan & Jackson, 2013).

In summary, the national water legislation must be articulated from the basis of the principle of responsibility, in such a way that both the use and management of water resources and the damage caused by their overexploitation and contamination, including compensation to victims, must be responsibility of the holders of economic projects and, lastly, of the State. Otherwise, the State would be failing to comply with its obligation to guarantee the full enjoyment and exercise of rights. In this order of ideas, the Water Law must empower the administration to limit the time for which the concessions or rights over the resource are granted, in such a way that a subsequent review of the conditions and priority uses against factors is made possible. external. As well as, to demand the payment of fees and taxes in accordance with the different uses that are given to water, whether domestic, agricultural, industrial, hydroelectric, among others; payment that, in any case, the administration must exempt when necessary to guarantee equal access to the right to water and sanitation to a certain community.

This last section deals with the particular principles of the IAD that guarantee both the human right to an adequate environment and the human right to water and sanitation.

Hydrographic Basin Unit Principle

The central axis of water resources planning must be constituted by the principle of hydrographic basin unit. The notion of hydrographic basin is used by geography to refer to the territorial area whose set of waters flow into the same river, lake or sea. Thus, the runoff of surface waters towards the sea is carried out through the basins of the main rivers or from slopes with intermittent waters85. This concept is aimed at ensuring an integrated and complete use of water resources, as well as protecting the quality of the waters and the ecosystems linked to the basin itself. For this reason, it is argued that its management must be carried out in a unitary and global manner, and should not be fragmented due to sections that serve river courses considered in isolation (Shupe, 1982).

Within the scope of the European Union, Directive 60/2000/CE, which establishes a community framework for action in the field of water policy, defines the hydrographic basin as "... the area of land whose surface runoff flows in its entirety through a series of streams, rivers and, eventually, lakes towards the sea through a single mouth, estuary or delta" (article 2.13).

At the universal level, the Convention on International Watercourses uses the term watercourse to refer to the "... system of surface and groundwater which, by virtue of their physical relationship, constitute a unitary whole and normally flow through a common mouth" (article 2 letter a).

In this way, the notion of international watercourse extends both to the States in whose territory surface waters flow and to those in which groundwater flows that flow into a river of another State, even affecting the private holders of exploitation rights. with obligations to preserve the river basin. Now, although the term international watercourses differs from that of hydrographic basin, in practice this difference is merely formal, since the concept of watercourses is made up of the same fundamental elements that make up that of hydrographic basin, such as material interdependence, geographic and economic unity, and the obligation of sustainable and integrated use (Baker, 2009).

Thus, in the case of shared basins, integrated management constitutes the cornerstone of an international strategy aimed at optimizing the distribution of water, thereby guaranteeing adequate social and economic development, leaving behind mere distribution. Likewise, it is understood that this principle requires the adoption of joint hydrological plans, as well as the relativization of the water sovereignty of the States, in order to deliver to a permanent international body the necessary powers to manage the ecosystem in a holistic manner. each shared basin, which would also imply the provision of management instruments and financial resources (Tarlock, 1990).

This preference of conventional International Law to favor organizational strengthening at the basin level would find its origin in a multiplicity of international treaties that, in turn, contemplate general and special mechanisms for conflict resolution on the matter. In this way, it would be possible to better ensure the protection of the environment, the application of the principle of equitable and reasonable use of water, as well as access to information and participation, and conflict resolution.

This principle involves the idea that only through a unitary management of the hydrographic basin can a balanced, integrated and global administration of all the interests that concur in the various existing uses be guaranteed. Therefore, the hydrographic basin unit principle requires the integration of all the sectors involved under a single institutional and operational framework (Naff & Dellapenna, 2017).

In short, any State that seeks sustainable management and planning of its water resources must structure its Water Law on the principle of basin unit. In this way, the water legislation must contemplate the existence of binding hydrological plans based on the characteristics of each basin, and giving special consideration to the various uses that are given to the water resources in the respective basin, as well as also to the population that inhabits said territorial surface. In turn, the bodies in charge of water management must be constituted by each hydrographic basin, endowed with powers that allow them to adopt decisions of a global and unitary nature in said territorial area, restricting only to exceptional and well-founded cases the management by sections or sections that are unaware of this principle. Likewise, basin organizations must be integrated by the administration, users and civil society, in such a way as to ensure the collaboration of all social actors tending to adequate water management, avoiding the simple distribution of water.

In summary, the water legislation must establish the legal provisions that strengthen this principle, so that it is the basis and measure of the management and planning of water resources.

Priority of Uses for Human Consumption Principle

In general terms, and in accordance with the principles and rules of International Law, States have, on the one hand, full sovereignty to exploit their natural resources in compliance with their own environmental policy and, on the other, the duty to guarantee that such exploitation does not harm the environment of other states or any area outside state jurisdiction (Tarlock, 1975).

The consecration of the rule of equitable and reasonable use in the Convention on International Watercourses constitutes a declarative effect of the customary principle of a general and abstract nature according to which shared fresh waters matter, on the one hand, the right of the riparian State to take advantage of the water resources that form part of its territorial sovereignty and, on the other, the obligation to exercise said right in such a way that the use of the waters does not affect the right of the other riparian States to use the shared waters in an equitable and reasonable manner.

In turn, said convention establishes that the equitable and reasonable use of watercourses will be determined on the basis of a series of factors of diverse nature -such as geographic, social, environmental, hydrographic, economic-, which will be analyzed jointly, and among which is "the population that depends on the water course in each State" (article 6).

Now, the aforementioned international instrument contemplates the general rule that no use of an international watercourse has in itself a priority over other uses, considering any agreement or custom to the contrary as an exception (article 10.1). However, it is provided that in case of conflict between various uses in an international watercourse, "vital human needs" must be taken into account in a special way (article 10.2).

Thus, the International Court of Justice, in the case of Costa Rica with Nicaragua, has recognized the existence of customary rights of the riverside populations over water resources, when these are necessary to cover the essential needs of daily life (Chikozho et al., 2018).

In this way, an interpretation of these rules in light of the principle of good faith, and especially in harmony with the ICESCR and with the aforementioned General Obligation No. 15, allows them to be understood as a preference for that use of the resources that has as its purpose guarantee vital human needs and thus the human right to water and sanitation (Elver, 2007).

CONCLUSION

The reason why international regulations on human rights and the environment must be incorporated into national water legislation finds its conceptual origin in the fact that these natural resources are part of the physical dimension of the legal right of the environment and this, in turn, of the global environment that serves as the basis for the full enjoyment and exercise of a series of rights of a substantive nature. Thus, there is a high degree of interdependence, interrelation and indivisibility between water, the environment and other human rights, so that adequate protection of water resources is not only essential for the respect and guarantee of human rights to water and an adequate environment, but also various rights such as life, health and self-determination, among others.

In this way, the State is obliged to configure its legislation based on the general guidelines of IHRL, having to enshrine and reinforce procedural human rights that, due to their own characteristics, allow for better collaboration of citizens in decision-making, strictly giving compliance with the principle of equality and non-discrimination. Otherwise, civil society will be able to activate all the international mechanisms that allow it to establish the responsibility of the State for human rights violations caused by environmental damage.

However, a correct normative integration requires that the national Laws implement the international norm in a complete and non-partialized manner, the Water Law being key as a sectoral branch, which must necessarily be configured based on the general obligations of IHRL, the rules particulars of the human right to water and sanitation, and to the principles of the DIMA and the DIA. In this way, the standard of International Law for a sustainable regulation and protection of water resources is revealed, which allows guaranteeing the different human rights involved.

In short, States enjoy broad autonomy to choose the measures that allow them to comply in the best possible way with their international obligations; however, it is necessary that they recognize each human right, apply the necessary means to respect and guarantee them, and establish the judicial and extrajudicial mechanisms that enable the effective protection of the same by citizens. Indeed, the Water Law must apply the transversal principle of equality and non-discrimination; to the principles of prevention, precaution, access to information, public participation in decision-making and access to justice; to the principle of hydrographic basin unit, and finally to the principle of priority of the use of water for human consumption.

REFERENCES

Adler, R. W. (2010). Climate change and the hegemony of state water law. *Stan. Envtl. LJ*, *29*, 1.

Alan, B. (2017). *Human rights and the environment: Where next?* Routledge.

Baker, L. A. (Ed.). (2009). *The water environment of cities*. Springer. doi:10.1007/978-0-387-84891-4

Ball, S., & Bell, S. (1994). *Environmental Law*. The Law and Policy Relating to the Protection of the Environment.

Benjamin, A. H., Marques, C. L., & Tinker, C. (2004). The water giant awakes: An overview of water law in Brazil. Tex L. *Rev.*, *83*, 2185.

Blumm, M. C. (1988). Public property and the democratization of western water law: A modern view of the public trust doctrine. *Envtl. L.*, *19*, 573.

Boyle, A., & Redgwell, C. (2021). *Birnie, Boyle, and Redgwell's International Law and the Environment*. Oxford University Press. doi:10.1093/he/9780199594016.001.0001

Chikozho, C., Saruchera, D., Danga, L., & da Silva, C. (2018). *A Compendium of the South African water law review post-1994*. Water Research Commission.

Eckstein, G. E. (2009). Water scarcity, conflict, and security in a climate change world: Challenges and opportunities for international law and policy. *Wis. Int'l LJ*, *27*, 409.

Elver, H. (2008). International environmental law, water and the future. In *International Law and the Third World* (pp. 191–208). Routledge-Cavendish.

Evans, M. D. (Ed.). (2014). *International law*. Oxford University Press. doi:10.1093/he/9780199654673.001.0001

Gleick, P. H. (1998). The human right to water. *Water Policy*, *1*(5), 487–503. doi:10.1016/S1366-7017(99)00008-2

Godden, L. (2005). Water law reform in Australia and South Africa: Sustainability, efficiency and social justice. *Journal of Environmental Law*, *17*(2), 181–205. doi:10.1093/envlaw/eqi016

Goldfarb, W. (2020). *Water law*. CRC Press. doi:10.1201/9781003069829

Klein, C. A., Angelo, M. J., & Hamann, R. (2009). Modernizing water law: The example of Florida. *Florida Law Review*, *61*, 403.

Koskenniemi, M. (2007). The fate of public international law: Between technique and politics. *The Modern Law Review*, *70*(1), 1–30. doi:10.1111/j.1468-2230.2006.00624.x

Malanczuk, P. (2002). *Akehurst's modern introduction to international law*. Routledge. doi:10.4324/9780203427712

McIntyre, O. (2010). The proceduralisation and growing maturity of international water law: Case concerning pulp mills on the river Uruguay (Argentina v Uruguay), International Court of Justice, 20 April 2010. *Journal of Environmental Law*, *22*(3), 475–497. doi:10.1093/jel/eqq019

McIntyre, O. (2016). *Environmental protection of international watercourses under international law*. Routledge. doi:10.4324/9781315580043

Naff, T., & Dellapenna, J. (2017). Can there be confluence? A comparative consideration of Western and Islamic fresh water law. In *International Law and Islamic Law* (pp. 281–305). Routledge. doi:10.4324/9781315092515-15

Orakhelashvili, A. (2018). *Akehurst's Modern Introduction to International Law*. Routledge. doi:10.4324/9780429439391

Rourke, J. T., & Boyer, M. A. (2008). *International politics on the world stage*. McGraw-Hill.

Sherk, G. W. (1990). Eastern water law: trends in state legislation. *Virginia Environmental Law Journal*, 287-321.

Shupe, S. J. (1982). Waste in Western Water Law: A Blueprint for Change. Or. *Law Review*, *61*, 483.

Tan, P., & Jackson, S. (2013). Impossible dreaming: Does Australia's water law and policy fulfil Indigenous aspirations. *Environment and Planning Law Journal*, *30*(2), 132–149.

Tarlock, A. D. (1975). Recent developments in the recognition of instream uses in western water law. *Utah L. Rev.*, 871.

Tarlock, A. D. (1990). Western Water Law, Global Warming, and Growth Limitations. *Loy. LAL Rev.*, *24*, 979.

van Rijswick, M. H. (2010). Interaction between European and Dutch water law. In *Water Policy in the Netherlands* (pp. 218–238). Routledge.

Weissbrodt, D., & Kruger, M. (2017). Norms on the responsibilities of transnational corporations and other business enterprises with regard to human rights. In *Globaization and International Investment* (pp. 199–220). Routledge.

Wilkinson, D. (2005). *Environment and law*. Routledge. doi:10.4324/9780203994443

ADDITIONAL READING

Malanczuk, P. (2002). *Akehurst's modern introduction to international law*. Routledge. doi:10.4324/9780203427712

Naff, T., & Dellapenna, J. (2017). Can there be confluence? A comparative consideration of Western and Islamic fresh water law. In *International Law and Islamic Law* (pp. 281–305). Routledge. doi:10.4324/9781315092515-15

Orakhelashvili, A. (2018). *Akehurst's Modern Introduction to International Law*. Routledge. doi:10.4324/9780429439391

Rourke, J. T., & Boyer, M. A. (2008). *International politics on the world stage*. McGraw-Hill.

Sherk, G. W. (1990). Eastern water law: Trends in state legislation. *Virginia Environmental Law Journal*, 287-321.

Shupe, S. J. (1982). Waste in Western Water Law: A Blueprint for Change. *Law Review*, *61*, 483.

Tan, P., & Jackson, S. (2013). Impossible dreaming: Does Australia's water law and policy fulfil Indigenous aspirations. *Environment and Planning Law Journal*, *30*(2), 132–149.

Tarlock, A. D. (1975). Recent developments in the recognition of instream uses in western water law. *Utah L. Rev.*, 871.

Tarlock, A. D. (1990). Western Water Law, Global Warming, and Growth Limitations. *Loy. LAL Rev.*, *24*, 979.

van Rijswick, M. H. (2010). Interaction between European and Dutch water law. In *Water Policy in the Netherlands* (pp. 218–238). Routledge.

Wilkinson, D. (2005). *Environment and law*. Routledge. doi:10.4324/9780203994443

KEY TERMS AND DEFINITIONS

Equity: Defined by UNEP to include intergenerational equity—"the right of future generations to enjoy a fair level of the common patrimony"—and intragenerational equity—"the right of all people within the current generation to fair access to the current generation's entitlement to the Earth's natural resources"—environmental equity considers the present generation under an obligation to account for long-term impacts of activities and to act to sustain the global environment and resource base for future generations. Pollution control and resource management laws may be assessed against this principle.

Polluter Pays Principle: The polluter pays principle stands for the idea that "the environmental costs of economic activities, including the cost of preventing potential harm, should be internalized rather than imposed upon society at large." All issues related to responsibility for environmental remediation costs and compliance with pollution control regulations involve this principle.

Precautionary Principle: One of the most commonly encountered and controversial principles of environmental law, the Rio Declaration formulated the precautionary principle: To protect the environment, the precautionary approach shall be widely applied by States according to their capabilities. Where there are threats of serious or irreversible damage, lack of complete scientific certainty shall not be used as a reason for postponing cost-effective measures to prevent environmental degradation. The principle may play a role in any debate over the need for environmental regulation.

Prevention: The concept of prevention can perhaps better be considered an overarching aim that gives rise to a multitude of legal mechanisms, including prior assessment of environmental harm, licensing or authorization that set out the conditions for operation and the consequences for violation of the conditions, as well as the adoption of strategies and policies. Emission limits and other product or process standards, the use of best available techniques, and similar techniques can all be seen as applications of the concept of prevention.

Public Participation and Transparency: Identified as necessary conditions for "accountable governments...industrial concerns," and organizations generally, public participation and transparency are presented by UNEP as requiring "effective protection of the human right to hold and express opinions and to seek, receive and impart ideas...a right of access to appropriate, comprehensible and timely information held by governments and industrial concerns on economic and social policies regarding the sustainable use of natural resources and the protection of the environment, without imposing undue financial burdens upon the applicants and with adequate protection of privacy and business confidentiality," and "effective judicial and administrative proceedings." These principles are present in environmental impact assessment, laws requiring publication and access to relevant environmental data, and administrative procedures.

Transboundary Responsibility: Defined in the international law context as an obligation to protect one's environment and prevent damage to neighboring environments, UNEP considers transboundary responsibility at the international level as a potential limitation on the sovereign state's rights. Laws that limit externalities imposed upon human health and the environment may be assessed against this principle.

Chapter 3
Innovations of Water Environmental Law

ABSTRACT

Human life and survival on Earth depend on the exploitation of diverse resources, including water. Improper use of environmental resources will lead to pollution and destruction. As one of the most sensitive areas of the environment to which human life depends, water is exposed to a variety of environmental pollutants. The protection of the health of water resources has created the need for intervention and the use of legal and criminal solutions in organizing their use. Domestic penal policy in the field of legislation, inspired by the provisions of Sharia law, along with local and national considerations for the protection of water resources, has directly and indirectly affected the requirements of accession to international instruments and has enacted regulations on the protection of small water resources.

INTRODUCTION

About seven-sevenths of the earth's surface is covered by water. In addition to being rich in food, water also has huge oil and gas resources (Gleick, 1998). For this reason, this vital resource is increasingly being exploited by humans. Extensive use has caused these resources as a large part of the environment to be exposed to the most diverse and serious environmental damage. With this industry, profound changes have taken place in various economic, social and cultural spheres. With the expansion of industries, there was a great need to exploit the natural resources that are used in the fields of energy supply and raw materials, it was felt that in proportion to increasing demand, the rate of human intervention in the natural environment, including water resources, has increased every year. The penetration of various pollutants into natural environments such as water sources increases. Extensive water resources such as springs, streams and small rivers as sources of drinking water have been exposed to many risks due to the influx of industrial and agricultural wastewater and the entry of solid waste from them. Extensive water resources from the tribes and oceans, which have been important since ancient times in terms of transportation and food reserves, are now exposed to flooding with the advancement of technology and as a result of the discovery of mineral and energy reserves (Alan, 2017).

DOI: 10.4018/978-1-6684-7188-3.ch003

The sources that have been identified as the cause and source of water pollution are divided into two categories: land-based and ship-based. Among the types of pollutants, oil and petroleum products have a very large role in creating water pollution. Crude oil is composed of a combination of saturated hydrocarbons, aromatics, resin and asphaltene (Alan, 2017) which is very dangerous for aquatic organisms and the marine environment and humans and causes damage. Serious illnesses become like cancer (Alan, 2017). Areas in which oil reserves are present are very limited and specific, and are usually far from the consumer markets of most industrialized countries, so every year in significant quantities of these materials, on the way to the consumer markets, the amount It is estimated at 320 million tons (Gleick, 1998), enters the water life cycle.

Our country, due to its location in the Persian Gulf oil basin, where most of the world's oil resources are located (Alan, 2017), is facing harmful damages due to the entry of these substances into the biological cycles. In this article, an attempt has been made to review the domestic and international documents to which our country has joined, to review the policies and policies adopted in relation to various types of pollutants, and to ask these two questions.

1. What are the criminal policies of the legislature at the national level in the face of various types of water pollution?
2. What measures have been taken regarding water pollution in the transnational documents to which the Islamic Republic of Iran has acceded?

In the protection of the aquatic environment, one of the most basic issues that must be determined is the concept and content of pollution. The Iranian legislature, in paragraph A of Article 1 of the Law on the Protection of Navigable Seas and Rivers against Contamination with Petroleum Products, approved on 6/31/2010, discharges or contaminates the discharge or the leakage of oil or oil or oil reservoirs or oil reserves. Has been stated. The disadvantage of the above definitions is that instead of a substantive definition that is comprehensive of the types of pollution, it is sufficient to express some examples of pollution (Alan, 2017).

Among the previous laws, we can mention Article 1 of the Law on the Protection of Marine and River Marziaz Contamination of Petroleum Products, approved on 11/14/1975, in which pollution is also mentioned in some cases, such as oil spills or leaks, or Any oil mixture is defined in the terms covered by this law.

This procedure is sometimes current among the regional documents to which the Government of the Islamic Republic of Iran has acceded. For example, in the Convention on the Protection of the Environment of the Caspian Sea (Tehran Convention adopted in 2003 and Implemented in 2006) concluded between the inland countries of the Caspian Sea, under the heading of the first part (general provisions) in Article Carcasses and examples of such are the criteria for defining the discharge of waste and the contamination of the water subject to the Convention (Gleick, 1998).

In terms of content among regional documents, the Kuwait 2 F1 Convention provides a more accurate definition of pollution. According to Article 1: "Marine pollution means the introduction of substances or energy into the marine environment by humans directly or indirectly, which have harmful effects such as damage to damaged resources and danger to human health and obstruction of activities. Marine, such as fishing and damage to quality, in terms of the use of seawater and the reduction of amenities, or the possibility of such risks. This definition is more appropriate in terms of the ways of causing pollution, whether direct or indirect, and also in terms of not paying attention to the intentional or unintentional nature of the pollution in the realization of responsibility, and is more than fair to others. In the definition of Article 4 of the Convention on the Law of the Sea, which was adopted on 10 December 1982 and entered into force on 16 November 1994, it is considered that adaptations have been made. Pursuant to Article 1 (4) of the Convention on Pollution "... the introduction of substances or energy into the environment directly or indirectly by humans, such as riverine tributaries, which may cause serious damage to the environment. And endangers marine life and human health and interferes with marine activities, including fishing and other legitimate uses of the sea, and undermines the quality of marine use and reduces its desirability) (Alan, 2017).

Feuer Bach, the originator of the term criminal policy, defines criminal policy as "the legislative wisdom of the state and the set of repressive methods by which the state demonstrates its reaction against crime" (Alan, 2017). From this perspective, criminal policy, in a limited sense, corresponds to the term "criminal policy" in terms of expressing responsibility. After presenting the definition of each of the schools, in accordance with the principles of thought and practical approaches, they have provided various and different definitions of this term. Among the latter, the definition of death of Delmars Martz seems more appropriate in terms of comprehensiveness and constraint. According to him, "criminal policy includes a set of methods by which the community body organizes the answers to the criminal phenomenon (crime and deviation)" (Alan, 2017). From a practical point of view, criminal policy is divided into categories such as legislative, judicial and participatory criminal policy (Gleick, 1998).

LEGAL MEASURES FOR WATER PROTECTION

Domestic Public Law

Iran is located in hot and dry regions in terms of climate and has faced various periods of drought and water shortage. This issue has become a ground for measures to be taken in Iran regarding water regulation. Due to the need of many people for water and lack of adequate access to surface water, canals have been built to access water. From a glacial point of view, the process of dehydration has led the legislature to require serious protection of water resources. The Civil Code of 1928 assigned articles to the provisions governing rivers, aqueducts, springs, and their sanctuaries, and in the regulation of these articles, the principles of individual ownership and exploitation or participation in private property were accepted. In particular, special attention has been paid to the harmless rule in the development of materials. Following the Civil Code, the Law on Canals was passed in 1930. This law also emphasizes the same principles of the civil law (principles of personal and non-harmful property) and Article 1 stipulates: The executor well will only be me in terms of ownership of the aqueduct and the executor and for the operations related to the aqueduct, and the owner of the property can build a well around the executor well or lands in the well up to the confines of the well and the executor. Do not be an executor. In permissible lands, the incident takes place around the executor's well or between two other wells (Alan, 2017).

Due to the growing importance of water during a gradual process, the need for increased government intervention in infrastructure such as water was felt. This was accelerated by the law establishing the Irrigation Company of 1943. Pursuant to the said article of the law, for the development and reform of the irrigation affairs of the country, an independent enterprise is established under the supervision of the Ministry of Agriculture called the Independent Irrigation Enterprise. The process of the government's strong presence in the field of water management was accelerated by the passage of the 1963

Law on the Establishment of the Ministry of Water and Electricity. In this law, the issue of protection and protection of the aquatic environment is explicitly mentioned and stated in paragraph 1 of Article 1, one of the main goals of the ministry is to monitor the use of the country's water resources, which in an inaccurate and Categories related to various aspects of the environment, such as water management, can be inferred. In 1966, the law on protection and protection of groundwater resources was approved. Article 1 stipulates that the protection and protection of groundwater resources and reserves and the supervision of all matters related to the Ministry of Energy (Alan, 2017).

With the enactment of the Water Law and the manner in which it was nationalized in 1968, water became public and national property, and the legislature, in its position of protection and protection of water resources, explicitly provided for legal mechanisms for water protection. According to Article 1, all irrigation water in rivers, natural streams, valleys, streams and any other natural route, including surface, groundwater, as well as floods, sewers, sewers, sewers, sewers, sewers, sewers, sewers, sewers, sewers, sewers, sewers Springs, mineral waters and groundwater resources belong to the national treasury and belong to the public, and the Ministry of Water and Electricity is responsible for maintaining and exploiting this national wealth and constructing and managing the water resources development facility.

In the seventh chapter, under the title of prevention of pollution of water resources, it was stipulated in Article 55 that water pollution is prohibited and institutions that supply water to municipal, industrial or mineral uses are obliged to submit a water purification and wastewater disposal plan to the Ministry of Water and Wastewater. And the Ministry of Health to prepare and implement. In order to provide an indicator for the detection of polluting actions, the legislature further defines in Article 56 (contamination of water) as "mixing of foreign substances into water to the extent that its physical or chemical or biologically harmful quality is harmful to humans." And be cattle, aquatic and plants, change and then consider foreign materials including petroleum, coal, acid and all kinds of carbon and sewage guilds(Boyd, 2012).

Among the other laws that have been approved in the field of water protection, the Law on Protection of the Sea and Border Rivers from Contamination with Petroleum Products approved in 1975. This law on increasing the criminal protection of the waters subject to the law mentioned in Articles 13, 10 and 14, stated the scope of duties and legal responsibilities of polluters.

Pursuant to Article 1, all ships covered by this law will be required to be insured against possible damage due to pollution of the sea while entering the Iranian coast. Ships that are not insured must carry a financial commitment to compensate. The purpose of the legislator in adopting these strategies was to ensure the payment of possible damages from oil pollution in the basins considered by the legislator. Article 13 stipulates that if the violation of the provisions of this law causes any damage to the ports and sea (beaches) or other coastal facilities of Iran or damage to aquatic and natural resources, the court shall Article 14 stipulated that pollution officials were required to pay all costs incurred by the competent authorities or on their orders and by other agents in order to limit and eliminate the effects of pollution.

The process of protecting water resources and protecting them continued after the victory of the Islamic Revolution. According to Article 45 of the Constitution, seas, lakes, rivers, and other public waters, except Anfal and public wealth, are considered and are at the disposal of the Islamic government to act in their public interest. It is also stated in the fiftieth principle that in the Islamic Republic of Iran, environmental protection, in which present and future generations should have a growing social life, is considered a public duty. Irreparable damage to it is prohibited. According to the principles and frameworks set out in the Constitution, the ordinary laws have been approved by the Islamic Consultative Assembly, which is in line with the objectives of the plan. One of these approvals is the Law on Fair Distribution of Water, which was approved in 1982. This law is one of the most progressive laws regarding the regulation of the use of water. Article 6, under the title of the second chapter of "Groundwater", stipulates that the owners and users of wells or aqueducts are responsible for preventing water pollution and are obliged to act in accordance with health regulations, if they have the power to do so. They are obliged to inform the Environmental Protection or Health Organization (Boyle & Redgwell, 2021).

Article 46 provides: "Pollution of water is prohibited, the responsibility for preventing and preventing the contamination of water resources is assigned to the Environmental Protection Organization."

Assignment of the mentioned matters to the Environmental Protection Organization means violation of Article 1 of the Water Law and the manner of nationalization approved in 1968, because under this article, the Ministry of Water and Electricity is responsible for water conservation and operation.

In order to prevent water pollution subject to the Law on Fair Distribution of Water, the Council of Ministers, based on the proposal of the Environmental Protection Organization and according to Article 46 of the Law on Fair Distribution of Water, approved bylaws with 22 articles in 1994. One of the points that has been approved in the by-laws but does not comply with the general principles of the rule of proper treatment of pollutants, especially water pollution, is paragraph 13 of Article 1, which has the right to: It is said that the management or management of polluting materials sources such as factories, workshops and other industrial facilities, either on its own, or on behalf of another person or natural or legal person, or is solely responsible. Normally, in such cases, the principle is to anticipate the responsibility of natural and legal persons and even the joint and several liability of individuals, so that none of the perpetrators of pollution, whether in the form of stewardship or causation; They are not relieved of any responsibility for their behavior. In addition, more guarantees were provided for the payment and compensation of certain individuals, both legal and natural (Boyle & Redgwell, 2021).

Pursuant to Article 22 of the said By-Laws, if violations of the regulations occur and cause damage to aquatic and natural resources, the court shall, at the request of the Environmental Protection Agency, sentence the authorities to pay and compensate the damages.

In terms of strategic and macro-legal solutions of the country in the field of environmental protection, we will need to adopt solutions in the protection of the natural environment of the country. Note 13 of the single article of the law of the first five-year plan of the Islamic Republic of Iran stipulates that factories and workshops are obliged to spend one per thousand of the sale of their products by recognizing and observing the environmental protection organization. Compensate for pollution and create green space.

In the first part, A (3-6-) Environmental sanitation is defined as one of the tasks of the government. The law of the second five-year plan stipulates in part A (2) (3) that during the implementation of the plan, all economic and social activities must be carried out with environmental considerations in mind, and Note 83 stipulates that in order to Prevention and remediation of water resources by industrial wastewaters and factories in cities and industrial towns, in relation to the creation and operation of water and the organization of subdivisions and network management (Evans, 2014).

In the law of the third five-year plan, in Article 104, in order to reduce the polluting factors of the environment, especially in the natural resources and water resources of the country, the production units are obliged to reduce the expenses incurred for adapting their technical specifications to the environmental standards. Consider biodegradation and pollution reduction as acceptable unit costs. Article 68, paragraph A of the Fourth Plan Law stipulates that the government is obliged to plan protection, restoration, reconstruction of reserves and elimination of pollution, and methods of sustainable exploitation of the country's marine environments until the end of the first year of the fourth economic and cultural development plan. Prepare and implement Iran. In the Fifth Development Plan, also in Article 189 (b), the Environmental Protection Agency is obliged to establish an environmental information system by the end of the fifth year of the Fifth Plan, and in accordance with the provisions of Jim and Dal, Article 193, is obliged to enforce Any industrial and mineral exploitation of water has been strictly prohibited from the second year of the program.

Benefiting from a favorable environment is also one of the priorities in the 20-year vision document.

It is expected by the Iranian society. In 2010, the last legislative policy of the legislator in the Law on Protection of Seas and Rivers capable of Shipping against Contamination of Petroleum Products is reflected in twenty-five articles and ten notes. Article 17 of this law stipulates: Owners, operators and officials of creating pollution subject to this law to compensate all damages caused by pollution and all costs of limiting and eliminating the effects of pollution and environmental monitoring, including the cost of equipment and equipment. Services provided by human agents are jointly and severally liable. Adopting this approach is in line with ensuring the payment of damages to the environment and related facilities (Koskenniemi, 2007).

According to this new environmental measure, the common and multiple responsibility for the total cost of cleaning is placed on the wrongdoers (Orakhelashvili, 2018).

The issue of responsibility is one of the most important and challenging areas of environmental law. In this area, we are witnessing the transition from the old doctrines of civil liability to modern systems of responsibility. The old theories were mostly based on the theory of fault, and the main goal of this approach is to return the situation to the previous one and compensate the damaged party(Orakhelashvili, 2018). Objections and criticisms of this theory due to the lack of complete protection of the environment; Led to the presentation of new theories. Risk-based civil liability theory is one of these theories. According to this theory, there is no need to prove the fault of the polluter in order to fulfill the responsibility, but as soon as the person engages in an activity that leads to environmental pollution and its destruction, he must be liable for the damage, whether or not. The reflection of this principle in the field of environmental responsibilities led to the formation of the "polluter, payer" principle, in which the wrongdoer must accept all the aspects and consequences of his actions and receive damages.

Transnational Documents

Chronologically, the beginning of the period of environmental protection in the international system dates back to half a century ago. The issue of a healthy and appropriate environment has become so important that it has been recognized as one of the examples of human rights from the beginning and has been mentioned in the Stockholm Declaration (Orakhelashvili, 2018). The conclusion of three hundred multilateral treaties and about one thousand bilateral treaties (Koskenniemi, 2007) and seventy international instruments are proof of this resistance. A significant portion of the documents are devoted to the issue of water protection.

TRANS-REGIONAL DOCUMENTS

Convention on the Law of the Sea

The emergence of some disputes between countries over the rights of the exploited powers has led the United Nations Commission of Inquiry to formulate principles for the use of the sea by governments. The actions of this commission led to developments in this field. Finally, on December 10, 1982, the

Convention on the Law of the Sea was ratified in 320 articles in seventeen sections and entered into force in 1994. The government of the Islamic Republic of Iran has signed this convention but has not ratified it (Orakhelashvili, 2018). The most recent achievement of this convention is the twelfth section, which deals with the commitment of states to the protection of the marine environment. Some of the basic provisions of the Convention on the Law of the Sea include Article 192 (Commitment of States to the protection and preservation of the environment); Article 193 (Rights of States in the Exploitation of Natural Resources); Article 194 (Obligation of states to take the necessary measures to prevent and protect the environment) and Article 195 (non-conversion of pollution from one type to another). In the twelfth chapter, it introduces a comprehensive framework of comprehensive measures for marine pollution control (Koskenniemi, 2007).

Ship Pollution Prevention Convention (MAPOL1F)

Contamination from ships can be one of the ways oil or chemicals and sewage enter the water. Oil pollution accounts for 71% of ship-based marine pollution. The annual rate of oil spill in the oceans is estimated at one million tons of oil discharged in standard oil activities and 200,000 tons of oil released as a result of oil tanker accidents per year (Klein et al., 2009). The International Maritime Organization had investigated these pollutions before the Convention on the Law of the Sea.

The mission of this organization is to promote healthy shipping as well as to apply and observe the rules of the environment in the watershed (Godden, 2005). The organization's efforts to develop comprehensive measures on marine pollution caused by oil led to the ratification of the International Convention for the Prevention of Pollution from Ships, known as the Marple; This convention was ratified in 1973 and amended in 1978 and finally entered into force in 1983. Iran has also formally signed and ratified the Convention (Benjamin et al., 2004).

The importance of the Convention is linked to its annexes, and the annexes to the Marple Convention deal with a variety of issues, including pollution by oil, toxic liquids, hazardous substances, and wastewater. In clauses 4 and 5 of this convention, the state is required to take appropriate steps to enforce the provisions of domestic law for the implementation of the provisions of the convention. Clause 6 allows Member States to inspect foreign vessels entering the territorial waters of a coastal State. This inspection is to control how hazardous materials are discharged by the ship. Clause 7 emphasizes that the inspection must be efficient in order to avoid undue and inappropriate delays and unnecessary seizures of the ship (Malanczuk, 2002).

Convention on Oil Pollution Preparedness and Response (OPRC)

This Convention is the result of the actions of the International Maritime Organization concerning shipwrecks and maritime emergencies that could lead to marine pollution and widespread environmental degradation(Goldfarb, 2020). This convention was ratified in 1990 and entered into force in 1995. In 1997, the Government of Iran acceded to the above-mentioned Convention. Increasingly, the national duty to protect the marine environment in the event of an accident has led to oil pollution. The territory covered by this national plan includes the coastal area and all waters under the supervision and rule of the Islamic Republic of Iran in the Persian Gulf region, the Oman Sea and the Caspian Sea. This plan has been ratified in accordance with paragraph 3 (2) of the Convention, which requires members to establish national systems and national plans to respond promptly to accidents leading to oil pollution (Malanczuk, 2002).

Kuwait Convention Regional Documents

The Persian Gulf region has been recognized as a semi-partisan sea by the International Maritime Organization and the United Nations Environment Program under the Convention on the Law of the Sea (Weissbrodt & Kruger, 2017). Article 60 (a) 211 of the Convention on the Law of the Sea, coastal States have the right to determine the specific and sensitive areas of their seas by providing details. In terms of convention; Special areas are points that, for reasons known to be associated with oceanic, ecological, and maritime traffic conditions, as well as due to the utilization and protection of their resources due to the need to avoid special measures in order to prevent them from being prevented. Widespread exploitation of oil tankers and ships is exposed to various types of environmental damage. The countries of Iran, Kuwait, Bahrain, Iraq, Qatar, and the United Arab Emirates ratified the Convention on April 24, 1978, with the aim of cooperating in the protection of marine habitat. It was implemented by the Ministry of Foreign Affairs of Kuwait (Rourke, 2008).

Following the ratification of this Convention, other protocols entitled "Protocol" on Marine Pollution from Exploration and Extraction from the Offshore Plateau "Protocol" on the Protection of the Marine Environment from Pollution from Land-Based and Red Crescent Resources . The government of the Islamic Republic of Iran has ratified all of them and only signed the maritime transport protocol.

According to Article 2, the provisions of the Convention do not apply to the inland waters of the members. According to Article 3, one of the key objectives of the Kuwait Convention is to reduce and combat pollution of the marine environment in the region. Pursuant to Article 16, a regional organization for the protection of the marine environment, called Rapmi, has been established for the implementation and implementation of the requirements of the Convention, which is based in Kuwait (Weissbrodt & Kruger, 2017).

Another center established under the Kuwaiti Convention is the Center for Mutual Cooperation at Sea (MEMAC), based in Bahrain. In addition to the two organizations, due to the importance of protecting the marine environment and taking timely action in the event of accidents leading to oil pollution, the necessary cooperation and coordination between the countries of the region in 2003 became the Center for Cooperation in Resolving Oil(ORC). The important task of this center will be to provide related measures to prevent and deal with oil spills at sea through cooperation between states, oil companies, and international competition organizations based on the principles of prudence and clean-up(Rourke, 2008).

Tehran Convention

The Caspian Sea, as the largest lake on earth, has a special environmental importance due to its special ecological location and is almost rectangular in shape (Rourke, 2008). The length of the Caspian Sea coastline is about 7000 km, of which about 1000 km from Astara to Atrak is one of the coasts of Iran (Boyd, 2012). Due to the closure of its aquatic ecosystem, their environment is always exposed to the dangers of environmental pollutants. Among the factors that severely threaten the health of the Caspian Sea are oil pollutants. In the latest assessments, the amount of oil reserves has reached more than 50 million networks (Rourke, 2008). In order to cooperate between the countries of the Caspian Sea basin in the field of environmental issues, on November 4, 2003, in Iran, the Tehran Convention was signed between the governments of Iran, Russia, Kazakhstan, Turkmenistan and Azerbaijan. The text of the Convention mentions eight protocols. One of the protocols is regional cooperation to deal with oil pollution in emergencies. In Article 2, the purpose of the Convention is to protect the environment of the Caspian Sea from all sources of pollution and to protect and maintain, rationally and sustainably use its living resources(Boyd, 2012).

Pursuant to Article 5, the Contracting Parties shall, in order to achieve the objectives of the Convention, adhere to principles such as the precautionary principle, the polluting principle, which is one of the new principles of environmental pollution, and the principle of access to information on environmental pollution. The types of contamination are also specified according to the source (Articles 7, 8, 9, 10, and 11).

CRIMINAL MEASURES

Human use of punishments to protect the values that are necessary to preserve the core resources of individuals and society dates back to several thousand years ago. However, it should be emphasized that the task of criminal law is not always to acknowledge, strengthen and protect the higher values of society, which are often observed by the people without resorting to punishment (Chikozho et al., 2018). This right can also control behaviors through criminalization and punishment and, as a result, produce values and norms that are valued and observed in accordance with the policies and strategies of the ruling party on social, political and economic issues of the society (Ball & Bell, 1994). At present, access to a healthy environment has been recognized as one of the collective values in many constitutions of the country. Due to the fact that water resources in terms of size and importance can be divided into two types of limited water resources such as springs and streams and small rivers and large water resources including rivers can be navigable and seas, therefore, in this regard. Related penalties are reviewed (Tan & Jackson, 2013).

Criminal Protection of Limited Water Resources

In 1968, with the passage of the Water Law and its nationalization, the legislature transferred ownership of water from individuals to the government, and the use and exploitation of water also became subject to new conditions. In the eighth chapter of the law, as offenses and crimes, some measures were criminalized and some punishments such as fines, blocking wells and disciplinary imprisonment were determined (Articles 59, 60, 61). Paragraph 5 of Article 60 stipulates that anyone who intentionally contaminates river water, public rivers, streams, reservoirs, springs, aqueducts and wells by adding foreign matter as provided in Article 56 of the law, in cases where the source of water is water (Tarlock, 1975). The perpetrator will be prosecuted under other laws. In this material, water contamination is the mixing of foreign substances with water to the extent that its physical, chemical or biological quality is harmful to humans, livestock, aquatic animals and plants. One of the other laws that has taken a criminal approach to the prevention and prevention of pollution, especially water, is; The Law on Environmental Protection and Improvement was approved in 1974. According to Article 1, protection, improvement and improvement of the environment, prevention and prevention of any pollution and any destructive action that causes the balance and appropriateness of the environment to be disturbed, as well as all matters related to the organization's environment, wildlife and conservation of aquatic life (van Rijswick, 2010).

Pursuant to Article 11, the Environmental Protection Agency, in compliance with the relevant regulations, identifies the factories and workshops that cause environmental pollution and notifies them in writing to eliminate the defects within a certain period of time (Naff & Dellapenna, 2017). If the persons concerned do not stop the notification of the prohibited work or activity, their continuation of their activity will be subject to the opinion of the court and in case of violation, they will be sentenced to imprisonment for one day to one year or a fine or both (van Rijswick, 2010).

At a glance, we can deduce the dispersion and incoherence of legislative legislative policies on water pollution. It is clear that one of the secrets of success in criminal legislation is coherence in the enactment of laws, which is not observed in the resolutions, and in some cases the legislature has only passively criminalized some matters. With the enactment of the Law on Fair Distribution of Water in 1982, despite the climatic situation of the country and the growing need for water resources and the growing problems that polluted water has created for the health of human beings (van Rijswick, 2010), only the case mentioned in Article 46 in It is only stipulated that "water pollution is prohibited, the responsibility for preventing and preventing the pollution of water resources is assigned to the Environmental Protection Organization". Unfortunately, this regulation has not improved compared to the previous laws. It should be noted that the legislature has not provided any enforcement guarantees for the actions of individuals who attempt to pollute water resources and has reduced the article to the level of moral advice. This legislative negligence becomes even more pronounced when in Article 45, for persons who seize another's water right without a license, the punishment of restitution and compensation and the punishment of 10 to 50 lashes or 15 days of disciplinary imprisonment Prediction; However, no metaphor has been envisaged for water contaminants (van Rijswick, 2010).

Finally, in the Islamic Penal Code adopted in 1996 on the criminalization of actions that cause environmental pollution in the watershed, we see the adoption of measures much heavier than the previous laws (Baker, 2009).

Article 688 of the Islamic Penal Code stipulates that any action that is considered a threat to public health, such as contaminating drinking water, or distributing contaminated drinking water, dumping poisonous substances in the river, is considered a major cause of death. They will be sentenced to one year. The diagnosis is made by the Ministry of Health and the Environment Organization. The legislature has enumerated some examples as an allegory, which can be useful in developing examples and similar cases that can be subject to the mentioned punishments (Sherk, 1990).

In Article 689, the legislator has also determined the damages to persons and if the actions mentioned in Article 688 that lead to murder or mutilation or other injuries, as the case may be, are subject to the permissions of the storytellers and the payment of diyat and total payment. In general, these two substances can be useful in combating pollution caused by limited water resources because they deal with both prevention and punishment and include both stages (Eckstein, 2009).

Criminal Protection of Large Water Resources

This law was approved in 1975 regarding the border rivers and inland waters and lands of Iran in the Oman Sea, the Persian Gulf and the Caspian Sea. According to Article 2 of the said law, contamination of the expressed water with oil or any kind of oil mixture by vessels and other oil facilities located on land or at sea is prohibited, and those who commit such crimes will be imprisoned and fined or fined. Two sentences are sentenced. If the contamination is due to negligence or carelessness, the contaminant will be fined. One of the notable points is the conviction of a person to the prescribed penalties in both cases of intentional and unintentional commission, which is one of the salient points considered and in line with the principles governing the determination of responsibility in environmental crimes (Adler, 2010).

A weakness that is considered in this law and in other environmental laws is the excessive use of imprisonment and fines by the legislature. Nowadays, environmental crimes are often committed by legal entities, so it seems more appropriate to use the legal and professional enforcement of alternatives to imprisonment. Today's criminal policy is, on the one hand, to reject the severity and, in particular, the subordination and deprivation of the criminal rights of convicts in the past, which is generally incompatible, and on the other hand, considering the economy. The circumstances of the offender and the social environment seek to establish a system quite different from prohibitions or restrictions that is capable of supporting both the potential victims of certain economic or professional activities and those of the perpetrators themselves (Wilkinson, 2005).

These punishments are usually complementary and optional, but can be foreseen in some cases as such (Adler, 2010). Environmental offenses include penalties such as commercial restrictions related to pollution or revocation of licenses, descriptive penalties that reduce the popularity of companies and corporations, and penalties such as those that increase social costs. And the job of these people can be used. Article 6 provides for exceptions to the absolute liability set out in Article 2. According to Article 6, the penalties prescribed in the article, in case the pollution is necessary in order to eliminate the danger from the ship or to save the lives of the persons; Provided that the proportion between the contamination with the hazard threatening the ship or its occupants is observed and immediate measures to eliminate the contamination are unimpeded. In the case of unintentional contamination that has not been predicted as a result of factors and events, immediate measures should be taken to prevent and eliminate its effects.

In Article 12, the officers of the Ports and Maritime Organization will be officers within the limits of their duties, provided that they are assigned by the respective organization and pass special courses for officers, under the supervision of the prosecutor and obtaining a certificate for passing these courses (Shupe, 1982). Pursuant to Article 17, the time of a claim for damages is six years from the date of the claim. This law has been repealed by the Law on the Protection of Navigable Seas and Rivers, which will be examined in the next section (Wilkinson, 2005).

Law on protection of navigable seas and rivers against pollution by petroleum products:

The mentioned law was approved in 2010. Certainly, the increasing spread of oil pollution in our country's marine basins has been motivated until after 35 years; A law similar to the Law on the Protection of Marine and Border Rivers from Oil Pollution, passed in 1975, should be enacted. The scope of application of both laws is the same and includes the law of Iran's maritime regions in the Persian Gulf and the Sea of Oman. It should be said that in the law mentioned in paragraph 6 of Article 6, the creation of any pollution of the marine environment, contrary to the provisions of the Islamic Republic of Iran, is considered a crime and subject to criminal punishment and civil liability (McIntyre, 2010).

Considering that most of the articles in the 2010 law are repetitions of the articles of the previous law, only those articles that are not mentioned in the previous law are analyzed (Tarlock, 1990). Article 4 criminalizes the misappropriation and non-performance of legal duties and the registration of false information in the oil registration book by the officials of the ship, tanker, platform and oil facilities in the case of criminalization and imposes fines and fines for it (Blumm, 1998). The crime of falsification and recording of untrue material is in accordance with the criminal titles of forgery of material and provisions in the Islamic Penal Code (Elver, 2007). Therefore, if a person commits these crimes, he will be fined for multiple crimes (spiritualizing an act and violating several laws) and according to the obligation of Article 46 of the Islamic Penal Code, if he has several titles in the punishable crimes of a single act. The punishment is more severe. Pursuant to Article 10, the above-mentioned officials, in case of contamination for any reason, should inform the competent authorities as soon as possible and be sentenced to administrative punishment and fines (Adler, 2010).

Among other innovations of the law is the provision of performance guarantees for the duties assigned to government officials in order to prevent the spread of infection and to deal with violators, which is a combination of administrative penalties and fines (Ball & Bell, 1994). One of the positive and novel points of the 2010 law is the prediction of a three-member expert board to diagnose and determine the appropriateness of actions with risks, the effectiveness of the actions taken with the materials that have resulted in the risk and the risk of harm and loss. which will play an effective role in making the decisions taken by the relevant authorities more objective.

CONCLUSION

The protection and preservation of large and small water resources has been going on for a long time and governments and international organizations due to the growing need for water and its scarcity, population growth and the threshold of high vulnerability of all kinds of water. National, regional and international levels have adopted solutions. In Iran, despite the severity and abundance of environmental hazards that threaten the country's water resources, so far only in some laws, legal and criminal measures have been taken to regulate and protect water resources. Certainly, passing legislation that is inspired by local and global strategies for the protection of the aquatic environment can be very effective. The existence of a specialized and comprehensive law in this regard, certainly in the degree of efficiency and increase the capability of the responsible agencies, in the protection of the above-mentioned objectives and the strengthening of inter-sectoral cooperation, which is one of the requirements of the restrictive watershed. One of the problems in the field of environmental protection is the non-institutionalization of environmental protection, as a fundamental and pivotal value in the relevant and responsible institutions, both legislative and judicial. Resolving this problem requires educational and cultural work at the sixteenth level to explain the position and role of the country's natural resources, especially the aquatic environment.

REFERENCES

Adler, R. W. (2010). Climate change and the hegemony of state water law. *Stan. Envtl. LJ*, *29*, 1.

Alan, B. (2017). *Human rights and the environment: where next?* Routledge.

Baker, L. A. (Ed.). (2009). *The water environment of cities*. Springer. doi:10.1007/978-0-387-84891-4

Ball, S., & Bell, S. (1994). *Environmental Law*. The Law and Policy Relating to the Protection of the Environment.

Benjamin, A. H., Marques, C. L., & Tinker, C. (2004). The water giant awakes: An overview of water law in Brazil. Tex L. *Rev.*, *83*, 2185.

Blumm, M. C. (1988). Public property and the democratization of western water law: A modern view of the public trust doctrine. *Envtl. L.*, *19*, 573.

Boyle, A., & Redgwell, C. (2021). *Birnie, Boyle, and Redgwell's International Law and the Environment*. Oxford University Press. doi:10.1093/he/9780199594016.001.0001

Chikozho, C., Saruchera, D., Danga, L., & da Silva, C. (2018). *A Compendium of the South African water law review post-1994*. Water Research Commission.

Eckstein, G. E. (2009). Water scarcity, conflict, and security in a climate change world: Challenges and opportunities for international law and policy. *Wis. Int'l LJ*, *27*, 409.

Elver, H. (2008). International environmental law, water and the future. In *International Law and the Third World* (pp. 191–208). Routledge-Cavendish.

Evans, M. D. (Ed.). (2014). *International law*. Oxford University Press. doi:10.1093/he/9780199654673.001.0001

Gleick, P. H. (1998). The human right to water. *Water Policy*, *1*(5), 487–503. doi:10.1016/S1366-7017(99)00008-2

Godden, L. (2005). Water law reform in Australia and South Africa: Sustainability, efficiency and social justice. *Journal of Environmental Law*, *17*(2), 181–205. doi:10.1093/envlaw/eqi016

Goldfarb, W. (2020). *Water law*. CRC Press. doi:10.1201/9781003069829

Klein, C. A., Angelo, M. J., & Hamann, R. (2009). Modernizing water law: The example of Florida. *Florida Law Review*, *61*, 403.

Koskenniemi, M. (2007). The fate of public international law: Between technique and politics. *The Modern Law Review*, *70*(1), 1–30. doi:10.1111/j.1468-2230.2006.00624.x

Malanczuk, P. (2002). *Akehurst's modern introduction to international law*. Routledge. doi:10.4324/9780203427712

McIntyre, O. (2010). The proceduralisation and growing maturity of international water law: Case concerning pulp mills on the river Uruguay (Argentina v Uruguay), International Court of Justice, 20 April 2010. *Journal of Environmental Law*, *22*(3), 475–497. doi:10.1093/jel/eqq019

McIntyre, O. (2016). *Environmental protection of international watercourses under international law*. Routledge. doi:10.4324/9781315580043

Naff, T., & Dellapenna, J. (2017). Can there be confluence? A comparative consideration of Western and Islamic fresh water law. In *International Law and Islamic Law* (pp. 281–305). Routledge. doi:10.4324/9781315092515-15

Orakhelashvili, A. (2018). *Akehurst's Modern Introduction to International Law*. Routledge. doi:10.4324/9780429439391

Rourke, J. T., & Boyer, M. A. (2008). *International politics on the world stage*. McGraw-Hill.

Sherk, G. W. (1990). Eastern water law: trends in state legislation. *Virginia Environmental Law Journal*, 287-321.

Shupe, S. J. (1982). Waste in Western Water Law: A Blueprint for Change. Or. *Law Review*, *61*, 483.

Tan, P., & Jackson, S. (2013). Impossible dreaming: Does Australia's water law and policy fulfil Indigenous aspirations. *Environment and Planning Law Journal*, *30*(2), 132–149.

Tarlock, A. D. (1975). Recent developments in the recognition of instream uses in western water law. *Utah L. Rev.*, 871.

Tarlock, A. D. (1990). Western Water Law, Global Warming, and Growth Limitations. *Loy. LAL Rev.*, *24*, 979.

van Rijswick, M. H. (2010). Interaction between European and Dutch water law. In *Water Policy in the Netherlands* (pp. 218–238). Routledge.

Weissbrodt, D., & Kruger, M. (2017). Norms on the responsibilities of transnational corporations and other business enterprises with regard to human rights. In *Globaization and International Investment* (pp. 199–220). Routledge.

Wilkinson, D. (2005). *Environment and law*. Routledge. doi:10.4324/9780203994443

ADDITIONAL READING

Malanczuk, P. (2002). *Akehurst's modern introduction to international law*. Routledge. doi:10.4324/9780203427712

Naff, T., & Dellapenna, J. (2017). Can there be confluence? A comparative consideration of Western and Islamic fresh water law. In *International Law and Islamic Law* (pp. 281–305). Routledge. doi:10.4324/9781315092515-15

Orakhelashvili, A. (2018). *Akehurst's Modern Introduction to International Law*. Routledge. doi:10.4324/9780429439391

Rourke, J. T., & Boyer, M. A. (2008). *International politics on the world stage*. McGraw-Hill.

Sherk, G. W. (1990). Eastern water law: trends in state legislation. *Virginia Environmental Law Journal*, 287-321.

Shupe, S. J. (1982). Waste in Western Water Law: A Blueprint for Change. Or. *Law Review*, *61*, 483.

Tan, P., & Jackson, S. (2013). Impossible dreaming: Does Australia's water law and policy fulfil Indigenous aspirations. *Environment and Planning Law Journal*, *30*(2), 132–149.

Tarlock, A. D. (1975). Recent developments in the recognition of instream uses in western water law. *Utah L. Rev.*, 871.

Tarlock, A. D. (1990). Western Water Law, Global Warming, and Growth Limitations. *Loy. LAL Rev.*, *24*, 979.

van Rijswick, M. H. (2010). Interaction between European and Dutch water law. In *Water Policy in the Netherlands* (pp. 218–238). Routledge.

Wilkinson, D. (2005). *Environment and law*. Routledge. doi:10.4324/9780203994443

KEY TERMS AND DEFINITIONS

Equity: Defined by UNEP to include intergenerational equity—"the right of future generations to enjoy a fair level of the common patrimony"—and intragenerational equity—"the right of all people within the current generation to fair access to the current generation's entitlement to the Earth's natural resources"—environmental equity considers the present generation under an obligation to account for long-term impacts of activities and to act to sustain the global environment and resource base for future generations. Pollution control and resource management laws may be assessed against this principle.

Polluter Pays Principle: The polluter pays principle stands for the idea that "the environmental costs of economic activities, including the cost of preventing potential harm, should be internalized rather than imposed upon society at large." All issues related to responsibility for environmental remediation costs and compliance with pollution control regulations involve this principle.

Precautionary Principle: One of the most commonly encountered and controversial principles of environmental law, the Rio Declaration formulated the precautionary principle: To protect the environment, the precautionary approach shall be widely applied by States according to their capabilities. Where there are threats of serious or irreversible damage, lack of complete scientific certainty shall not be used as a reason for postponing cost-effective measures to prevent environmental degradation. The principle may play a role in any debate over the need for environmental regulation.

Prevention: The concept of prevention can perhaps better be considered an overarching aim that gives rise to a multitude of legal mechanisms, including prior assessment of environmental harm, licensing or authorization that set out the conditions for operation and the consequences for violation of the conditions, as well as the adoption of strategies and policies. Emission limits and other product or process standards, the use of best available techniques, and similar techniques can all be seen as applications of the concept of prevention.

Public Participation and Transparency: identified as necessary conditions for "accountable governments...industrial concerns," and organizations generally, public participation and transparency are presented by UNEP as requiring "effective protection of the human right to hold and express opinions and to seek, receive and impart ideas...a right of access to appropriate, comprehensible and timely information held by governments and industrial concerns on economic and social policies regarding the sustainable use of natural resources and the protection of the environment, without imposing undue financial burdens upon the applicants and with adequate protection of privacy and business confidentiality," and "effective judicial and administrative proceedings." These principles are present in environmental impact assessment, laws requiring publication and access to relevant environmental data, and administrative procedures.

Transboundary Responsibility: Defined in the international law context as an obligation to protect one's environment and prevent damage to neighboring environments, UNEP considers transboundary responsibility at the international level as a potential limitation on the sovereign state's rights. Laws that limit externalities imposed upon human health and the environment may be assessed against this principle.

Chapter 4
Globalization and Environmental Justice

ABSTRACT

Globalization is a growing and unstoppable process that began with integrating the global economy and free trade. It also affects various fields besides economics and is one of the most important sectors affected by the environment. Some believe that economic activities may pose a serious threat to the environment. On the other hand, some believe that economic growth is needed to achieve a healthier environment with less pollution. This chapter aims to examine some of the opportunities and threats of globalization to the environment and the possible positive or negative effects of economic growth resulting from the globalized economy. Also, solutions to reduce the negative effects of globalization on the environment include paying special attention to sustainable development, supporting environmentally friendly technologies, reforming energy consumption patterns, encouraging renewable and clean energy, imposing carbon taxes, and environmental labels suggested.

INTRODUCTION

Today, globalization has become one of the most important issues facing human beings. Although many years ago, it began its growing trend in human life, following the world's industrialization and expanding communications and related technologies, rapidly Has found impressive. Since capitalism entered the world as a sustainable form of human society, globalization has been going on for four or five centuries (Sze, 2006).

There are various definitions of globalization, some of which are mentioned. Larsen sees globalization as a kind of shrinking world and shortening distances. A phenomenon that is conducive to the expansion of welfare causes one person on one side of the world to interact with the other on the other side of a mutually beneficial relationship. According to Giddens, globalization is the expansion and strengthening of global relations, which will cause internal events in one area to affect the situation in other areas and vice versa. In his research, Samian points to three other definitions: Waters sees globalization as a social process in which the geographical constraints that overshadow social and cultural relations are

DOI: 10.4018/978-1-6684-7188-3.ch004

removed, and people increasingly reduce this constraint. And the clauses become aware; Harvey sees globalization as the density of time and space so that every event in every corner of the globe is quickly communicated to everyone, and practice time and space have lost their meaning. Finally, Henderson sees globalization as the integration of markets and defines conditions for it, including the free movement of goods, services, labor, capital, and creating a single market. Economically, no one is a foreigner, and everyone is equal in the market.

Globalization is the compression of the world and the intensification of awareness of the world as a unit. Globalization can be understood in terms such as internationalization, liberalization - in the sense of removing the restrictions imposed by governments on activities between countries to create a free and borderless world economy, globalization - in the sense of dissemination of various experiences and goals for people in the four corners of the world interpreted Western-style modernity in terms of Westernization, especially in its Americanization form, as well as territorialization. Globalization is a multifaceted phenomenon so that it has found its way into various structures of social, economic, political, cultural, military, technological, and environmental action.

The historical stages of globalization have been the movement from large to small. In the early stages of globalization, nations had to think global to flourish or at least survive. In the second phase of globalization, the companies had to think globally to thrive or survive. Finally, in stage three, globalization, people must think global to flourish or at least stay. In other words, globalization has reached the globalization of the individual from the stage of globalization of industry.

One of the most important areas of influence in globalization is the field of the environment. The environment refers to the set of living and non-living elements that comprise an organism or group of organisms, and the environment is virtual all the conditions, components, and factors that affect the growth and development of an organism - whether living or non-living, such as light, temperature, and climate.

The process of globalization began to increase international trade and reduce trade tariffs. As a result, global studies show that global trade has increased significantly during this period, while this growth is followed by carbon dioxide emissions, which play an important role in global warming and climate change.

From an economic point of view, this process can harm the environment in several ways (Aukusti Lehtinen, 2006):

Elimination of tariffs will allow goods to be traded at a lower cost, and demand for goods produced in the country with a greater comparative advantage will increase. In this case, the producing country, by increasing its production, will try to export the maximum of the above product in the competitive market. However, high production puts additional strain on the country's raw materials, and in some cases, improper harvesting of forests, the sea, agricultural land, rivers, and mineral resources will have irreversible effects on the environment.

The enticing benefits of increased production and free trade in the globalized space encourage manufacturers to produce more using any technology. However, a completely economical and unequivocal approach will lead to using methods and technologies that emit greenhouse gases and air pollutants, discharge industrial effluents into rivers, and ignore human health's adverse effects on producing goods noise pollution the like.

If not global consensus, there is at least a relative consensus that globalization directly affects the economy. Concerns arise when it comes to "non-economic effects", of which the environment is the most important.

All the inhabitants of the earth have an equal share in the benefit of the blue planet. Therefore, the earth and everything in it is considered a kind of human property. However, some argue that unless we are personally confronted with the costs of our type of performance, we will continue to pollute shared climate resources and overuse natural resources.

The question is, is globalization effective in achieving a balance between economic and environmental goals? Does it damage or improve this process? Will international trade and investment allow countries to achieve higher economic growth while maintaining the environment at the appropriate level, or will they have to destroy the environment to achieve a level of economic growth?

In the present study, the problem is first stated. Then, in the following, the Kuznets environmental curve and its discussion are presented. Then the effects of globalization on the environment and, finally, some appropriate strategies in adopting appropriate policies on the environment and globalization are mentioned.

BACKGROUND

According to their supporters, if we review the history of forces opposed to globalization, environmental activists have always played a significant role. Proponents of the environment have long thought that the economy's globalization is a threat to the environment. Economists have defended it. This disagreement is mainly due to the completely different views and lifestyles of economists and environmental activists. In other words, economists generally believe that nature is at the service of humanity; However, if not all environmentalists, at least a significant part of them reject this view and emphasize the independence of nature (Agyeman et al., 2016).

Meanwhile, some critics of the globalization of economics believe that the World Trade Organization (WTO), in the name of free trade, hinders countries' protection and simultaneously harms the global environment and countries' sovereignty. Even today, in the best of circumstances, humans are destroying their little planet with more garbage, more heat, and more smoke in a global invasion of natural resources faster than ever before in history. The important issue here is the number of inhabitants of the planet and the impact of their particular way of life on the environment. It is clear that if the human population were still only a few million people migrating and their occupation was nothing but hunting and fishing and they still knew nothing about agriculture, the level of pollution of human origin, at least or with a little negligence, would be nothing. However, this is not the case. Today, human beings face many problems in the field of the environment. For example, in some cases, global warming - it is possible that the time will come when it is too late for any compensatory action.

At the beginning of the third millennium AD, human societies faced new and irreplaceable challenges other than globalization; For example, no one can definitively deny the greenhouse effect. It may not be an exaggeration to say that the greatest human challenge of the present century - at least in the field of the environment - is global warming. Global warming will continue, and we will see a huge change in world climate. The level of energy consumption in the world has increased, and if no effort is made to reduce dependence on fossil fuels and energy efficiency is not increased, greenhouse gas emissions will be higher than they already are, which results in global warming and the devastating consequences that follow will not happen. In the meantime, each part of human activities has a different share in greenhouse gas emissions.

Increasing acid rain due to sulfur dioxide and nitrogen oxide are other problems that threaten the global environment's health; It is predicted that by 2030, the emission of gases that are the source of these rains in Asia will double. Increasing the amount of nitrogen from fertilizers and wastewater can disrupt the nitrogen cycle and affect fertile lands, lakes, rivers, and coastal waters is another environmental challenge facing human beings today. Between 1960 and 2015, almost one-fifth of the world's tropical forests were destroyed; Although in rich countries, the level of forest cover has remained constant or increasing, in underdeveloped countries, the process of deforestation to create agricultural land continues. It is important to note that deforestation, deforestation, and pollution can threaten biodiversity in many areas.

The growing trend of coral reef degradation and marine aquatic species' threat following widespread fishing methods threatens the seas (Pellow & Brulle, 2005).

Another crisis facing humanity is the water crisis. The use of freshwater resources in the world is increasing rapidly. To the extent that some predict competition for access to water resources will even be the source of wars in the 21st century. Currently, on average, one out of every three people in the world lives with a water shortage, and this figure will change to two out of every three people within the next twenty years, i.e., by 2040.

This raises the question of the extent to which globalization plays a role in degrading the environment. How many environmental problems are internal and local, and what are these problems due to globalization. Environmental degradation is not just a local phenomenon within borders or regionally but occurs globally. Some environmental problems, especially in the field of pollutants and their consequences, such as rising global temperatures, ozone depletion, and acid rain, are considered external influences for some countries; Although these effects are the result of the actions of one country within its sovereignty, or occur in global commonalities where no country is sovereign, on a large scale, they involve several countries and in some cases the whole world. Obviously, in transboundary pollution, the harmful effects of these pollutants and the costs of controlling them do not belong to the citizens of a particular region or a particular government.

International law has also highlighted several international conflicts. One of the biggest controversies is managing globally shared resources such as the oceans, the atmosphere, the ozone layer, the climate, and so on. These conflicts are reflected in issues such as fishing in open waters, migration and population movement, pest control, and the exploitation of other natural resources.

High consumption due to rising incomes is itself a potential threat to the environment. Intensification of competition also leads to a "race that continues to the brink of extinction" and creates a "haven" for pollution. Governments may even lower their environmental standards to achieve a relative advantage. Therefore, industrialization is very fast in an environment created by globalization, but incomes are still low, environmental degradation is possible.

The growing use of human resources will prevent species and ecosystems from adapting rapidly to these conditions; Thus, much of the earth's biodiversity will be lost. The result will be transforming natural ecosystems into agricultural and urban lands, and eventually, global warming. Even many species that are currently under threat will become extinct (Faber & McCarthy, 2012).

Undoubtedly, the economic growth resulting from a free and global economy affects the environment. There are two main views between the proponents and the opponents. First, some believe that according to the Kuznets "green" or "environmental" curve, more capital inflows and increased economic growth in southern countries will ultimately result in environmental benefits. This chart shows the relationship between economic development and the severity of environmental pollution. Although environmental pollution is relatively low in developing countries, the higher the economic growth, the higher the pollution level until it reaches a maximum. In contrast, in industrialized countries, the environment's costs are reduced. In other words, although the environmental situation worsens at first, over time, following economic growth and improving the quality of life of the people, the quality of the environment also improves, and the demand for the favorable environmental situation increases.

In contrast, another group - critics of the Kuznets green curve - believe that this U-shaped diagram will eventually change to an N-shaped diagram. In this way, following the increase in economic growth, any improvement in the environment's state is neutralized (Walker, 2009).

Besides, some believe that strict environmental regimes in industrialized countries are causing industries that are increasingly polluting, especially dirty ones, to relocate from rich, industrialized countries to neighboring and developing countries. This hypothesis states that manufacturing industries are more concentrated wherever environmental laws or enforcement are weak. In other words, these areas become a haven for pollution. But, on the other hand, this issue confirms the impact of environmental standards on these industries' location.

DISCUSSION

Water, land, forests, and biodiversity are vital human assets that can be catastrophic if they are not adequately protected internationally. An important part of these disasters will be on the countries of the tropics. Climate change and environmental degradation, along with the damage to agricultural production following global warming and the eventual spread of certain tropical diseases, will be the consequences of this neglect.

It is clear that both incomes are important in the economy, and the environment's desired quality can not be ignored. Although the advocates of each differ on how important they are and their share, the result should not be a political stalemate in which both the environment and the economy are worse off.

Environmentalists have the right to argue that economic growth should consider the extent of the damage done to the environment. This is where the importance of environmental valuation comes into play. However, it is important to note that extreme valuations should be avoided in valuations, except in a few cases where it seems reasonable to overemphasize the preservation of the environment. In other words, the optimal situation is a situation in which some commercial benefits are obtained, and some environmental damage occurs.

As mentioned earlier - based on the Kuznets green curve - some believe that economic growth primarily harms the environment in developing countries. Still, over time, as the economic situation improves and people become richer, people themselves tend to clean up the pollution they have caused. London can be mentioned as an example. Nearly half a century ago, London's air was very heavy due to the smoke from burning coal. But now that the city is much richer than before, the air is cleaner than before. Despite the heavy traffic and the large number of cars that travel in the city every day, the air is far more suitable for breathing due to unleaded gasoline in the cars. This situation can be explained by the Kuznets environmental curve. Assuming a positive relationship between globalization and economic growth, according to some indicators in the early stages of economic growth, the environmental situation will first deteriorate; But in the next stage (Carruthers, 2008).

"In backward areas—like Nepal—the only fuel available to the people is the wood of trees. The use of trees for fuel has destroyed forests and farmland," says Dasgupta Parta of Cambridge University. Erodes ultimately lead to environmental degradation, which will lead to increasing poverty. In other words, poverty can be a major cause of environmental pollution, and on the other hand, if the environment is degraded, it will itself increase poverty (Stiglitosis); briefly, it is the decrease in per capita income that leads to the deterioration of the environment and not the improvement in per capita income.

Given these studies, it is possible to reject the extremist view that pollution will increase with increasing revenue; Simultaneously, it should not be assumed that economic growth will take care of the environment without implementing environmental policies.

In the meantime, it should be noted that democratic governments are more likely than authoritarian governments to accept the demand of the Greens and implement environmental policies. However, respect for agreements and treaties is often difficult for countries that typically have sovereign governments. Therefore, international law should focus more on agreements that have been or will be voluntarily accepted by governments or international organizations.

It is also important to note that decisions are not made to ignore business or environmental goals. Sometimes trade barriers aimed at helping the environment do more harm than good. For example, boycotting the Amazon timber trade with the motive of protecting forests may seem like a good idea at first glance; But there is direct trade loss and indirect environmental damage because if loggers are unable to export timber, they may turn to farm for a living, which in turn will have far more negative consequences. Besides, it can be said that if Brazil's forests can prevent global warming by absorbing carbon dioxide, CO2-producing countries will help the country manage, preserve and rehabilitate its forests while exploiting them.

According to developing countries, developed countries have had a high share of global pollution during and after the Industrial Revolution and have benefited from the fruits of their products since then. Therefore, based on the principle of environmental compensation, developing countries seek to increase the share of payments of developed countries to control carbon dioxide (Faber, 2005).

Despite all this, advocates of free trade and globalization of trade believe that the most preventive environmental laws are in place in developed countries today. In comparison, these countries have freer trade. The group also emphasizes that many global companies, despite the lack of codified environmental laws in the host countries, prefer to continue using environmentally friendly technologies rather than good reputations. Spend themselves. Although there are some exceptions, the prevailing attitude is towards environmentally friendly technologies. To the extent that today some countries have adopted less growth to protect the environment. At the same time, countries that have higher standards often innovate. Many of these innovations have led to fundamental technical changes that are often environmentally friendly and efficient. Therefore, their use is beneficial.

The positive and negative effects of globalization on the environment:

Examining the views of the proponents and opponents of globalization, its positive and negative effects on the environment can be enumerated as follows:

With the expansion of globalization of the economy and countries' specialization in producing goods in which they have a relative advantage, industries' geographical location changes. As a result, some countries absorb dirtier industries such as steel and chemical plants, while others absorb cleaner sectors such as computer programming or call centers. Therefore, in some cases, pollution problems are transmitted around the world.

The process of globalization can boost trade and economic growth. Under these circumstances, if there is no desire for clean production and recycling of waste, this economic growth will be to the detriment of the environment; Because economic growth leads to revenue growth and consequently aggregate demand, thereby intensifying production under any circumstances and payment from primary sources without adhering to the principle of substitution.

Globalization can lead to "downward competition" because governments may ignore or lower their environmental standards to attract foreign investment.

In addition to the negative effects that may occur following the process of globalization on the environment, some positive effects can also be listed:

Globalization expands green technologies. Advanced companies that set up overseas factories often bring environmentally friendly technologies with them, forcing local competitors to use clean technology in their production operations to maintain high productivity; At the same time, exporters are increasingly forced to raise their standards to meet the demands of their strict customers.

Economic growth can ultimately lead to a cleaner environment. It is clear that if governments are forced to respond to stricter environmental laws, the final cost of implementing them will increase for the people. Globalization can cover this cost by boosting economic growth and increasing per capita income; Because when people get richer, they tend to a society where people demand a cleaner environment; in these circumstances, they will be able to pay for such a demand.

Globalization has created a situation where environmental pressures and protests are no longer confined to countries' borders. In other words, the possibility of cooperation and unity of procedure and the joint decision has been created for environmental activists. Furthermore, due to international non-governmental organizations' activities and the global dissemination of information, environmental activities are expanding worldwide (Mohai et al., 2009).

"Consumer power" can be used as one of the consequences of globalized trade as a tool to help the environment. Because of the tendency towards environmental labeling of goods globally, consumers will be able to use the tools of their desire to buy environmentally friendly goods to express their desires. The environmental label indicates the environmental aspects of a product or service. At the same time, for these strategies to be more effective, laws and standards need to be enacted in the absence of which manufacturers can not present their products as environmentally friendly, or a country unfairly labels imports of competing goods to get more market share of that commodity.

Another positive effect of globalization on the environment is the creation of a system by which countries can interact with each other under certain rules established by multilateral negotiations and at the same time their actions by some multilateral organizations. Be monitored. The need for such a system is well felt, especially in global environmental problems and issues involving more than one or more countries. It can act as the role of the World Trade Organization in economics as a global watchdog and organizer in the field of environment and provide the ground for the implementation of previous agreements such as the Kyoto Protocol.

Globalization makes it possible for countries to draw on each other's experiences based on their statistical documentation in the field of the environment, and ultimately with a proper understanding of how globalization and growth can or can affect environmental goals and helps them to make the right decisions.

STRATEGIES AND RECOMMENDATIONS

Wealthy countries have played a major role in creating environmental problems due to their time and industry growth. These are the countries that are the main consumers of energy and make the most use of natural resources and, of course, create the major part of pollution. In contrast, developing and underdeveloped countries have a different role in environmental degradation due to their lower access to advanced technology and industry. Although the average resident of a developed country consumes resources several times as much as the developing countries and leaves behind pollution, it must also be borne in mind that today's less influential people are gradually becoming Elements are influential.

Even though there is no complete consensus on solving environmental problems and adopting appropriate policies to achieve the goals of globalization and at the same time minimizing the adverse environmental effects, but on the issue that multilateral solutions to transboundary environmental problems - There is a kind of consensus, both regionally and globally - over unilateral (one-dimensional) solutions. At the same time, it should not be overlooked that any global action would not be complete without environmental policies at the national and local levels. It can be hoped that the situation will improve when the global environmental laws are properly implemented at the national level and countries work together to solve transboundary problems such as global warming (Brulle, 2006).

Despite the biased views of hardliners on the one hand and environmentalists on the other, it can be said that the prevailing opinion among the public is to try not to lose either and benefit from both, namely economic growth and improving global welfare. To be and also to have the right to a clean environment. None of them can completely improve the standard of living without the other. An optimal combination of both goals is needed to absorb the positive effects to the maximum and reduce the minimum's negative effects. This optimal combination can be achieved in the concept of "sustainable development" as one of the millennium's most important goals. Sustainable development means that policies are adopted to meet the present aspirations without compromising the future's wants and needs.

Various solutions can be proposed to reduce the effects of human abuse on the environment while paying attention to globalization.

Abolish government subsidies in agriculture, especially subsidies for pesticides and chemical fertilizers, impose a tax on pesticides to reduce their consumption. The abolition of subsidies and the imposition of taxes on fertilizers and pesticides will strengthen trade and give people access to cheap food and help the environment.

Abolish government subsidies in the fisheries sector and further monitor fishing in coastal waters and at the same time take coordinated decisions on the exploitation of offshore resources. This will improve free trade while allowing fishing to continue, and as people pay lower taxes in this sector, it will also be possible for people to increase their per capita fish consumption.

Reduce or eliminate energy sector subsidies and impose taxes on oil, coal, and, in a word, carbon in general (carbon tax). In this case, companies and industries are at least forced to consider the damage they cause to the environment. If governments cancel energy subsidies, the trade will be freer, and only fuels will be traded whose economic benefits outweigh their environmental costs. As a result, less damage is likely to be done to the environment. In some cases, closing coal mines and paying miners to work will be cheaper and much better for the environment.

Liberalizing the environmental services trade is another solution in this area. Barriers to investment and trade are obstacles to developing useful technologies if their release can improve the environment while facilitating conditions for environmental consultants and providers of green solutions at the international level.

Pollution control by the polluting industries and the development of new, smaller, and cleaner technologies will reduce the industries' environmental costs. Besides, if companies anticipate that environmental laws will become stricter over time, using the latest scientific advances to spend more money later makes sense. It should be noted that in Iran, according to environmental laws and regulations, the environmental assessment takes precedence over economic assessment. According to all developers, they are committed to conducting an environmental assessment before the project (Temper et al., 2015).

International agreements to solve cross-border and environmental trade problems, especially in cases where the destructive factor is related to more than one or more countries or affect the whole world, are other solutions. Examples include greenhouse gas emissions that result in global warming and consequent changes in seasons, rising sea levels due to the melting of polar ice caps, climate change, and even the spread of some diseases such as malaria. Under these agreements, economic sanctions can be imposed on countries with a high share of greenhouse gas emissions.

Considering "sustainable development" along with correct national and international policies for the use of energy and resources and population control can prevent the destruction of ecosystems.

Combining free trade with an appropriate environmental policy, in other words, having an appropriate environmental policy to protect the environment and then pursuing free trade to obtain benefits from it.

Prohibiting the sale of second-hand and out-of-category machines that are not environmentally friendly, following new and environmentally friendly technologies by rich countries to developing countries with low environmental standards.

Supporting companies that produce environmentally friendly technologies so that they can set and implement certain standards themselves. In this case, these groups will establish a favorable coexistence relationship with environmental groups. However, from the point of view of the first group, making a profit and from the point of view of the second group, friendship with nature is the motive for this cooperation; But the result is that pushing the world to move towards higher standards is ultimately in the interest of the environment and the sectors that benefit from its health.

Licensing for environmental labeling, by doing so, consumers can freely choose which of the products produced following environmental standards and compatible with it and products without this standard.

Developed countries that, due to the exploitation of sufficient wealth, have the opportunity to benefit from environmentally friendly technologies, and by changing public attitudes or under pressure from pro-environmental groups, tighten trade and imports from poor or developing countries. Recognizing that developing and emerging countries face resource shortages, they provide technical assistance to ensure both parties' views (Gould, 2010).

Establish an environmental information network to encourage the economic community towards an environmental information system and require developers to publish and disseminate the environmental impact of their projects to the public.

Reducing energy consumption as a national priority and changing consumption patterns. Using and encouraging others to use renewable and clean energy such as wind, solar, and biomass alongside fossil fuels. Imposing a tax on gasoline and using long-term incentives to produce renewable energy can help achieve this goal.

Collaborate with large firms and encourage them to reduce the impact of their work on the environment by adopting environmentally friendly mechanisms while maintaining profitability for firms and, more importantly, the practicality of these mechanisms. The result is the protection of water resources, energy, and, in a word, better management of these resources. These collaborations do not conflict with government regulations and oversight and can even serve as executive support for government laws; Because the strict rules that are often imposed on users are often not enforced properly and, in some cases, are not enforced at all. This collaboration also helps to remove the barrier between stakeholders; In other words, it eliminates the gap between the two sides of this wall, i.e., the environmentalists on the one hand and the exploiters on the other.

An appropriate role for international environmental groups is to help local groups, in turn, work through domestic political channels and legal and legitimate means to shift the balance of political forces towards greater value for the environment.

Finally, it is important to note that severe austerity measures are more effective in those sectors of the business that have the greater global impact or in which different parts of the world are involved, and in other local cases, more balanced measures are more effective (Schlosberg & Carruthers, 2010).

CONCLUSION

Globalization itself is a complex phenomenon that has many forces and effects. Naturally, not all of these effects or forces will benefit the environment if it is equally illogical to consider them all harmful to the environment. Environmental degradation is a global issue today that is noteworthy from all three ecological, political, and economic perspectives. In the meantime, global environmental problems need global solutions. Experimental studies on various contaminants have reached a bell-shaped curve. In this way, the level of pollution initially increases with increasing income and then decreases. Development destroys the environment in the first place; But eventually, it improves. Therefore, it can be said that development threatens the environment and can effectively improve its situation. In this way, as income increases, concern for the environment increases, and this concern encourages a response that improves the environment. Environmental degradation and the increasing growth of trade have led to increased public awareness and response to the effects of destructive economic misconduct and increased environmental demands. Simultaneously, in adopting trade policies and reforms of the World Trade Organization, the balance of interests between countries should be considered, and issues other than trade, such as the environment, should also be addressed. We should not forget that we are the guests of this planet and the test of our generation is to pass it on to the next generation in the same way or better. Without paying special attention to the environment, justice is practically neglected to benefit different generations of nature. So it can be said that the real supporters of democracy in the world are environmentally friendly and live green.

REFERENCES

Agyeman, J., Schlosberg, D., Craven, L., & Matthews, C. (2016). Trends and directions in environmental justice: From inequity to everyday life, community, and just sustainabilities. *Annual Review of Environment and Resources*, *41*(1), 321–340. doi:10.1146/annurev-environ-110615-090052

Aukusti Lehtinen, A. (2006). 'Green waves' and globalization: A Nordic view on environmental justice. *Norsk Geografisk Tidsskrift-Norwegian Journal of Geography*, *60*(1), 46–56. doi:10.1080/00291950600548881

Brulle, R. J., & Pellow, D. N. (2006). Environmental justice. *Annual Review of Public Health*, *27*, 103–124. doi:10.1146/annurev.publhealth.27.021405.102124 PMID:16533111

Carruthers, D. V. (2008). The globalization of environmental justice: Lessons from the US-Mexico border. *Society & Natural Resources*, *21*(7), 556–568. doi:10.1080/08941920701648812

Faber, D. (2005). Building a transnational environmental justice movement: Obstacles and opportunities in the age of globalization. *Coalitions across borders: Transnational protest and the neoliberal order*, 43-68.

Faber, D. R., & McCarthy, D. (2012). Neo-liberalism, globalization and the struggle for ecological democracy: linking sustainability and environmental justice. In *Just sustainabilities* (pp. 55–80). Routledge.

Gould, C. C. (2010). Moral issues in globalization. The Oxford handbook of business ethics, 305-334.

Mohai, P., Pellow, D., & Roberts, J. T. (2009). Environmental justice. *Annual Review of Environment and Resources*, *34*(1), 405–430. doi:10.1146/annurev-environ-082508-094348

Pellow, D. N., & Brulle, R. J. (2005). *Power, justice, and the environment: toward critical environmental justice studies*. Academic Press.

Schlosberg, D., & Carruthers, D. (2010). Indigenous struggles, environmental justice, and community capabilities. *Global Environmental Politics*, *10*(4), 12–35. doi:10.1162/GLEP_a_00029

Sze, J. (2006). *Noxious New York: The racial politics of urban health and environmental justice*. MIT Press. doi:10.7551/mitpress/5055.001.0001

Temper, L., Del Bene, D., & Martinez-Alier, J. (2015). Mapping the frontiers and front lines of global environmental justice: The EJAtlas. *Journal of Political Ecology*, *22*(1), 255–278. doi:10.2458/v22i1.21108

Walker, G. (2009). Globalizing environmental justice: The geography and politics of frame contextualization and evolution. *Global Social Policy*, *9*(3), 355–382. doi:10.1177/1468018109343640

ADDITIONAL READING

Agyeman, J., Schlosberg, D., Craven, L., & Matthews, C. (2016). Trends and directions in environmental justice: From inequity to everyday life, community, and just sustainabilities. *Annual Review of Environment and Resources*, *41*(1), 321–340. doi:10.1146/annurev-environ-110615-090052

Brulle, R. J., & Pellow, D. N. (2006). Environmental justice. *Annual Review of Public Health*, *27*, 103–124. doi:10.1146/annurev.publhealth.27.021405.102124 PMID:16533111

Carruthers, D. V. (2008). The globalization of environmental justice: Lessons from the US-Mexico border. *Society & Natural Resources*, *21*(7), 556–568. doi:10.1080/08941920701648812

Faber, D. (2005). Building a transnational environmental justice movement: Obstacles and opportunities in the age of globalization. *Coalitions across borders: Transnational protest and the neoliberal order*, 43-68.

Faber, D. R., & McCarthy, D. (2012). Neo-liberalism, globalization and the struggle for ecological democracy: linking sustainability and environmental justice. In *Just sustainabilities* (pp. 55–80). Routledge.

Gould, C. C. (2010). Moral issues in globalization. The Oxford handbook of business ethics, 305-334.

Mohai, P., Pellow, D., & Roberts, J. T. (2009). Environmental justice. *Annual Review of Environment and Resources*, *34*(1), 405–430. doi:10.1146/annurev-environ-082508-094348

Schlosberg, D., & Carruthers, D. (2010). Indigenous struggles, environmental justice, and community capabilities. *Global Environmental Politics*, *10*(4), 12–35. doi:10.1162/GLEP_a_00029

Sze, J. (2006). *Noxious New York: The racial politics of urban health and environmental justice*. MIT Press. doi:10.7551/mitpress/5055.001.0001

KEY TERMS AND DEFINITIONS

Equity: Defined by UNEP to include intergenerational equity—"the right of future generations to enjoy a fair level of the common patrimony"—and intragenerational equity—"the right of all people within the current generation to fair access to the current generation's entitlement to the Earth's natural resources"—environmental equity considers the present generation under an obligation to account for long-term impacts of activities and to act to sustain the global environment and resource base for future generations.

Polluter Pays Principle: The polluter pays principle stands for the idea that "the environmental costs of economic activities, including the cost of preventing potential harm, should be internalized rather than imposed upon society at large."

Precautionary Principle: One of the most commonly encountered and controversial principles of environmental law, the Rio Declaration formulated the precautionary principle: To protect the environment, the precautionary approach shall be widely applied by States according to their capabilities.

Prevention: The concept of prevention can perhaps better be considered an overarching aim that gives rise to a multitude of legal mechanisms, including prior assessment of environmental harm, licensing or authorization that set out the conditions for operation and the consequences for violation of the conditions, as well as the adoption of strategies and policies.

Transboundary Responsibility: Defined in the international law context as an obligation to protect one's environment and prevent damage to neighboring environments, UNEP considers transboundary responsibility at the international level as a potential limitation on the sovereign state's rights.

Chapter 5
Green Energy Economy Legislation

ABSTRACT

The legal basis for renewables is the environmental and human rights obligations of governments under international instruments and treaties. Despite this, and given that WTO members do about 97% of world trade, the implementation of programs and policies must also be consistent with supporting the development of new energy must be following the obligations of governments following WTO rules and regulations. Therefore, the organization's member states must balance their international environmental and trade commitments in this regard. Recent WTO jurisprudence has provided criteria for balancing these two categories of obligations. Since many countries join the World Trade Organization, regulation in energy, especially new energy, and investment guarantee contracts and renewable electricity purchase must be done under the organization's rules and jurisprudence. In this chapter, considering the World Trade Organization's recent practice, criteria will be presented to balance governments' environmental and commercial commitments supporting renewable energy.

INTRODUCTION

Human-related climate change is affecting the energy sector and undermining the foundations of energy security. Similarly, climate change mitigation measures accelerate energy efficiency and sustainable energy sector policies (Adib et al., 2015). "Accepting this fact, in climate regimes, the approaches adopted in some of the initial environmental treaties, which required members to use trade restrictions against non-member countries, were abandoned (Aichele & Felbermayr, 2013). Today, reducing greenhouse gas emissions in the light of precautionary measures is considered a strategic environmental goal and stabilizing climate regimes. Therefore, most energy policymakers acknowledge the need to develop new energies through policies and support programs. This is being done in developed and developing countries, and in Iran, in addition to bylaws and contracts to support the production of renewable electricity, it is also reflected in the 2020 budget law.

DOI: 10.4018/978-1-6684-7188-3.ch005

GATF, the World Trade Organization's predecessor and the organization itself, has not been indifferent to environmental issues since its inception. Preservation of the environment is currently accepted in the jurisprudence of the organization as an exception to the rules of the organization, provided that the general principles of non-discrimination are observed, and the review body, by providing a broad interpretation of the condition of environmental protection measures(United Nations, 1988). Nevertheless, environmental issues are multifaceted, and paying attention to them as an exception to the organization's agreements and regulations does not seem sufficient and effective. For example, the cornerstone of dispute resolution in the US case - Migo - is that Article 20 of the GAT states that the exceptions are "limited" and "conditional." This is even though in GATS, environmental exceptions are only limited and unconditional. Quantitative resource constraints to generate electricity from fossil fuels, greenhouse gas emissions, and environmental damage to soil, air, and water have shifted energy policies to greater focus and support for clean and endless electricity sources for power generation. Supports that pursue two major goals: security of supply and protection of the environment (Alston, 2020).

On the other hand, government support for producing some goods following international trade rules is contrary to the principle of freedom of trade and increased competition. It is a challenge that pits energy and environmental policies against trade policies, and one must inevitably be balanced in favor of the other. Policies to support the production of renewable electricity are one of the points of gravity of this challenge.

The challenge, like an iceberg, has many mysterious aspects that do not seem at first glance (Bradbrook, 1996). Also, the pre-litigation of cases pending or pending at the ICSID Arbitration Center over electricity and renewable energy over other common issues in international arbitration such as oil and gas in 2015 indicates the growing importance of renewable energy and renewable energy.

POLICIES TO SUPPORT THE PRODUCTION OF GREEN ENERGY

Within the framework of the rules and regulations of the World Trade Organization, which is the Charter of International Trade, the lack of negotiations to establish an agreement on the environment and trade or support for renewable energy, tensions over the organization's licensing subsidies has been created to promote the production of energy from renewable sources. This is evident in the three cases of international trade disputes involving renewable energy policies. The last two major cases relating to guaranteed electricity purchase from renewable energy sources by the State of Ontario, Canada. These support programs include policies adopted by the public or quasi-government sector to expedite investment in renewable energy development projects, subject to specific criteria (Alston, 2020) . These policies often pursue two main goals: to ensure the electricity supply's security and protect the environment. Accordingly, the Government of Canada and the State of Ontario plan to generate electricity from small and large-scale renewable power plants using solar, wind, water, and biomass sources for 20 years at a minimum fixed base price for general electric injection. It is guaranteed to be purchased. This rate can be increased annually for more desirable projects and every few years for other projects. Similar policies are in place in most developed countries, even in the field of renewable energy. For example, 18 of the 27 member states of the European Union have adopted programs that guarantee a minimum price for the resale of electricity generated from renewable sources (Birnie & Boyle, 2002).

China, Australia, New Zealand, South Africa, India, Switzerland, and the United States have implemented such programs at the state, local, and municipal levels in 37 states, including California, Colorado, and Florida. These programs are already very successful and have led to the growth of investments and electricity production from renewable energy sources .

These programs have also led to a 40% increase in solar panels in the UK and a 15% increase in Germany's renewable electricity production (Brownlie, 2008). In Iran, according to Article 62 of the Law on Regulation of State and Government Finance "B" Article 133 of the Law on the Fifth Development Plan and approved by the Economic Council, renewable electricity with a threefold increase in 2020, at ten times the sale price with a guaranteed purchase of up to 10 years it is possible. Meanwhile, the electricity generated by thermal power plants is bought simultaneously as the selling price, and small-scale power plants are bought at 1.5 times the selling price. Although this huge price difference is also related to the huge costs of setting up renewable power plants, part of this price difference is an incentive policy to grow investments in this sector.

Note 2 of the 2020 Budget Law also deals exclusively with developing new energies and the guarantee of investments made in this field. Renewable energy protection programs are often based on two environmental and economic considerations: greenhouse gas emissions, which do not directly affect the final costs of goods and services and do not affect producers. The failure of the market to calculate the costs of emitting greenhouse gases at final prices means that consumers and producers do not have the financial incentives and incentives to reduce emissions pursued in renewable power generation plans (Brownlie, 2008).

The economic rationale is that the lack of such policies supporting renewable energy leads to a reduction in electricity generated from fossil fuels relative to renewable energy. With this trend, investors are not encouraged to invest in new energy development projects because, without supportive policies, they will not be able to calculate and take into account the benefits they bring to the public from the implementation of new energy development projects at the final price (Brownlie, 2008). Therefore, nationally adopted measures in this regard do not seem to be anything more than complementary policies to increase the security of supply and protect the environment (Bradbrook, 2008) . The benefits of these projects, which can benefit national economies and globally and for humanity in general, suggest that governments have vested interests in pursuing policies to mitigate the effects of global climate change and global warming . Moreover, in the absence of global standards and comprehensive international mechanisms to reduce greenhouse gas emissions and climate change and the economic and environmental foundations of government intervention in the market, it supports and promotes renewable energy systems and renewable energy systems. (Bradbrook, 2008). In this regard, the policies and laws adopted to pursue these goals by the World Trade Organization members will be a potential factor for their separate design in the form of an agreement.

REGULATION OF PROTECTION SUBSIDIES WITHIN THE WTO REGULATIONS

Given that subsidies that increased exports and restricted imports were seen as a barrier to international trade, the Tokyo Round negotiations were of great importance and led to an agreement's ratification in this regard. Agreements that did not clearly define what subsidy was one of its main weaknesses were considered a significant success (Bradbrook, 2013). In the Uruguay Round, the Subsidy Agreement was discussed as a separate issue from non-tariff barriers, and the shortcomings of the Tokyo Round Agreement were amended. Accordingly, subsidies were classified into three categories: non-traceable, traceable, and prohibited. Negotiations on the definitions and scope of each of these classifications, the Subsidies Agreement, and Compensation Measures were finalized in 1991. Providing a clear and acceptable definition of the subsidy to all members, this agreement was adopted in 1994 as one of the multilateral documents under Annex A of the organization's founding document, which was required of all members. Given the need for this agreement for all members of the organization and taking into account current considerations, renewable energy protection policies should be: In the first step, they should be made to not fall into the definition of subsidies according to the agreement. In the next step, if this is not possible, they should be implemented so that they are not considered prohibited subsidies. Finally, if the policies' scope is such that the adoption of them is considered a prohibited subsidy because of these measures' benefits, they should be justified under the exceptions of Article 20 GATT 1994 (Bradbrook, 2013). For this purpose, two main criteria should be considered in this regard.

Rule of Non-Specificity

Subsidies are defined in Article 1 of the Agreement and refer to assistance provided by the government or public institutions to the private sector (Barboza, 2010). In other words, the government transfers or directs aid to the private sector; For example, it provides the conditions for the private sector to supply electricity to generate electricity from renewable sources instead of fossil fuels and to implement these projects. Therefore, the first condition for assistance to be considered a subsidy is its grant from the government or public sector. The second condition is that the benefit is transferred to the private sector.

The cornerstone of the Canada-Passenger Airplane case is that the transfer of benefits means that the private sector is better positioned to receive the aid than when it did not receive it. The third and necessary condition is the specificity of the aid. The specificity or validity of the assistance provided to certain institutions is assessed based on their type of production or based on the activity's geographical location. It also subsidizes the creation of restrictions for a group of companies that have virtually the same effect on subsidies (Barboza, 2010).

Energy policymakers can make it possible for renewable energy to be provided in general and for all sectors by providing conditions for access to aid and eliminating these aids (Biswas, 1981). But it is challenging in simple terms and practice: first, that subsidies to all sectors of the economy are impossible given the limited resources of countries, especially developing countries, while these countries may have a comparative advantage. More than other developed countries in the production of renewable electricity. Second, extensive administrative action is needed to ensure that some companies may be denied access to subsidies. According to the members' discussions in the implementation of Article 31 of the Agreement, all non-specific (non-traceable) subsidies were included in traceable subsidies. Practically, the subsidies were restricted to two groups: These include government grants for R&D projects, development of less deprived areas, and grants for environmental protection projects that are currently subject to special conditions: Prohibited subsidies are prohibited only because of their effects, and traceable subsidies are identified based on the analysis of their effects on other member states (Barboza, 2010). It is useful to note that all renewable energy subsidies dealt with in the context of an organization's negotiation or dispute settlement mechanism are inherently discriminatory and do not have an inherently supportive approach. For example, the Ontario State Plan made the guaranteed purchase of electricity conditional on the supply of 50 and 60 percent of the solar and wind farm designs' project equipment, respectively, from state production. Also, China's subsidies for wind projects, which led to a lawsuit filed by the United States in 2010, were subject to export performance and the use of domestic products (Barboza, 2010).

Rule of Non-Discrimination

The absence of discrimination can be considered the key principle of the World Trade Organization's rights, according to which subsidies should not legally or practically lead to discrimination between domestic and imported goods (Borowy, 2014). This principle is also reflected in many of the organization's provisions, including Article 4 of the GATT and Article 2 of the Trims Agreement. The standard of national behavior and the complete government principle are also a reflection of this general principle. Accordingly, any action that causes foreign and imported goods to be less fortunate than domestic products is prohibited. Therefore, even assuming the justification of renewable energy subsidies under Article 20 (g) of GATTA, the allocation of these subsidies and aids should not be considered arbitrary, discriminatory, or a hidden restriction on international trade (Bruce, 2013). In some cases, the organization's jurisdiction is that exceptions to the 1994 GATT Rule 20 may also apply to other organization agreements, including the Subsidy Agreement and the Compensation Measures . Article 20 refers to specific and vital strategic policies adopted to protect human, plant, and animal life, public health and ethics, the conservation of finite natural resources, and so on. Inciting this article, two issues need to be considered to justify subsidies and aids to renewable energy. The first is that measures and protection should be included in defining one of the exceptions to Article 20; the criteria at the beginning of the article must be observed to adopt these measures to justify deviations from the organization's rules. Paragraph (g) of Article 20 refers to the exception of measures related to the protection of non-renewable resources, provided that it is accompanied by measures to limit the production and consumption of these resources within the country (Bruce, 2013). There are two distinct provisions in this article: First, protection measures must protect non-renewable resources. Contrary to some paragraphs of Article 20, which state that measures taken to achieve the objectives outlined in paragraphs of that article must be "necessary and necessary," paragraph "g" refers only to measures "in connection" with the protection of natural resources. Thus, they not only include the necessary measures for protection but also cover a wider range (Bruce, 2013). The two criteria of "primary purpose" and "the existence of a logical connection between the intended purpose and the action were taken" have been presented in the salmon fish case - Canada's case - and the new formulation gasoline, respectively, to maintain the relationship between the actions taken.

Pillar of Appeal in the US-Shrimp Case This issue is inferred from Article 20 (g) of protection of non-renewable living resources, such as fossil fuels, such as related measures for the conservation of non-living natural resources (Bruce, 2013). As a result, measures to support renewable energy are also included. In the light of declining fossil fuel resources and the damage that climate change is doing to the global economy, catalyst measures for the development of renewable projects are related to the conservation of finite natural resources. The second point is that restrictions should accompany these measures on the production and consumption of fossil resources. In other words, restrictions on the protection of non-renewable natural resources should be imposed not only on imports but also on domestic products (Bruce, 2013).

The compatibility of new energy protection policies with this criterion depends on the structure of policies and actions and each country's situation. Still, it is important to evaluate this coordination whether the implementing member of supportive policies, measures, and other measures in other areas is still important. Has it adopted the effects of climate change or not? (Bruce, 2013). The second issue is how to take protective measures that should not be discriminatory, arbitrary, or biased. The US-Shrimp Dispute Settlement Pillar seeks a broader interpretation of discrimination and strictly enforces even examples such as not consulting and objecting to this article's provisions, which at first glance may not constitute discrimination. Interpreted the concept of discrimination at the beginning of Article 20. In this case, the review pillar described the lack of transparency in domestic processes and the lack of authorization by some exporting countries as arbitrary and biased. It seems that considering the records of the organization and GATT investigations, especially in the cases of gasoline and shrimp, negotiating and involving other member countries in the adoption of policies to support renewable electricity, to prove that this action was taken, Not biased and discriminatory, plays a key role (Bruce, 2013) .

TRADE AND ENVIRONMENTAL COMMITMENTS TOWARD RENEWABLE ENERGY

The description of the grants and subsidies granted to the renewable energy sector under Article 20 has not been necessary but sufficient to deviate from the organization's regulations, but the rule at the top of the article must also be observed. Another point is that renewable electricity subsidies are linked to international environmental agreements' goals and objectives, including the United Nations Framework Convention on Climate Change and the Kyoto Protocol. This relationship has two purposes concerning the rule of Article 20: first, and more importantly, this relationship represents the goal of preserving the finite natural resources mentioned in the article, and second, this relationship indicates the objective of being an objective criterion for the legitimacy of subsidies. Renewable energy development projects deny the claim that policies to support new energy are a hidden constraint on international trade (Bruce, 2013).

following Articles 26, 30, and 31 of the Convention on the Law of Treaties and the purpose of environmental protection, which was noted in the preamble to the Moroccan agreement, and the membership of most of the members of the UN Convention The intention of subsidizing members to develop renewable energy based on meeting international environmental commitments will strike an acceptable balance between countries' international trade and environmental commitments (Bruce, 2013). Accordingly, the environmental exceptions to Article 20 GATT in this regard should be in the light of paragraph 4 of Article 2 of the Kyoto Protocol, which encourages research, development, and expansion of new and renewable energy and carbon dioxide carbon monoxide emission control technologies. The risk to the environment is interpreted in paragraph 5 of the same article, which refers to reducing and eliminating market deficiencies in the greenhouse gas emitting industries (Bruce, 2013). Article 20 of the Rules of Procedure for the Non-Discriminatory Implementation of these Rules is also in line with paragraph 5 of article 3 of the UN Climate Change Convention. According to this clause: "Measures commonly used to combat climate change, including unilateral measures, shall not in international trade cause unintentional or intentional means or implicit restrictions on international trade."

However, given the key role of the environment in sustainable development and governments' international environmental commitments, the Subsidies and Compensation Agreement does not consider environmental considerations, especially subsidies, especially projects. The environment has been a traceable subsidy since 2001 (Bruce, 2013). The next issue is the burden of proof that the organization's agreement and jurisprudence have imposed on renewable energy development projects to subsidize members, which hinders the adoption and implementation of policies to protect reproduction. Environmental considerations should be more prominent and subject to international regulation in the form of a comprehensive trade and environment agreement.

Table 1. Most important raised renewable energy legal cases in the WTO jurisprudence court

File ID	Caller	Called	Issue	Result
Ds437/2012	PRC	USA	China protests US retaliatory measures for importing solar panels from China with Chinese state-owned companies' help to panel manufacturers.	The pillar of the dispute with the United States recommended that it coordinate these actions following its obligations to the organization.
Ds419/2010	USA	PRC	The United States is protesting against Chinese government loans and grants to wind turbine equipment manufacturers.	Negotiations are ongoing, but no request has been made for a panel.
Ds412/2010	Japan	Canada	Japan protests against Ontario's support programs for the guaranteed purchase of renewable electricity on the condition that domestic manufacturers supply equipment.	The pillar recommended that the retaliation be coordinated with the organization following its obligations to the organization.
Ds452/2012	China	EU	China protests against EU support programs and its members for using renewable energy to force domestic production to benefit from the support provided in these programs.	Negotiations are ongoing, but no request has been made for a panel so far.
Ds426/2011	EU	Canada	European Union protests against Ontario's support programs for guaranteed renewable electricity purchase based on domestic manufacturers' supply equipment.	The Reconciliation Section of Canada recommended that the measures be coordinated following its obligations to the organization.
Ds449/2012	PRC	USA	China protests against compensatory measures by the United States regarding importing some products from China, including renewable energy equipment.	Pillaring the dispute resolution with the United States recommended that the measures be coordinated following its obligations to the organization.
Ds473/2013	Argentina	EU	Argentina protests against the European Union's anti-dumping measures regarding the import of biodiesel from Argentina to the Argentine government's subsidy to produce this product.	Due to the lack of qualified legal staff in this field, the panel was formed with a delay and is scheduled to issue its report soon.
Ds456/2013	USA	India	The United States protests against the Indian government's demand for domestic products to manufacture solar panels in the program Jawaharlal Nehru National, Solar Mission - Wikipedia.	The members of the review panel have been selected, but the review has not started yet.
DS 216/2000	Brazil	Mexico	Brazil protests against the Mexican government for taking anti-dumping measures based on a subsidy granted by Brazil to produce transformers with a capacity of more than 1,000,000 kilowatts.	Negotiations are underway, but no request has been made for a panel.
Ds480/2014	Indonesia	EU	Indonesia protests the European Union's anti-dumping measures regarding the import of biodiesel from Argentina with the Indonesian government subsidy's help to produce this product.	Negotiations are underway, but no request has been made for a panel.

EFFECTIVE MEASURES IN THE FIELD OF RENEWABLE ENERGY PRODUCTION

The rules under which the European Union has filed a lawsuit against Canada are:

Paragraphs 4 and 8 of Article 3 GATT 1994: Clause 4 requires members to comply with the national code of conduct concerning goods imported from other members' territory. Therefore, members should not take any action to reduce sales or harm to imported goods compared to similar domestic goods.

Clause 8 also excludes the application of the provisions of Article 3 in the case of government procurement which is for non-commercial purposes.

Clause 1 Article 2 of the Investment-Related Measures Agreement (TRIMS): This article obliges members to bring their investment-related measures in line with the National Code of Conduct, and Article 11 of the 1994 GATT, which calls for the removal of any quantitative restrictions other than tariff and tax restrictions is not contradictory .

Articles 1 and 3 of the Subsidies and Compensation Agreement: Article 1 defines subsidies and explains the conditions under which government aid and the public and public sectors are considered subsidies. Article 3 also refers to the explanation of prohibited and follow-up subsidies, which according to paragraphs 1 and 2 of this article, subsidies granted by states subject to export performance or the use of domestic products are prohibited.

In this case, the main challenge is to ensure domestic production supply and use in renewable electricity generation projects, which puts domestic production in a better position than similar imported products. Consider tariff and tax restrictions. After reviewing the panel, the panel accepts the EU's claim that Ontario Canada's plan to purchase guaranteed renewable electricity complies with the Canadian government's obligations under the National Code of Conduct and the Trims Agreement . However, concerning this program's inconsistency with the Subsidies and Compensation Agreement, the panel considers that Article 1 of the agreement's condition is considered a subsidy for the guaranteed purchase of electricity, which does not constitute a transfer of benefits to this part of the Union.

The panel argues that given the return rate on investment in similar projects and the average capital rate in Canada, the program offers nothing more than the usual rate for renewable electricity generators and only guarantees it. Therefore, the price set in this program for the guaranteed purchase of electricity has been determined according to the fair rate of return on investment and the risk contained in renewable electricity generation plans .

In a case filed by Japan v. Canada, the panel also stated that the structure of electricity supply and demand in Ontario is such that the competitive wholesale electricity market cannot achieve the goals set in this regard. It is also argued that such markets' economic structure rarely attracts the capital needed to provide a sustainable electricity grid, so supportive action is necessary(Bruce, 2013).

Following the Canadian appeal, a review body is formed. The revision column confirms the panel's findings in all cases except subsidies. According to the Board of Appeals, the panel erred in its reasoning in assessing the concept of interest and market assessment in this regard. However, the revision pillar suffices with the same argument in this regard and does not examine the circumstances under which it leads to the transfer of benefits or how it conflicts with the subsidy agreement .

Canada's conviction in these two cases is not only for supporting the production of renewable electricity through guaranteed purchase but also for making such protection conditional on the use of domestic products in renewable projects that are contrary to national standards of conduct. Canada, of course, referred to Article 8 1994 3 of the 1994 GATT in this regard, but it was not agreed upon by the panel and the pillar. According to this paragraph, the provisions of Article 3, particularly the National Code of Conduct, apply to laws, regulations, and requirements that apply to purchases of government entities solely for government purposes and not with a commercial approach or to use these purchases to produce other goods or resell. It is their business, and it is not applied. In both cases, the panel and the review panel acknowledged that Canada had failed to prove that the Ontario Bureau had no business approach to the program and the purchase of guaranteed electricity.

The challenge of the organization's jurisprudence regarding renewable energy is mainly related to the national standard of conduct and is rarely related to explaining the concept of subsidized or non-guaranteed purchase of electricity in these projects. In particular, the technical characteristics of wholesale electricity markets complicate the analysis of the concept and transfer of benefits that are the pillars of subsidies.

Besides, the panels and review bodies relied on the security of energy supply in justifying programs and policies to support the production of renewable electricity. It was not directly enshrined in the organization's rules and hardly referred to in Article 20. These policies must bear the brunt of proving their good faith in these programs and fulfilling their obligations under the organization's rules. Given that many countries today pursue various programs to support renewable energy, the lack of adjustment of the organization's regulations will gradually become a major challenge to implementing these programs. Establishing a comprehensive agreement addressing trade, particularly in the energy and environment trade, and removing the restrictions placed on the subsidy agreement by these programs, is an effective step towards protecting the environment and protecting the environment. Global support for renewable energy will be within the framework of WTO rules.

INTERNATIONAL LAW AND RENEWABLE ENERGY

Given the transboundary effects of climate change, the need for energy supply, and ultimately sustainable development, international law, albeit scattered, has addressed the use of alternative energy sources. Of course, in this context, our aim is not to examine in detail the existing provisions but merely to reach a general conclusion on the status of renewable energy in the light of international law is sufficient for the present article. In this section, the existing international documents and the principles of international law will be examined, respectively. The legal hierarchy will first examine the applicable documents, public international law, and the existing soft law in this field (Cherp et al., 2011).

Renewables and Hard Law

Two binding instruments have regulations in renewable energy at the international level: the Energy Charter Treaty and the Kyoto Protocol. The Energy Charter Treaty is the first and only international treaty on international cooperation exclusively in energy. The main purpose of drafting the Energy Charter Treaty is to facilitate investment in the energy sector by defining the necessary provisions to reduce the non-commercial investment risks, removing trade and transit barriers. It is worth mentioning that Iran joined the Energy Charter Treaty in 2002 as a supervisory member. Therefore, because the authorities are considering accession to the Energy Charter Treaty, it will be important to pay attention to the Energy Charter Treaty provisions on renewable energy development. Article 1 (5) refers to the definition of economic activity in the energy sector. According to this article, electricity generators' production and operation are considered one form of economic activity in the contract. However, according to the final law of the European Energy Charter Conference, there is no reference in this article to the production of electricity through renewable energy sources, which serves as an interpreter of the Energy Charter Treaty, which includes electricity from renewable energy sources. Therefore, the production of energy from renewable sources falls within the scope of the Energy Charter Treaty.

According to Article 19 (D) of this Treaty, which deals with environmental aspects, States Parties shall pay particular attention to the development and use of renewable energy sources. However, as authors such as Bruce and Bradbrook have pointed out, the wording of this article does not reflect the binding commitment of member states to develop and strengthen the use of renewable energy sources (Bradbrook, 2013; Bruce, 2013). However, there is no denying that governments are explicitly obliged to develop renewable energy in the Energy Charter Treaty. But, of course, since no specific quantitative goals have been mentioned in the development of renewable energy sources, the guarantee of the implementation of this article will remain in an aura of ambiguity.

Climate change and its effects on the environment led to drafting the United Nations Convention on Climate Change in 1992. This internationally binding document is very much related to renewable energy. In this document, member states agreed to reduce greenhouse gas concentrations in the atmosphere to some extent possible environmental hazards. However, this document does not mention any specific commitment to developing the use of renewable energy sources to reduce greenhouse gas emissions. Therefore, governments are free to determine the methods needed to reduce greenhouse gas emissions "through human activities." In this regard, according to the Kyoto Protocol, which was drafted within the framework of the Convention, governments were divided into two general categories, and different obligations were set for each of these two categories (Cherp, 2011).

Under the protocol, developed countries (Annex A countries) must reduce greenhouse gas emissions individually or in partnership by at least 5% lower than in 1990 emissions by 2008 and 2012. Of course, some members, such as the European Union (15 members at the time), set higher targets (8%) for reducing their greenhouse gas emissions. On the other hand, developing countries (non-annexed countries "A") must also develop a program to reduce greenhouse gas emissions. Of course, the text of the Kyoto Protocol does not include any binding commitment to determine the types of tools to reduce greenhouse gas emissions, and governments are fully free to determine the mechanisms to meet their obligations. Nevertheless, Article 2 (1) obliges developed countries (Appendix (a) to develop and use renewable energy sources following their internal conditions, which is not in fact of a binding nature. However, one of the most important mechanisms implemented to achieve the set goals has been using renewable resources in practice. For example, in the European Union(EU), in addition to the economic crisis that led to a decrease in energy demand, the use of renewable energy has played a significant role in reducing greenhouse gas emissions in the realization of the ceiling of reducing greenhouse gas emissions. In addition to the above, Article 12 of the Kyoto Protocol refers to the Clean Development Mechanism (CDM). Following paragraph 2 of the said article, the purpose of this mechanism is to assist on the one hand the non-annexed countries "A" to achieve sustainable development and ultimately to achieve a reduction in greenhouse gas emissions and, on the other hand, to "help the countries." Their commitment is to reduce greenhouse gas emissions. One of the most important projects that have been considered and implemented under CDM is renewable energy projects (Cottier et al., 2011).

In another study examining the effects of the Kyoto Protocol on reducing global emissions, the authors concluded that the adoption and implementation of the Kyoto Protocol would increase the use of renewable energy sources by up to 3% (Aichele & Felbermayr, 2013). Therefore, what is clear is the importance of implementing the Kyoto Protocol in increasing the rate of development and utilization of renewable energy. However, because of the issues raised, it can be stated that in the applicable documents, while the capacity for the development of renewable energy has been considered, no obligation has been assumed to take into account the sources.

Renewables and Soft Law

Apart from the issues that may arise about the nature and guarantee of soft law enforcement, international instruments have already been developed in the form of soft law in many areas, especially in the environment. Thus, as one author defines, soft law refers to those rules that do not guarantee legal enforcement but have enforcement effects (Drake, 2016). Of course, as another author points out, soft law, although not inherently binding, in practice guarantees political performance that may lead to legal effects (Drake, 2016). The authors have expressed various views on nature and guarantee the implementation of soft rights beyond this study's scope. However, given the growing importance of soft law in the development of international law, it will not be useless to review soft law documents on renewable energy.

The first international document to be adopted on renewable energy sources is the United Nations Conference on New and Renewable Energy. The most important outcome of this international conference is the need to develop the use of renewable energy, and secondly, the formation of a committee to develop and use renewable energy sources. Of course, this conference does not include implementing executive policies on renewable energy development, and even the committee was eventually merged into the UN Commission on Sustainable Development. In principle, the development of renewable energy sources was pursued within the framework of sustainable development, until in 1997, the General Assembly of the United Nations adopted a plan to implement Agenda 21 further. Article 45 of the resolution emphasizes the need for international participation to promote the use of renewable energy. In paragraph 46, the need to transfer technical knowledge from developed countries to developing countries is considered to such an extent that it will enable developing countries to increase energy production from renewable sources. The need for developing countries to systematize the use of renewable energy was also emphasized. However, there is no obligation on the part of the states to comply with the provisions of the said resolution, which in principle is not binding (Drake, 2016).

International efforts to reduce greenhouse gas emissions and explain sustainable development in 2002 adopted the Johannesburg Executive Plan. As mentioned, this document is one of the most comprehensive international instruments ever adopted on renewable energy. Chapter 3 of the Johannesburg Executive Plan is devoted to changing unsustainable patterns of production and consumption. Clause 20 (c) of the plan encourages governments and other international actors to use other types of energy to increase the share of renewable energy. Section 20(e) emphasizes the diversification of various energy sources, including fossil fuels, hydroelectric sources, and renewable energy sources. Besides, increasing the global share of renewable energy in total global energy production is emphasized. The importance of investing in renewable energy development by developed countries and international financial organizations in developing countries has also been emphasized on several other occasions.

Therefore, in the Johannesburg Executive Plan, paying attention to the development of renewable energy in the energy portfolio of countries is more explicitly emphasized. To the extent that the need to increase the share of these energies in total global energy production is also mentioned.

In 2005, the UN resolution on the World Summit outcome emphasized the need to take more effective steps to develop and support renewable energy. Finally, in 2011, the UN General Assembly adopted a resolution on sustainable development (United Nations, 2010).

The resolution designated 2012 as the "Year of Sustainable Energy for All." Of course, this document does not directly reference governments' commitments to developing the use of renewable energy sources. However, the document calls on the UN Secretary-General to report on implementing the resolution, in particular, to develop and promote the use of renewable energy. The report, entitled "Promoting New and Renewable Energy Sources," seeks to diversify programs at the national and international levels to promote sustainable development by considering different policies for developing and developing countries. The most recent document adopted on renewable energy development is the 2013 UN General Assembly Resolution on the "Promotion of New and Renewable Energy Sources."

Examining the existing soft law documents on renewable energy, it can be concluded that the emphasis on the use and development of renewable energy has grown exponentially in recent years and has been emphasized in many international instruments. Contrary to these emphases, however, no specific goals have been set for increasing the global share of renewable energy in the global portfolio. There can be several reasons for this; According to the authors, the most important factor is the difference in energy supply sources in different countries and technological and economic capacities for the development of this group of energy sources (Gunningham, 2012); This is because these distinctions, especially considering the very high costs (at least in the short term) for the operation of such projects, have led to the difficulty of regulating comprehensive global regulations in this field. At the same time, the effects of non-binding documents in supporting the development of renewable energy sources cannot be ignored. For example, as mentioned at the end of the Rio + 20 conference, the document "The Future We Want" specifically mentions the need to develop renewable energy. In this regard, we can mention the steps taken by UNESCO to implement the "Renewable Energy for Us" program. The program will use renewable energy sources to expand renewable energy sources to reduce greenhouse gas emissions to provide energy in biosphere reserves and World Heritage Sources to renewable energy sources (Gunningham, 2012).

Public International Law

As one of the main sources of international law, public international law plays an important role in regulating international relations between states. The rules of public international law become enforceable once States have recognized them in practice. The authors have expressed different views on how to form and the elements that constitute the rules of international custom (Gielen et al., 2019). However, this chapter examines those rules of international custom that are recognized in international law. Although the rules on the development of renewable energy are not currently explicitly accepted within the framework of international custom, some of the accepted rules can be referred to as public international law. For example, the two non-damage and sustainable development principles are among the most relevant principles regarding renewable energy development. The principle of non-damages, which is now an integral part of public international law, first appeared in the arbitration of The Trail Smelter Case in 1941 in British Columbia, Canada, and then in the case of the Carrefour Canal in The International Court of Justice has been emphasized (Heffron & Talus, 2016). This principle is also recognized in the draft of the United Nations Commission on International Law on the Prevention of Cross-Border Damage Caused by Dangerous Activities in 2001.

Furthermore, this principle has been used in several cases in the International Court of Justice. However, numerous opinions have been expressed about the definition and limits of this principle. For example, Article 3 of the draft law of the United Nations Commission on the Prevention of Cross-Border

Damage resulting from Dangerous Activities defines: Significant cross-border damage or in the event of an accident to reduce the risks.

On the other hand, Article 21 of the Stockholm Declaration states: "States, following the Charter of the United Nations and the principles of international law, have the right to use their resources under their respective environmental policies." Activities carried out within their sovereignty or control shall not damage the environment of other countries or regions outside the national sovereignty. With a slight change, the original Rio Declaration provides the same definition as the Stockholm Declaration of this principle. In its Advisory on Nuclear Weapons, the International Court of Justice defines these principles and the obligations of States about the prevention of transboundary damage as follows: The environment of other countries or the sphere of the sovereignty of countries is now part of the body of international law" (Holden et al., 2014).

Examining the rulings of the International Court of Justice and the rulings of the International Court of Justice in this regard, it can be concluded that the principle of non-damage is currently one of the main principles of international environmental law. And governments are now obliged to comply with it. The question that comes to mind now is the relation of this principle to renewable energies. For example, RK. In the case of Nauru-Australia (1992), some authors argue that the principle of non-harm includes protecting the earth's atmosphere; they state that the earth's atmosphere, like the oceans and seas, is part of its common heritage (Birnie & Boyle, 2002). Therefore, the same protection of the oceans in international law should apply to the atmosphere. According to the principles of customary international law, governments have international responsibility for greenhouse gas emissions.

However, as some authors have pointed out, there is no standard for determining the due diligence of states in this area (Lyster & Bradbrook, 2006); Because the standards set out in the Long-Range Transboundary Air Pollution Convention or the Ozone Layer Convention are not analogous to determining the appropriate effort to emit greenhouse gases. The standards contained in the Kyoto Protocol are also not like international custom; Because they are not subject to universal consensus. Some authors also argue that the principle of non-damage can be applied if the error or intention of governments to cause cross-border damage is established. Otherwise, the damage can not be considered a violation of international law and responsible for governments (Brownlie, 2008). However, as another author acknowledges, the principle of non-damage does not necessarily imply governments' intent or error; this principle applies to non-prohibited acts and is common in international law (Borowy, 2014). In this regard, numerous lawsuits have been filed nationally and internationally against countries and companies that emit greenhouse gases (Gunningham, 2012).

Many authors also emphasize the international responsibility of states for damages that have been introduced in terms of greenhouse gas emissions (Cherp, 2011; Cottier, 2011). But so far, no lawsuit has been filed that has resulted in a ruling on reducing pollutants or compensation. Therefore, although it can be argued that governments are obliged to prevent damage to the earth's atmosphere and thus reduce greenhouse gases in the main shadow of no damage, the limits of damage and the commitment of governments to prevent it have not been determined. At the same time, For governments to be held accountable in this regard, there are numerous problems, such as standard, appropriate efforts, proving the causal relationship, the competent authority, etc., that have a long way to go to solve them, therefore, in the light of this principle, governments can not be expected to develop renewable energy, which will reduce greenhouse gas emissions.

Renewable Energy and International Institutions

The role of international organizations in shaping various fields of international law is well known. Different international organizations are working hard to establish an appropriate legal framework for energy and energy governance systems.

United Nations agencies are actively working to promote renewable energy worldwide. UNEP is the leading United Nations organization supporting renewable energy advertising. UNEP's major contribution to the Renewable Energy Law is the publication of the "Handbook" (2007) and "Guide" (2016) to assist designers in developing countries, and the annual report "Global Renewable Energy Investment Trends," which involves various themes of renewable energy investment cases. UNDP has been active in transforming the global energy system for more than two decades by improving energy efficiency and using renewable energy systems. Recently, UNDP released the "Sustainable Energy Strategy Note 2017-2020: Sustainable Energy Supply in Climate Change", the first UNDP program, mission, method, guidance, principle, and focus on sustainable energy. To this end, it puts renewable energy in its vision and mission (United Nations, 2010). This is an important initiative of the UNDP program because the organization has played an effective role in development activities mainly from countries in Asia and sub-Saharan Africa, where disadvantaged and marginalized people cannot afford electricity.

The fact is that despite the active participation of United Nations agencies in global development activities, their energy activities are relatively ambiguous (Gunningham, 2012). There is a lack of coordination and trust among these organizations; that is, "the regime's system is ineffective for renewable energy to achieve sustainable energy" (Heffron, 2016). Therefore, the United Nations Department of Energy, composed of 21 member organizations, was established in 2004 to respond to energy Challenges; the United Nations system has established institutional mechanisms to promote coordination and coherence (United nations, 2010) The International Energy Agency (IEA) was established in 1974 after the oil crisis of 1973-79. It aims to coordinate the response of oil-importing countries to major oil supply disruptions. It is the most famous international organization dedicated to energy security, environmental awareness, and economic development. IEA was established following the decision of the OECD Council and is an autonomous organization within the framework of the OECD (International Energy Planning Agreement). But this is separate from the OECD and has been recognized by the "United Nations Framework Convention on Climate Change." Historically, the main focus of the IEA is neither renewable energy nor conventional fuels. The Renewable Energy Working Group (REWP) was recently established. Its task is to provide renewable-relevant technology and policy recommendations and projects to various IEA organizations. A very authoritative global energy outlook is published every year, which is the main reference for energy-related stakeholders by IEA. Founded in 2011, International Renewable Energy Agency (IRENA) is a leading international organization focusing on renewable energy. As of September 2017, it has 152 members and 28 future members. Through the 2009 IRENA Charter, the parties expressed their willingness to promote the recognition of renewable energy in terms of sustainable development and firmly believe that renewable energy provides a huge opportunity to solve and gradually reduce energy security issues and unsustainable energy prices. Therefore, IRENA enjoys the same privileges and immunities as the United Nations (Gielen, 2019), although IRENA has no explicit or implicit authority to set binding RE targets.

At the International Renewable Energy Conference held in Bonn in 2004, the International Geothermal Association, International Hydropower Association, International Solar Energy Association, World Biotechnology Association, and World Wind Energy Association formed an international alliance called

Renewable Energy Alliance. It was later renamed to the 21st Century Renewable Energy Policy Network or REN21. The network connects a wide range of stakeholders and provides international leadership in promoting renewable, focusing on the annual "Report on the Global State of Renewable." In addition, various development banks such as the World Bank, Asian Development Bank, African Development Bank, Inter-American Development Bank, and New Development Bank play an important role in promoting renewable energy through financing projects. However, these banks did not have a comprehensive strategy to incorporate renewable energy into their overall loan plans (Cottier, 2011). Therefore, it is clear that many non-governmental actors have been involved in promoting and developing renewable energy; although their activities do not overlap in some cases, IRENA is in a leading position.

SOLUTIONS AND RECOMMENDATIONS

Besides, although policies such as the transfer of energy production technology through renewable energy sources or the creation of financial incentives for investment in this sector have been proposed in the international instruments, it still seems relevant. And it is certainly easier to formulate and implement these policies in international law than to create international obligations for renewable energy development.

FUTURE RESEARCH DIRECTIONS

Nowadays, more researchers study the renewable energy industry, market, etc., and there is a significant interest and attraction to increasing the share of renewables in national energy portfolios. This global convergent ambient for renewable energy development owes much to international conventions, legal principles, and public international laws. This chapter suggests that the current successful trend continues, and a deep energy transition occurs in the decades to come.

CONCLUSION

Policies to protect the production of renewable electricity do not conflict with the World Trade Organization rules. These policies, if they violate the principle of non-discrimination, national conduct, and other regulations of the organization and are not justifiable under the exceptions of General Gat 1994 and in particular paragraph (g) of Article 20, will violate the commitment of the government taking protective measures. It is essential to observe goodwill and that such programs do not impose a hidden limit on international trade. However, Article 20 (g) alone does not appear to support renewable electricity generation policies and does not address requirements such as the electricity supply's security. It should be a comprehensive agreement on energy and the environment and within the framework of other organization regulations. Until then, and following the current regulations of the organization, governments such as Iran (if it becomes a member of the organization), which in their laws and regulations protecting the production of renewable electricity, are subject to the use of domestic products or the application of protection only to domestic companies. They shall be responsible for justifying it and invoking Article 20 (g) (g) under the heading of the protection of non - renewable resources. Discounts for developing countries to bring their trade policies into line with the organization's regulations will include countries about to join, such as Iran, but its government appears to be very close to using domestic production in energy and energy. It has taken other parts, and it will not be easy to leave it. Therefore, before accession, it is necessary to formulate trade policies in line with the organization's regulations, and through alternative measures such as rules of origin, to support domestic production so that other sectors also follow the process of compliance with the requirements of the electricity supply and facilitation of the renewable electricity.

REFERENCES

Adib, R., Murdock, H. E., Appavou, F., Brown, A., Epp, B., Leidreiter, A. & Farrell, T. C. (2015). *Renewables 2015 global status report*. REN21 Secretariat Press.

Aichele, R., & Felbermayr, G. (2013). The Effect of the Kyoto Protocol on Carbon Emissions. *Journal of Policy Analysis and Management, 32*(4), 731–757. doi:10.1002/pam.21720

Alston, P. (2020). The Committee on Economic, Social and Cultural Rights. *NYU Law and Economics Research Paper, 15*, 20-24.

Barboza, J. (2010). *The Environment, Risk and Liability in International law*. Martinus Nijhoff Publishers.

Birnie, P. W., & Boyle, A. E. (2002). *International Law and the Environment*. Oxford University Press.

Biswas, M. R. (1981). The United Nations Conference on New and Renewable Sources of Energy: A review. *Mazingira, 5*, 11–16.

Borowy, I. (2014). *Defining sustainable development for our common future. A history of the world commission on environment and development*. Routledge Press.

Bradbrook, A. (2008). *The development of renewable energy technologies and energy efficiency measures through public international law*. Oxford University Press. doi:10.1093/acprof:oso/9780199532698.003.0006

Bradbrook, A. J. (1996). Energy law as an academic discipline. *Journal of Energy & Natural Resources Law, 14*(2), 193–217. doi:10.1080/02646811.1996.11433062

Bradbrook, A. J. (2013). *International Law and Renewable Energy: Filling the Void*. Duncker & Humblot Press.

Brownlie, I. (2008). *Principles of Public International Law*. Oxford University Press.

Bruce, S. (2013). International law and renewable energy: Facilitating sustainable energy for all. *Melbourne Journal of International Law, 14*, 18.

Cherp, A., Jewell, J., & Goldthau, A. (2011). Governing global energy: Systems, transitions, complexity. *Global Policy, 2*(1), 75–88. doi:10.1111/j.1758-5899.2010.00059.x

Cottier, T., Malumfashi, G., Matteotti-Berkutova, S., Nartova, O., De Sepibus, J., & Bigdeli, S. Z. (2011). *Energy in WTO law and policy: The prospects of international trade regulation from fragmentation to coherence*. WTO Press. doi:10.1017/CBO9780511792496

Drake, L. (2016). International law and the renewable energy sector. The Oxford Handbook of International Climate Change Law, 1, 357.

Gielen, D., Boshell, F., Saygin, D., Bazilian, M. D., Wagner, N., & Gorini, R. (2019). The role of renewable energy in the global energy transformation. *Energy Strategy Reviews*, *24*, 38–50. doi:10.1016/j.esr.2019.01.006

Gunningham, N. (2012). Confronting the challenge of energy governance. *Transnational Environmental Law*, *1*(1), 119–135. doi:10.1017/S2047102511000124

Heffron, R. J., & Talus, K. (2016). The evolution of energy law and energy jurisprudence: Insights for energy analysts and researchers. *Energy Research & Social Science*, *19*, 1–10. doi:10.1016/j.erss.2016.05.004

Holden, E., Linnerud, K., & Banister, D. (2014). Sustainable development: Our common future revisited. *Global Environmental Change*, *26*, 130–139. doi:10.1016/j.gloenvcha.2014.04.006

United nations. (1988). *Protection of global climate for present and future generations of mankind.* General Assembly Press.

United Nations (2010). *Delivering on Energy: An Overview of Activities of UN Energy and Its Members.* United Nations Press.

ADDITIONAL READING

Barboza, J. (2010). *The Environment, Risk and Liability in International law*. Martinus Nijhoff Publishers.

Birnie, P. W., & Boyle, A. E. (2002). *International Law and the Environment.* Oxford University Press.

Biswas, M. R. (1981). The United Nations Conference on New and Renewable Sources of Energy: A review. *Mazingira*, *5*, 11–16.

Borowy, I. (2014). *Defining sustainable development for our common future. A history of the world commission on environment and development.* Routledge Press.

Bradbrook, A. (2008). *The development of renewable energy technologies and energy efficiency measures through public international law.* Oxford University Press. doi:10.1093/acprof:oso/9780199532698.003.0006

Bradbrook, A. J. (1996). Energy law as an academic discipline. *Journal of Energy & Natural Resources Law*, *14*(2), 193–217. doi:10.1080/02646811.1996.11433062

Bradbrook, A. J. (2013). *International Law and Renewable Energy: Filling the Void.* Duncker & Humblot Press.

Brownlie, I. (2008). *Principles of Public International Law.* Oxford University Press.

Bruce, S. (2013). International law and renewable energy: Facilitating sustainable energy for all. *Melbourne Journal of International Law*, *14*, 18.

Cherp, A., Jewell, J., & Goldthau, A. (2011). Governing global energy: Systems, transitions, complexity. *Global Policy*, *2*(1), 75–88. doi:10.1111/j.1758-5899.2010.00059.x

Cottier, T., Malumfashi, G., Matteotti-Berkutova, S., Nartova, O., De Sepibus, J., & Bigdeli, S. Z. (2011). *Energy in WTO law and policy: The prospects of international trade regulation from fragmentation to coherence*. WTO Press. doi:10.1017/CBO9780511792496

Drake, L. (2016). International law and the renewable energy sector. The Oxford Handbook of International Climate Change Law, 1, 357.

Gielen, D., Boshell, F., Saygin, D., Bazilian, M. D., Wagner, N., & Gorini, R. (2019). The role of renewable energy in the global energy transformation. *Energy Strategy Reviews*, *24*, 38–50. doi:10.1016/j.esr.2019.01.006

Gunningham, N. (2012). Confronting the challenge of energy governance. *Transnational Environmental Law*, *1*(1), 119–135. doi:10.1017/S2047102511000124

Heffron, R. J., & Talus, K. (2016). The evolution of energy law and energy jurisprudence: Insights for energy analysts and researchers. *Energy Research & Social Science*, *19*, 1–10. doi:10.1016/j.erss.2016.05.004

Holden, E., Linnerud, K., & Banister, D. (2014). Sustainable development: Our common future revisited. *Global Environmental Change*, *26*, 130–139. doi:10.1016/j.gloenvcha.2014.04.006

United Nations. (2010). *Delivering on Energy: An Overview of Activities of UN Energy and Its Members*. United Nations Press.

KEY TERMS AND DEFINITIONS

Equity: Defined by UNEP to include intergenerational equity—"the right of future generations to enjoy a fair level of the common patrimony"—and intragenerational equity—"the right of all people within the current generation to fair access to the current generation's entitlement to the Earth's natural resources"—environmental equity considers the present generation under an obligation to account for long-term impacts of activities and to act to sustain the global environment and resource base for future generations.

Polluter Pays Principle: The polluter pays principle stands for the idea that "the environmental costs of economic activities, including the cost of preventing potential harm, should be internalized rather than imposed upon society at large."

Precautionary Principle: One of the most commonly encountered and controversial principles of environmental law, the Rio Declaration formulated the precautionary principle: To protect the environment, the precautionary approach shall be widely applied by States according to their capabilities.

Prevention: The concept of prevention can perhaps better be considered an overarching aim that gives rise to a multitude of legal mechanisms, including prior assessment of environmental harm, licensing or authorization that set out the conditions for operation and the consequences for violation of the conditions, as well as the adoption of strategies and policies.

Transboundary Responsibility: Defined in the international law context as an obligation to protect one's environment and prevent damage to neighboring environments, UNEP considers transboundary responsibility at the international level as a potential limitation on the sovereign state's rights.

Chapter 6
Sustainable Energy Economic Development Law

ABSTRACT

The development of renewable energy sources has grown significantly in the current trend of international law. Several international instruments have been adopted to regulate countries' environmental and energy policies. A wide range of binding and soft law instruments are used to formulate these regulations. However, the proliferation of international instruments regarding renewable energy sources, and the uncertainty regarding the principles and rules of international law with regard to renewable energy, necessitate the consideration of an international law approach to renewable energy. This chapter will examine the principles and rules of international law in this field, considering the importance of developing renewable energy as a global issue.

INTRODUCTION

Globally, the states have used various energy sources in response to their socio-economic potentials to meet their demand and development. Since the second millennium BC, PR China has used coal as an energy fuel for its thermal needs and natural gas since 200 BC (Spataru, 2017). Since the late nineteenth century, the global society has been dependent on conventional fossil fuels, such as coal, oil, natural gas, etc., to provide around 75% of its energy needs. These fuels, however, have some unavoidable challenges and problems. Studies have shown that conventional and unconventional oil and gas resources could last for another few decades with current exploration and extraction technologies (maybe five decades for oil and fifteen for natural gas) (Ottinger, 2005).

In addition, based on the uneven geographical distribution of fossil fuel carriers, more than 90 percent of the proven oil reserves are located in only 15 countries (Ruta & Venables, 2012). Although countries with oil reserves should theoretically be able to access the fruits of modern scientific developments and innovations, such a situation creates a serious barrier for sustainable and inclusive development in other parts of the world. Around 1.2 billion people, or 16.5% of the global population, mostly in sub-Saharan Africa and developing Asia, do not have access to electricity (IEA, 2013). The countries with fossil fuel

DOI: 10.4018/978-1-6684-7188-3.ch006

reserves are blessed as finite sources because those people cannot afford the energy price, but new energies are theoretically infinite. Hence, switching the focus from non-renewable fossil fuels to renewables so as to keep the stream of overall sustainable and clean development activities is known as a better and viable alternative, even though there are some inherent initial concerns, which are not unique to renewable energy generation but are present whenever there are new technological developments.

The threat of climate change is a shared concern of humanity (United Nations, 1998) and threatens sustainable and equitable development. Nearly two-thirds of the world's greenhouse-gas emissions come from fossil fuels, which contribute to global warming leading to climate change (Nhamo et al., 2020; IEA, 2015). It is therefore recommended that 75% of all fossil fuel reserves, i.e., 35% of petroleum, 52% of natural gas, and 88% of coal reserves worldwide, be left unused until 2050 in order to keep the temperature variation relative to 1900s under 2°C (McGlade & Ekins, 2015; IEA, 2014). In the world energy industry, it is recommended to change these sources or at least increase the share of renewable energy carriers, including solar, wind, hydropower, ocean, biomass, and hydrogen (Adib et al., 2015). Although fossil fuel prices have declined dramatically in the last decade, clean energy still produced an estimated 23.7% of the world's electricity in 2017. Hydropower, wind, bioenergy, solar, and geothermal energy, respectively, contributed 70%, 15%, 8%, 5%, and 1%, and marine energy contributed the remaining 1% (IRENA, 2017).

Having collaborated successfully on the Millennium Development Goals in 2015, the United Nations members conducted the 2030 Agenda for Sustainable Development and Sustainable Development Goals to end poverty, protect the environment, and promote sustainable development. Although it is difficult to define precisely (Holden et al., 2014), it requires energy savings on the demand side, efficient energy production, continuous flow of energy with fewer environmental impacts (Lund, 2007), and requires the percentage of renewable energy to be more than 27% (IRENA, 2017).

Anyone who is aware of the nature and global effects of climate change should expect that "hard" international law provisions will play an active role. Realistically, there is no direct or specific international instrument that is binding or 'hard' to promote renewable energy. Several international environmental law instruments, for example, the Convention on Biological Diversity, 1993, the Convention on the Conservation of Migratory Species of Wild Animals, 1983, and the Ramsar Convention on Wetlands of International Importance, 1975, etc. contain isolated provisions that should be considered before undertaking any wind energy projects. However, the paucity of international legal provisions addressing renewable energy does not negate the growing importance of its implications to combat climate change. Consequently, normative international law and the positive initiatives of global actors, regional players, and international trade investment systems may play an instrumental role in promoting renewable energy.

Some scholars have already considered the position of renewable energy under international law without any hard instruments (Omorogbe, 2008; Bruce, 2013; Drake, 2016; Mulyana, 2016). The nature of international law has changed since their work. In accordance with the United Nations Framework Convention on Climate Change, the historic Paris Agreement entered into force on 4 November 2016. A 197-country United Nations Framework Convention on Climate Change (UNFCCC) agreement has been signed, whereas 178 countries have already ratified the agreement to save the human world from the effects of climate change. In this agreement, these countries committed to limit global temperature rise to 1.5 °C over pre-industrial levels and to keep global warming below 2 °C during this century. This will be possible due to the efficient exploitation of renewables, a clean and green source of energy, which will aid member countries in fulfilling their national commitments made under the Paris Agreement. In such a changing environment, discussing renewables under international law will be very timely. For decision-makers regarding renewable energy, the current international law approach provides a clear picture. By examining international documents and procedures, the authors intend to examine the commitment of governments to develop the use of renewable energy. We first discuss renewable energy sources, then the principle of state sovereignty over natural resources, as well as the challenges to international commitments on renewable energy development. Finally, we have carefully considered the existing international standards and regulations. First, the binding documents of international law are explained, followed by public international law, and finally the development of renewable energies will be discussed from the perspective of soft law (Warschauer, 2010).

LITERATURE REVIEW

From various events in the past few decades, the international movement can transfer traditional energy to renewable energy. In the legal field and under the auspices of the United Nations, the use and promotion of renewable energy can be indirectly mentioned in the 1972 Stockholm Declaration on the Human Environment (1972), and the risk of erosion of non-renewable land resources has been verified (United Nations, 1972). Subsequently, the 1987 report of the World Commission on Environment and Development emphasized that renewable energy should be used to lay the foundation for the global energy structure in the 21st century. Next, the "Nairobi Action Plan for the Development and Use of New Energy and Renewable Energy" emphasizes the importance of developing new and renewable energy sources to help meet sustained economic and economic needs, especially in developing countries (Boroway, 2013). Even if these measures have no direct and significant impact on promoting renewable energy, it should be understood that since the importance of environmental protection as a global concern in the 1960s, the international community has regarded it as a reality almost immediately (Biswas, 1981).

In international law, the establishment of scientific institutions such as the Intergovernmental Panel on Climate Change (IPCC) by the United Nations Environment Programme (UNEP) and the World Meteorological Organization (WMO) in 1988 may be a turning point. Forget it. Designed to reduce the impact of climate change. Then, at the Rio Earth Summit in June 1992, an international agreement was approved is the UNFCCC. However, the UNFCCC does not explicitly mention renewable energy. Finally, the Third Conference of the Parties (COP3), held in December 1997, approved the "Kyoto Protocol" of the "United Nations Framework Convention on Climate Change." The "Kyoto Protocol" introduced some innovative mechanisms to reduce greenhouse gas emissions, and these mechanisms have an indirect impact on the promotion of global renewable energy (United Nations, 1988).

In 2011, the UN Secretary-General launched the "Sustainable Energy for All Initiative" (SE4ALL) to achieve 30% of the global renewable energy target by 2030 and emphasized that these targets should be achieved mainly through internal actions and specific targets. Resolution 151/65 of the United Nations General Assembly (UNGA) declared 2012 as the "International Year of Sustainable Energy for All" and subsequently declared 20-20-2014 as the "Decade of Sustainable Energy for All" (United Nations, 2012). In 2015, the Sustainable Development Goals were adopted through Resolution A/RES/7-7 on 25 September 2015, including 17 global and 169 regional goals. The goal is to realize the transformation of a sustainable planet by 2030. Of all the goals set, goals 7 and 13 are particularly important for the discussion in this article. Goal 7 points out that to ensure affordable, reliable, sustainable, and modern energy sources for all and indicates that large amounts of renewable energy and fossil fuels are needed. By 2030, the share of renewable energy in the global energy structure will increase significantly (United

Nations, 2015). In addition, Goal 13 calls on the international community to take immediate action to address climate change and its impact. Understandably, as renewable promised, green, pollution-free energy can be used as a tool to combat climate change. Nevertheless, despite the ambitions of these SDGs, the survey of the SDG progress report shows that this progress is not satisfactory, and most of the activities related to renewable energy are electricity, not the final achievement. Agreement. Paris was ratified in 2016. The United Nations Framework Convention on Climate Change (UNFCCC) is not legally binding on materials but is binding on reports.

Regarding whether to adopt binding or non-binding legal documents, the international community is divided into different platforms. It is important to decide on this issue because the provisions of the legal documents necessary to reduce greenhouse gas emissions are used indirectly to promote renewable energy development. Developing countries are more willing to seek legal instruments because they have the largest greenhouse gas emissions, imposing a more pronounced burden on developing countries' climate change. In addition, because national legal systems have different effects on the signing and ratifying of international instruments at the municipal level, some countries may prefer to adopt them. Therefore, these countries tend to be flexible in this regard. However, the international community has already faced it. In this regard, the "Paris Agreement" seems to mix the needs of the two countries. Even though not all terms of the agreement are binding on the signatory, it does include some terms that can ensure transparency, accountability, and accuracy, and these terms should be finally considered when assessing the long-term importance of the agreement.

RENEWABLE ENERGY LAW

International law mainly concerns the rights and responsibilities of more than one country, while energy issues are more regarded as municipal issues(United Nations, 2016). As a result, energy law and energy law issues are being developed as international disciplines (Omorogbe, 2008; Bradbrook, 1996). However, a comprehensive set of laws, norms, and norms on energy and supply systems is difficult within international law and regulations. Although some attempts have been made to define the scope of energy law, these disciplines are considered underdeveloped (Gunningham, 2012) and maybe one of the most complex areas of law due to their conflict with many other laws (Heffron, 2016). The Energy Law stipulates rules and regulations related to the exploration, exploitation, distribution, exploitation, development, and supply of coal, oil, and natural gas reserves and nuclear energy to a certain extent. In the case of transportation, supply, and sale of energy, this situation considers that the role of international energy law and renewable energy very fragmented and largely incompatible (Bradbrook, 2008; Cottier, 2009) and complex (Cherp, 2011).

The Renewable Energy Law is a part of the Energy Law, which solves various issues related to the development, implementation, and commercialization of energy generated by renewable energy. Furthermore, since this type of law encourages renewable energy development in renewable energy development, land use, housing, and financial issues faced by entrepreneurs, projects such as food tariffs (FiTs), etc., other economic law incentives are also carried out following context. Due to these factors, at the international level, renewable legal issues have become a major challenge in public international law and may be used as case studies in legal "divisions" (Leal-Arcas, 2016).

It is important to discuss renewable energy from the perspective of international law. As an important tool for mitigating the impact of climate change, international law helps regulate the government's normative behavior; However, energy is mainly regarded as an internal issue, cooperation between countries and a compilation of technology transfer commitments are vital in energy security and stability of the states (Cherp, 2011). In the absence of any specific and direct international instruments on renewable energy, the existing literature in this field discusses international environmental law, especially sustainable international development law, and international economics (trade and investment), environment (climate, water, and biodiversity), and social rights (human rights and social development) are considered important components of the International Sustainable Development Law. Based on this concept, renewable energy law can be derived that from all aspects, the right to development will be regarded as human rights, climate change, trade, economics, and investment law.

Necessity to Develop and Use Renewable Energy

As mentioned, in recent years, special attention has been paid to the development of the use of renewable energy sources. Now the question that needs to be answered is why this is the case. Of course, since this issue itself requires a detailed discussion, a detailed description of it is not included in this space and will be provided only for the sake of a brief overview of the need to use renewable sources. Over the years, the most important reason for the increase in the average temperature on the ground has been the increase in greenhouse gas emissions (IPCC, 2007). Fossil fuels have played the most important role in greenhouse gas emissions. At the same time, climate change has been described as the most serious and long-term environmental challenge that can potentially affect all vital aspects of the planet. "Scientists believe that to reduce the harmful effects of global warming, and the average temperature should not exceed 2 degrees Celsius by 2050" (Watkins et al., 2007).

This has led to the adoption and implementation of several international instruments to reduce greenhouse gas emissions. Although applicable international instruments do not make direct reference to renewable energy substitution in this area, at present, the reduction of greenhouse gas emissions is largely achieved by the replacement of renewable energy sources. Therefore, the first reason raised in this context is the destructive effects of fossil fuels in increasing greenhouse gas emissions.

In this regard, it should be said that Iran currently has a significant share in the emission of greenhouse gases in the region and the world with a growing trend (Moradi & Aminian, 2010). In particular, as the most important source of greenhouse gas production, the energy sector has a major share in greenhouse gas emissions.

The second proposition that justifies the need to accelerate the use of renewable energy is the issue of limited fossil energy resources. Global energy consumption has increased by 45% over the past 25 years and will increase by 39% over the next 20 years.

The amount of natural gas consumption in 2000 was equal to 11724Bcm (BP, 2001) which reached 31690Bcm in 2010" (BP, 2011). Crude oil consumption is also on a completely upward trend, albeit with a lower slope than natural gas. Crude oil consumption in 2000 was 606000 barrels per day. This amount reached 873000 in 2010. Of course, the number of resources discovered and the amount of fossil energy production are also rising. But for now, it is predicted that with the same energy consumption, fossil energy sources can be exploited for the next 50 years, after which it is in a state of ambiguity. Of course, the predictions made in this regard are not definitive; "Because several factors, such as the discovery of new resources, the exploitation of unconventional resources, the increasing rate of exploitation of new reservoirs, the environmental costs and the high costs of exploiting unconventional resources, will be exacerbated by the current fossil fuels (Richter, 2010).

The third issue that needs to be addressed is the issue of energy supply. This issue can be examined nationally and internationally. In general, fossil energy sources are widely distributed around the world. The geography of resource distribution has led to many costs being incurred to transfer fossil fuels and some of the events that occur during the transfer. Also, significant costs are usually imposed due to the waste of resources along the transmission route.

These factors, along with other energy supply factors that are beyond the scope of this discussion, have led to the "decentralization" of "energy sources" to the attention of national and international actors (Smith & Taylor, 2008).

STATE SOVEREIGNTY AND REGULATIONS ON RENEWABLE ENERGY

With a brief description of renewable energy globally, the main topic can be dealt with, studying international law and renewable energy principles and rules. First, one of the most important challenges is addressed: governments' sovereignty over their natural resources. If there be no legislation on renewable energy in national regulations, international regulations will face certain difficulties.

The sovereignty of states over natural resources has been accepted as one of the principles of international law. The principle of sovereignty over natural resources means that states can regulate energy resources within their territory. Except in cases prohibited by international law. This principle has not only been emphasized in the declarations of the UN General Assembly but has been widely recognized in the "practice of states" (Schrijver, 1997). Therefore, there is no doubt that this principle is respected in international law, to the extent that some authors believe that this principle is now considered part of "public international law." As Shamsaii (2006) points out, this principle entered the field of international law from the 1960s onwards, following the independence-seeking approach of formerly colonized countries. This principle was primarily based on the human rights instruments (Lewis, 2020; Alston, 2020) and the recognition of the right to self-determination for nations, especially the newly independent states. This principle is also explicitly mentioned in the Stockholm Declarations. Thus, the Stockholm Declaration limited state sovereignty over natural resources to environmentally friendly use. Following the Stockholm Declaration, the focus on environmental protection in the exploitation of natural resources accelerated. To the extent that the concept of sustainable development was explicitly mentioned in the Rio Declaration (Mohseni, 2013). According to the principle of the Rio Declaration, first, the sovereignty of governments over their resources, concerning environmental policies and national development, and second, the responsibility of governments to control the actual activities in their territories are referred to as the concept of "preventing environmental damage." Thus, the emphasis on the sovereignty of states over their resources has, of course, been taken into account in the context of national environmental regulations (Najafifard, 2014).

But the question that comes to mind here is whether this principle is considered as one of the rules of international law or not? The need to answer this question arises from the fact that if that principle of the rule of thumb is considered, its observance will bind all subjects of international law. Therefore, this principle will prevail over other principles of international law. As a result, all international treaties and instruments in conflict will be considered null and void. As a result, it will be difficult to impose restrictions on governments' domestic policies and laws by creating international obligations to require the development and use of renewable resources(Wilkins, 2010).

On the other hand, if this principle is not considered a rule of thumb, it does not take precedence over other principles of international law and can even be modified in the shadow of other international rules. In answer to this question, Schrijver argues that, contrary to the strong position of this principle in international law, it cannot be considered part of the rules of jurisprudence. The most important part of Eschzaifer's argument is based on the fact that if this principle is taken from the rules of *jus cogens*, all international treaties contrary to it will automatically be annulled. At the same time, many international treaties use the usual contractual terms, such as stability, international arbitration, etc., limiting the principle of state sovereignty over resources(Spataru, 2017). Governments also have freedom of action in negotiating and concluding international agreements. Therefore, it is very difficult to take this principle into account as a rule of thumb; Because the current practice of international law shows the opposite.

This view is endorsed by the growing trend of global economic development through foreign investment and international regulations approved and implemented by international organizations in this field; Because governments have in many cases been required to enforce international law instead of their domestic law. Besides, in many cases, governments are required to cooperate to protect the common environment. Thus, in practice, governments' principle of sovereignty of natural resources is significantly limited and applicable in the light of other international regulations. For example, Article 18 of the Energy Charter Treaty refers to the sovereignty of states over energy sources. In this article, the sovereignty of states over resources is recognized if it is exercised following international law. Moreover, in this article, the exercise of sovereignty over resources should also not affect the treaty's purpose to develop access to various types of energy(Leal-Arcas, 2016).

INTERNATIONAL LAW AND RENEWABLE ENERGY

Given the transboundary effects of climate change, the need for energy supply, and ultimately sustainable development, international law, albeit scattered, has addressed the use of alternative energy sources. Of course, in this context, our aim is not to examine in detail the existing provisions but merely to reach a general conclusion on the status of renewable energy in the light of international law is sufficient for the present article. In this section, the existing international documents and the principles of international law will be examined, respectively. The legal hierarchy will first examine the applicable documents, public international law, and the existing soft law in this field.

Renewables and Hard Law

Two binding instruments have regulations in renewable energy at the international level: the Energy Charter Treaty and the Kyoto Protocol. The Energy Charter Treaty is the first and only international treaty on international cooperation exclusively in energy. The main purpose of drafting the Energy Charter Treaty is to facilitate investment in the energy sector by defining the necessary provisions to reduce the non-commercial investment risks, removing trade and transit barriers. It is worth mentioning that Iran joined the Energy Charter Treaty in 2002 as a supervisory member. Therefore, because the authorities are considering accession to the Energy Charter Treaty, it will be important to pay attention to the Energy Charter Treaty provisions on renewable energy development. Article 1 (5) refers to the definition of economic activity in the energy sector. According to this article, electricity generators' production and operation are considered one form of economic activity in the contract. However, according to the final law of the European Energy Charter Conference, there is no reference in this article to the production of electricity through renewable energy sources, which serves as an interpreter of the Energy Charter Treaty, which includes electricity from renewable energy sources. Therefore, the production of energy from renewable sources falls within the scope of the Energy Charter Treaty.

According to Article 19 (D) of this Treaty, which deals with environmental aspects, States Parties shall pay particular attention to the development and use of renewable energy sources. However, as authors such as Bruce and Bradbrook have pointed out, the wording of this article does not reflect the binding commitment of member states to develop and strengthen the use of renewable energy sources (Bradbrook, 2013: 242; Bruce, 2013). However, there is no denying that governments are explicitly obliged to develop renewable energy in the Energy Charter Treaty. But, of course, since no specific quantitative goals have been mentioned in the development of renewable energy sources, the guarantee of the implementation of this article will remain in an aura of ambiguity.

Climate change and its effects on the environment led to drafting the United Nations Convention on Climate Change in 1992. This internationally binding document is very much related to renewable energy. In this document, member states agreed to reduce greenhouse gas concentrations in the atmosphere to some extent possible environmental hazards. However, this document does not mention any specific commitment to developing the use of renewable energy sources to reduce greenhouse gas emissions. Therefore, governments are free to determine the methods needed to reduce greenhouse gas emissions "through human activities." In this regard, according to the Kyoto Protocol, which was drafted within the framework of the Convention, governments were divided into two general categories, and different obligations were set for each of these two categories (Telesetsky, 1999).

Under the protocol, developed countries (Annex A countries) must reduce greenhouse gas emissions individually or in partnership by at least 5% lower than in 1990 emissions by 2008 and 2012. Of course, some members, such as the European Union (15 members at the time), set higher targets (8%) for reducing their greenhouse gas emissions. On the other hand, developing countries (non-annexed countries "A") must also develop a program to reduce greenhouse gas emissions. Of course, the text of the Kyoto Protocol does not include any binding commitment to determine the types of tools to reduce greenhouse gas emissions, and governments are fully free to determine the mechanisms to meet their obligations. Nevertheless, Article 2 (1) obliges developed countries (Appendix (a) to develop and use renewable energy sources following their internal conditions, which is not in fact of a binding nature. However, one of the most important mechanisms implemented to achieve the set goals has been using renewable resources in practice. For example, in the European Union(EU), in addition to the economic crisis that led to a decrease in energy demand, the use of renewable energy has played a significant role in reducing greenhouse gas emissions in the realization of the ceiling of reducing greenhouse gas emissions. In addition to the above, Article 12 of the Kyoto Protocol refers to the Clean Development Mechanism (CDM). Following paragraph 2 of the said article, the purpose of this mechanism is to assist on the one hand the non-annexed countries "A" to achieve sustainable development and ultimately to achieve a reduction in greenhouse gas emissions and, on the other hand, to "help the countries." Their commitment is to reduce greenhouse gas emissions. One of the most important projects that have been considered and implemented under CDM is renewable energy projects (Rahimi et al., 2004).

In another study examining the effects of the Kyoto Protocol on reducing global emissions, the authors concluded that the adoption and implementation of the Kyoto Protocol would increase the use of renewable energy sources by up to 3% (Aichele & Felbermayr, 2013). Therefore, what is clear is the importance of implementing the Kyoto Protocol in increasing the rate of development and utilization of renewable energy. However, because of the issues raised, it can be stated that in the applicable documents, while the capacity for the development of renewable energy has been considered, no obligation has been assumed to take into account the sources.

Renewables and Soft Law

Apart from the issues that may arise about the nature and guarantee of soft law enforcement, international instruments have already been developed in the form of soft law in many areas, especially in the environment. Thus, as one author defines, soft law refers to those rules that do not guarantee legal enforcement but have enforcement effects (Yanfang, 2011). Of course, as another author points out, soft law, although not inherently binding, in practice guarantees political performance that may lead to legal effects (Omorogbe, 2008). The authors have expressed various views on nature and guarantee the implementation of soft rights beyond this study's scope. However, given the growing importance of soft law in the development of international law, it will not be useless to review soft law documents on renewable energy.

The first international document to be adopted on renewable energy sources is the United Nations Conference on New and Renewable Energy. The most important outcome of this international conference is the need to develop the use of renewable energy, and secondly, the formation of a committee to develop and use renewable energy sources. Of course, this conference does not include implementing executive policies on renewable energy development, and even the committee was eventually merged into the UN Commission on Sustainable Development. In principle, the development of renewable energy sources was pursued within the framework of sustainable development, until in 1997, the General Assembly of the United Nations adopted a plan to implement Agenda 21 further. Article 45 of the resolution emphasizes the need for international participation to promote the use of renewable energy. In paragraph 46, the need to transfer technical knowledge from developed countries to developing countries is considered to such an extent that it will enable developing countries to increase energy production from renewable sources. The need for developing countries to systematize the use of renewable energy was also emphasized. However, there is no obligation on the part of the states to comply with the provisions of the said resolution, which in principle is not binding(Ruta, 2012).

International efforts to reduce greenhouse gas emissions and explain sustainable development in 2002 adopted the Johannesburg Executive Plan. As mentioned, this document is one of the most comprehensive international instruments ever adopted on renewable energy. Chapter 3 of the Johannesburg Executive Plan is devoted to changing unsustainable patterns of production and consumption. Clause 20 (c) of the plan encourages governments and other international actors to use other types of energy to increase the share of renewable energy. Section 20(e) emphasizes the diversification of various energy sources, including fossil fuels, hydroelectric sources, and renewable energy sources. Besides, increasing the global share of renewable energy in total global energy production is emphasized. The importance of investing in renewable energy development by developed countries and international financial organizations in developing countries has also been emphasized on several other occasions.

Therefore, in the Johannesburg Executive Plan, paying attention to the development of renewable energy in the energy portfolio of countries is more explicitly emphasized. To the extent that the need to increase the share of these energies in total global energy production is also mentioned.

In 2005, the UN resolution on the World Summit outcome emphasized the need to take more effective steps to develop and support renewable energy. Finally, in 2011, the UN General Assembly adopted a resolution on sustainable development (Spataru, 2017).

The resolution designated 2012 as the "Year of Sustainable Energy for All." Of course, this document does not directly reference governments' commitments to developing the use of renewable energy sources. However, the document calls on the UN Secretary-General to report on implementing the resolution, in particular, to develop and promote the use of renewable energy. The report, entitled "Promoting New and Renewable Energy Sources," seeks to diversify programs at the national and international levels to promote sustainable development by considering different policies for developing and developing countries. The most recent document adopted on renewable energy development is the 2013 UN General Assembly Resolution on the "Promotion of New and Renewable Energy Sources."

Examining the existing soft law documents on renewable energy, it can be concluded that the emphasis on the use and development of renewable energy has grown exponentially in recent years and has been emphasized in many international instruments. Contrary to these emphases, however, no specific goals have been set for increasing the global share of renewable energy in the global portfolio. There can be several reasons for this; According to the authors, the most important factor is the difference in energy supply sources in different countries and technological and economic capacities for the development of this group of energy sources (Nhamo, 2020); This is because these distinctions, especially considering the very high costs (at least in the short term) for the operation of such projects, have led to the difficulty of regulating comprehensive global regulations in this field. At the same time, the effects of non-binding documents in supporting the development of renewable energy sources cannot be ignored. For example, as mentioned at the end of the Rio + 20 conference, the document "The Future We Want" specifically mentions the need to develop renewable energy. In this regard, we can mention the steps taken by UNESCO to implement the "Renewable Energy for Us" program. The program will use renewable energy sources to expand renewable energy sources to reduce greenhouse gas emissions to provide energy in biosphere reserves and World Heritage Sources to renewable energy sources(Najafifard, 2014).

Public International Law

As one of the main sources of international law, public international law plays an important role in regulating international relations between states. The rules of public international law become enforceable once States have recognized them in practice. The authors have expressed different views on how to form and the elements that constitute the rules of international custom (Lepard, 2010). However, this chapter examines those rules of international custom that are recognized in international law. Although the rules on the development of renewable energy are not currently explicitly accepted within the framework of international custom, some of the accepted rules can be referred to as public international law. For example, the two non-damage and sustainable development principles are among the most relevant principles regarding renewable energy development. The principle of non-damages, which is now an integral part of public international law, first appeared in the arbitration of The Trail Smelter Case in 1941 in British Columbia, Canada, and then in the case of the Carrefour Canal in The International Court of Justice has been emphasized (Barboza, 2010). This principle is also recognized in the draft of the United Nations Commission on International Law on the Prevention of Cross-Border Damage Caused by Dangerous Activities in 2001.

Furthermore, this principle has been used in several cases in the International Court of Justice. However, numerous opinions have been expressed about the definition and limits of this principle. For example, Article 3 of the draft law of the United Nations Commission on the Prevention of Cross-Border Damage resulting from Dangerous Activities defines: Significant cross-border damage or in the event of an accident to reduce the risks.

On the other hand, Article 21 of the Stockholm Declaration states: "States, following the Charter of the United Nations and the principles of international law, have the right to use their resources under their respective environmental policies." Activities carried out within their sovereignty or control shall not damage the environment of other countries or regions outside the national sovereignty. With a slight change, the original Rio Declaration provides the same definition as the Stockholm Declaration of this principle. In its Advisory on Nuclear Weapons, the International Court of Justice defines these principles and the obligations of States about the prevention of transboundary damage as follows: The environment of other countries or the sphere of the sovereignty of countries is now part of the body of international law".

Examining the rulings of the International Court of Justice and the rulings of the International Court of Justice in this regard, it can be concluded that the principle of non-damage is currently one of the main principles of international environmental law. And governments are now obliged to comply with it. The question that comes to mind now is the relation of this principle to renewable energies. For example, RK. In the case of Nauru-Australia (1992), some authors argue that the principle of non-harm includes protecting the earth's atmosphere; they state that the earth's atmosphere, like the oceans and seas, is part of its common heritage (Birnie & Boyle, 2002). Therefore, the same protection of the oceans in international law should apply to the atmosphere. According to the principles of customary international law, governments have international responsibility for greenhouse gas emissions.

However, as some authors have pointed out, there is no standard for determining the due diligence of states in this area (Lyster & Bradbrook, 2006); Because the standards set out in the Long-Range Transboundary Air Pollution Convention or the Ozone Layer Convention are not analogous to determining the appropriate effort to emit greenhouse gases. The standards contained in the Kyoto Protocol are also not like international custom; Because they are not subject to universal consensus. Some authors also argue that the principle of non-damage can be applied if the error or intention of governments to cause cross-border damage is established. Otherwise, the damage can not be considered a violation of international law and responsible for governments (Brownlie, 2008). However, as another author acknowledges, the principle of non-damage does not necessarily imply governments' intent or error; this principle applies to non-prohibited acts and is common in international law (Borowy, 2014). In this regard, numerous lawsuits have been filed nationally and internationally against countries and companies that emit greenhouse gases (Gunningham, 2012).

Many authors also emphasize the international responsibility of states for damages that have been introduced in terms of greenhouse gas emissions (Cherp, 2011; Cottier, 2011). But so far, no lawsuit has been filed that has resulted in a ruling on reducing pollutants or compensation. Therefore, although it can be argued that governments are obliged to prevent damage to the earth's atmosphere and thus reduce greenhouse gases in the main shadow of no damage, the limits of damage and the commitment of governments to prevent it have not been determined. At the same time, For governments to be held accountable in this regard, there are numerous problems, such as standard, appropriate efforts, proving the causal relationship, the competent authority, etc., that have a long way to go to solve them, therefore, in the light of this principle, governments can not be expected to develop renewable energy, which will reduce greenhouse gas emissions.

Renewable Energy and International Institutions

The role of international organizations in shaping various fields of international law is well known. Different international organizations are working hard to establish an appropriate legal framework for energy and energy governance systems.

United Nations agencies are actively working to promote renewable energy worldwide. UNEP is the leading United Nations organization supporting renewable energy advertising. UNEP's major contribution to the Renewable Energy Law is the publication of the "Handbook" (2007) and "Guide" (2016) to assist designers in developing countries, and the annual report "Global Renewable Energy Investment Trends," which involves various themes of renewable energy investment cases. UNDP has been active in transforming the global energy system for more than two decades by improving energy efficiency and using renewable energy systems. Recently, UNDP released the "Sustainable Energy Strategy Note 2017-2020: Sustainable Energy Supply in Climate Change", the first UNDP program, mission, method, guidance, principle, and focus on sustainable energy. To this end, it puts renewable energy in its vision and mission (United Nations, 2016). This is an important initiative of the UNDP program because the organization has played an effective role in development activities mainly from countries in Asia and sub-Saharan Africa, where disadvantaged and marginalized people cannot afford electricity.

The fact is that despite the active participation of United Nations agencies in global development activities, their energy activities are relatively ambiguous (Gunningham, 2012). There is a lack of coordination and trust among these organizations; that is, "the regime's system is ineffective for renewable energy to achieve sustainable energy" (Heffron, 2016). Therefore, the United Nations Department of Energy, composed of 21 member organizations, was established in 2004 to respond to energy Challenges; the United Nations system has established institutional mechanisms to promote coordination and coherence (United nations, 2015) The International Energy Agency (IEA) was established in 1974 after the oil crisis of 1973-79. It aims to coordinate the response of oil-importing countries to major oil supply disruptions. It is the most famous international organization dedicated to energy security, environmental awareness, and economic development. IEA was established following the decision of the OECD Council and is an autonomous organization within the framework of the OECD (International Energy Planning Agreement). But this is separate from the OECD and has been recognized by the "United Nations Framework Convention on Climate Change." Historically, the main focus of the IEA is neither renewable energy nor conventional fuels. The Renewable Energy Working Group (REWP) was recently established. Its task is to provide renewable-relevant technology and policy recommendations and projects to various IEA organizations. A very authoritative global energy outlook is published every year, which is the main reference for energy-related stakeholders by IEA. Founded in 2011, International Renewable Energy Agency (IRENA) is a leading international organization focusing on renewable energy. As of September 2017, it has 152 members and 28 future members. Through the 2009 IRENA Charter, the parties expressed their willingness to promote the recognition of renewable energy in terms of sustainable development and firmly believe that renewable energy provides a huge opportunity to solve and gradually reduce energy security issues and unsustainable energy prices. Therefore, IRENA enjoys the same privileges and immunities as the United Nations (Gielen, 2019), although IRENA has no explicit or implicit authority to set binding RE targets.

At the International Renewable Energy Conference held in Bonn in 2004, the International Geothermal Association, International Hydropower Association, International Solar Energy Association, World Biotechnology Association, and World Wind Energy Association formed an international alliance called Renewable Energy Alliance. It was later renamed to the 21st Century Renewable Energy Policy Network or REN21. The network connects a wide range of stakeholders and provides international leadership in promoting renewable, focusing on the annual "Report on the Global State of Renewable." In addition, various development banks such as the World Bank, Asian Development Bank, African Development Bank, Inter-American Development Bank, and New Development Bank play an important role in promoting renewable energy through financing projects. However, these banks did not have a comprehensive strategy to incorporate renewable energy into their overall loan plans (Wilkins, 2010). Therefore, it is clear that many non-governmental actors have been involved in promoting and developing renewable energy; although their activities do not overlap in some cases, IRENA is in a leading position.

SOLUTIONS AND RECOMMENDATIONS

Besides, although policies such as the transfer of energy production technology through renewable energy sources or the creation of financial incentives for investment in this sector have been proposed in the international instruments, it still seems relevant. And it is certainly easier to formulate and implement these policies in international law than to create international obligations for renewable energy development.

FUTURE RESEARCH DIRECTIONS

Nowadays, more researchers study the renewable energy industry, market, etc., and there is a significant interest and attraction to increasing the share of renewables in national energy portfolios. This global convergent ambient for renewable energy development owes much to international conventions, legal principles, and public international laws. This chapter suggests that the current successful trend continues, and a deep energy transition occurs in the decades to come.

CONCLUSION

The need to develop these sources of energy production is undeniable by examining the documents and principles of international law on renewable energy. This can be achieved by examining existing international instruments and in the light of the principles of international law, such as the principle of non-damage and the principle of sustainable development. China, which is not primarily responsible for reducing greenhouse gas emissions under the Kyoto Protocol, is currently one of the largest energy-producing countries through renewable sources. On the other hand, the right to a healthy environment has been mentioned as one example of the third generation of human rights in several international documents, and renewable energy is one of the most important sources of energy production. They reduce environmental pollution, especially air pollution. The authors believe that international law does not currently impose a specific obligation on governments to develop renewable energy for sustainable development and reduce the effects of greenhouse gases. At the same time, establishing such obligations is possible through international treaties and is not considered contrary to the sovereignty of states over their resources. For example, we can refer to EU laws in this regard. EU countries agreed in 2009 to supply 20% of total energy production through renewable energy sources by 2020, which has largely achieved this goal.

It is important to note that energy governance is a complex issue, and there is no suitable "best model" for renewable energy governance. At the national level, there is a gap between what is needed and what the government is doing, while at the regional and global levels, collective action challenges often create insurmountable obstacles. Several bills can be considered, including the promulgation and promulgation of special renewable energy laws, including a series of policies that support the introduction of renewable energy power generation prices. Providing incentives and subsidies can help bring about positive changes in Europe. International investment treaties and binding legal instruments can guarantee universal access to energy for all, although such plans require political commitment and excessive cross-border tendencies. Regional and institutional frameworks indicate that renewable energy has become an important issue for further developing cross-border countries. From Africa, South America to Asia, and other regions in developed countries, indirect measures (if not direct measures) taken in most parts of the world should be attributed to using and promoting renewable energy for sustainable change. Global and regional institutions are also actively promoting renewable energy and its potential contribution to sustainable development for all. Therefore, the use and consumption of renewable energy have promoted sustainable development in many ways. The United Nations has approved the Sustainable Development Goals and recognizes the importance of renewable energy in providing clean and cost-effective energy for all. Generally, in the context of the 21st century, the preference and dependence on rare piles of the earth is an internal, regional, and global reality, which cannot be denied or denied in any country.

ACKNOWLEDGMENT

This research received no specific grant from any funding agency in the public, commercial, or not-for-profit sectors.

REFERENCES

Adib, R., Murdock, H. E., Appavou, F., Brown, A., Epp, B., Leidreiter, A. & Farrell, T. C. (2015). *Renewables 2015 global status report.* REN21 Secretariat Press.

Aichele, R., & Felbermayr, G. (2013). The Effect of the Kyoto Protocol on Carbon Emissions. *Journal of Policy Analysis and Management, 32*(4), 731–757. doi:10.1002/pam.21720

Alston, P. (2020). The Committee on Economic, Social and Cultural Rights. *NYU Law and Economics Research Paper, 15*, 20-24.

Barboza, J. (2010). *The Environment, Risk and Liability in International law*. Martinus Nijhoff Publishers.

Birnie, P. W., & Boyle, A. E. (2002). *International Law and the Environment.* Oxford University Press.

Biswas, M. R. (1981). The United Nations Conference on New and Renewable Sources of Energy: A review. *Mazingira, 5*, 11–16.

Borowy, I. (2014). *Defining sustainable development for our common future. A history of the world commission on environment and development.* Routledge Press.

Bradbrook, A. (2008). *The development of renewable energy technologies and energy efficiency measures through public international law.* Oxford University Press. doi:10.1093/acprof:oso/9780199532698.003.0006

Bradbrook, A. J. (1996). Energy law as an academic discipline. *Journal of Energy & Natural Resources Law, 14*(2), 193–217. doi:10.1080/02646811.1996.11433062

Bradbrook, A. J. (2013). *International Law and Renewable Energy: Filling the Void.* Duncker & Humblot Press.

British Petroleum. (2001). *Statistical Review of World Energy.* British Petroleum press.

British Petroleum. (2011). *Statistical Review of World Energy.* British Petroleum press.

Brownlie, I. (2008). *Principles of Public International Law.* Oxford University Press.

Bruce, S. (2013). International law and renewable energy: Facilitating sustainable energy for all. *Melbourne Journal of International Law, 14*, 18.

Cherp, A., Jewell, J., & Goldthau, A. (2011). Governing global energy: Systems, transitions, complexity. *Global Policy, 2*(1), 75–88. doi:10.1111/j.1758-5899.2010.00059.x

Cordonier Segger, M. C., & Khalfan, A. (2004). *Sustainable development law: principles, practices and prospects.* Oxford university press.

Cottier, T., Malumfashi, G., Matteotti-Berkutova, S., Nartova, O., De Sepibus, J., & Bigdeli, S. Z. (2011). *Energy in WTO law and policy: The prospects of international trade regulation from fragmentation to coherence.* WTO Press. doi:10.1017/CBO9780511792496

Drake, L. (2016). International law and the renewable energy sector. The Oxford Handbook of International Climate Change Law, 1, 357.

Gielen, D., Boshell, F., Saygin, D., Bazilian, M. D., Wagner, N., & Gorini, R. (2019). The role of renewable energy in the global energy transformation. *Energy Strategy Reviews, 24*, 38–50. doi:10.1016/j.esr.2019.01.006

Gunningham, N. (2012). Confronting the challenge of energy governance. *Transnational Environmental Law, 1*(1), 119–135. doi:10.1017/S2047102511000124

Heffron, R. J., & Talus, K. (2016). The evolution of energy law and energy jurisprudence: Insights for energy analysts and researchers. *Energy Research & Social Science, 19*, 1–10. doi:10.1016/j.erss.2016.05.004

Holden, E., Linnerud, K., & Banister, D. (2014). Sustainable development: Our common future revisited. *Global Environmental Change, 26*, 130–139. doi:10.1016/j.gloenvcha.2014.04.006

Intergovernmental Panel on Climate Change. (2007). *Climate change: The physical science basis: Summary for policymakers.* IPCC Press.

International Energy Agency. (2014). *World energy outlook 2014: Executive summary.* International Energy Agency Press.

International Energy Agency. (2015). *Energy and climate change: World energy outlook special report.* International Energy Agency Press.

International Energy Agency & Birol, F. (2013). World energy outlook 2013. International Energy Agency Press.

International Renewable Energy Agency. (2017). *Renewable Energy Highlights.* International Renewable Energy Agency Press.

Leal-Arcas, R., & Minas, S. (2016). The micro level: Insights from specific policy areas: Mapping the international and European governance of renewable energy. *Yearbook of European Law, 35*(1), 621–666. doi:10.1093/yel/yew022

Lepard, B. D. (2010). *Customary International Law: A New Theory with Practical Applications.* Cambridge University Press. doi:10.1017/CBO9780511804717

Lewis, M. K. (2020). Why China Should Unsign the International Covenant on Civil and Political Rights. *Vanderbilt Journal of Transnational Law, 53*, 131.

Lund, H. (2007). Renewable energy strategies for sustainable development. *Energy, 32*(6), 912–919. doi:10.1016/j.energy.2006.10.017

McGlade, C., & Ekins, P. (2015). The geographical distribution of fossil fuels unused when limiting global warming to 2 C. *Nature, 517*(7533), 187–190. doi:10.1038/nature14016 PMID:25567285

Michalena, E., & Hills, J. M. (2013). *Introduction: Renewable Energy Governance: Is it Blocking the Technically Feasible in Renewable Energy Governance.* Springer Press.

Mohseni, R., & Shokri, M. (2013). Study of carbon dioxide emissions in Iran with a fuzzy approach. *Iranian Journal of Energy, 16*(1), 1–16.

Moradi, A., & Aminian, M. (2009). Iran's greenhouse gas emissions in 2009. *Nesha Alam Magazine, 13*, 55–59.

Mulyana, I. (2016). The Development of International Law in the Field of Renewable Energy. *Hasanuddin Law Review, 1*(1), 38–60. doi:10.20956/halrev.v1i1.213

Najafifard, M., & Mashhadi, A. (2014). *Green Economy Based on Sustainable Development in the Light of the Rio+20 Declaration. International Conference and Online Green Economy*, Tehran, Iran.

Nhamo, G., Nhemachena, C., Nhamo, S., Mjimba, V., & Savić, I. (2020). *SDG7-Ensure Access to Affordable, Reliable, Sustainable, and Modern Energy.* Emerald Group Publishing. doi:10.1108/9781789737998

Omorogbe, Y. O. (2008). *Promoting sustainable development through the use of renewable energy: The role of the law.* Oxford University Press. doi:10.1093/acprof:oso/9780199532698.003.0003

Ottinger, R. L., Robinson, N., & Tafur, V. (2005). *Compendium of sustainable energy laws*. Cambridge University Press. doi:10.1017/CBO9780511664885

Rahimi, N., Kargari, N., & Khodi, M. (2004). A Study of the PAC Development Mechanism in the Kyoto Protocol and the Financing of Projects. *Iranian Journal of Energy*, *21*, 57–71.

Richter, B. (2010). *Beyond Smoke and Mirrors: Climate Change and Energy in the 21st century*. Cambridge University Press. doi:10.1017/CBO9780511802638

Ruta, M., & Venables, A. J. (2012). International trade in natural resources: Practice and policy. *Annual Review of Resource Economics*, *4*(1), 331–352. doi:10.1146/annurev-resource-110811-114526

Schrijver, N. (1997). *Sovereignty Over Natural Resources: Balancing Rights and Duties*. Cambridge University Press. doi:10.1017/CBO9780511560118

Shamsaii, M. (2006). International Law and Sustainable Development. *Law and Politics Research*, *19*, 7–24.

Smith, Z.A. & Taylor, K.D. (2008). *Renewable and Alternative Energy Resources: A Reference Handbook*. ABC-CLIO Press.

Spataru, C. (2017). *Whole energy system dynamics: Theory, modelling and policy*. Taylor & Francis Press. doi:10.4324/9781315755809

Telesetsky, A. (1999). Kyoto Protocol. *Ecology Law Quarterly*, *26*, 797–813.

United Nations. (1972). *United Nations Conference on the Human Environment*. United Nations Press.

United Nations. (1988). *Protection of global climate for present and future generations of mankind*. General Assembly Press.

United Nations. (2010). *Delivering on Energy: An Overview of Activities of UN Energy and Its Members*. United Nations Press.

United Nations. (2012). *United Nations General Assembly Declares 2014–2024 decade of Sustainable Energy for All*. United Nations Press.

United Nations. (2015a). *Transforming our world: The 2023 agenda for sustainable development*. United nation press.

United Nations. (2015b). *Transforming our world: The 2030 agenda for sustainable development*. Seventieth General Assembly Press.

United Nations. (2016). *Sustainable development goals report 2016*. United nation press.

Warschauer, M., & Liaw, M. L. (2010). *Emerging Technologies in Adult Literacy and Language Education*. National Institute for Literacy Press. doi:10.1037/e529982011-001

Watkins, K. (2007). *Human Development Report 2007/2008 - Fighting Climate Change: Human solidarity in a divided world*. United Nations Press.

Wilkins, G. (2010). *Technology transfer for renewable energy*. Taylor & Francis. doi:10.4324/9781849776288

Yanfang, L., & Wei, C. (2011). Framework of Laws and Policies on Renewable Energy and Relevant Systems in China under the Background of Climate Change. *Vermont Journal of Environmental Law, 13*(4), 823–865. doi:10.2307/vermjenvilaw.13.4.823

ADDITIONAL READING

Bruce, S. (2013). International law and renewable energy: Facilitating sustainable energy for all. *Melbourne Journal of International Law, 14*, 18.

Dursun, B., & Gokcol, C. (2014). Impacts of the renewable energy law on the developments of wind energy in Turkey. *Renewable & Sustainable Energy Reviews, 40*, 318–325. doi:10.1016/j.rser.2014.07.185

Farrell, A. E., Plevin, R. J., Turner, B. T., Jones, A. D., O'hare, M., & Kammen, D. M. (2006). Ethanol can contribute to energy and environmental goals. *Science, 311*(5760), 506–508. doi:10.1126cience.1121416 PMID:16439656

Liu, L. Q., Liu, C. X., & Wang, J. S. (2013). Deliberating on renewable and sustainable energy policies in China. *Renewable & Sustainable Energy Reviews, 17*, 191–198. doi:10.1016/j.rser.2012.09.018

Obeng-Darko, N. A. (2019). Renewable energy and power: A review of the power sector reform and renewable energy law and policy nexus in Ghana. *African Review (Dar Es Salaam, Tanzania), 11*(1), 17–33. doi:10.1080/09744053.2018.1538677

Painuly, J. P. (2001). Barriers to renewable energy penetration; a framework for analysis. *Renewable Energy, 24*(1), 73–89. doi:10.1016/S0960-1481(00)00186-5

Peeters, M., & Schomerus, T. (2014). *Renewable energy law in the EU: legal perspectives on bottom-up approaches*. Edward Elgar Publishing. doi:10.4337/9781783473199

Schuman, S., & Lin, A. (2012). China's Renewable Energy Law and its impact on renewable power in China: Progress, challenges and recommendations for improving implementation. *Energy Policy, 51*, 89–109. doi:10.1016/j.enpol.2012.06.066

Wagner, A. (2000). *Set for the 21st century-Germany's new Renewable Energy Law*. Renewable Energy World Press.

Zhao, Z. Y., Zuo, J., Fan, L. L., & Zillante, G. (2011). Impacts of renewable energy regulations on the structure of power generation in china–a critical analysis. *Renewable Energy, 36*(1), 24–30. doi:10.1016/j.renene.2010.05.015

KEY TERMS AND DEFINITIONS

Energy Charter Conference: Article 33 of the ETC establishes the Energy Charter Conference, which is the governing and decision-making body of the Organisation and has United Nations General Assembly observers status in the resolution 62/75 adopted by the General Assembly on 6 December

2007. Members consist of Countries, and Regional Economic Integration Organisations signed or acceded to the treaty and represented in the conference and its subsidiary bodies. The conference meets regularly to discuss issues affecting energy cooperation among Members, review the implementation of the treaty and PEEREA provisions, and consider new activities within the Energy Charter framework.

Energy Charter Treaty: The Energy Charter Treaty (ECT) is an international agreement that establishes a multilateral framework for cross-border cooperation in the energy industry. The treaty covers commercial energy activities, including trade, transit, investments, and energy efficiency. In addition, the treaty contains dispute resolution procedures both for States Parties to the treaty (vis-a-vis other States) and between States and the investors of other States, who have made investments in the former territory.

Energy Community: The Energy Community, also referred to as the Energy Community of southeast Europe, is an international organization established between the European Union (EU) and several third countries to extend the EU internal energy market to Southeast Europe and beyond. With their signatures, the Contracting Parties commit themselves to implement the relevant EU energy acquis communautaire, develop an adequate regulatory framework, and liberalize their energy markets in line with the acquis's acquis under the treaty.

EurObserv'ER: EurObserv'ER is a consortium dedicated to monitoring the development of the various sectors of renewable energies in the European Union. Created in 1999 by Observ'ER, the Observatory of renewable energies in France, it is composed of five other partners: ECN (The Energy research Centre of the Netherlands), IEO (EC BREC Institute of Renewable Energetic Ltd), RENAC (Renewables Academy AG), FS (Frankfurt School of finance and management) and IJS (Institut *Jozef Stefan*).

European Integrated Hydrogen Project: The European integrated Hydrogen Project (EIHP) was a European Union project to integrate United Nations Economic Commission for Europe (UNECE or ECE) guidelines and create a basis for ECE regulation of hydrogen vehicles the necessary infrastructure replacing national legislation and regulations. This project aimed to enhance the safety of hydrogen vehicles and harmonize their licensing and approval process.

International Energy Charter: The International Energy Charter is a non-binding political declaration underpinning key principles for international energy cooperation. The declaration attempts to reflect the changes in the energy world that have emerged since the development of the original Energy Charter Treaty in the early 1990s. The International Energy Charter was signed on 20 May 2015 by 72 Countries plus the EU, Euratom, and ECOWAS at a Ministerial conference hosted by the government of the Netherlands.

Jus Cogens: Jus cogens, also known as the peremptory norm, is a fundamental and overriding principle of international law. It is a Latin phrase that translates to 'compelling law.' It is absolute, which means that there can be no defense for the commission of any act prohibited by jus cogens.

Chapter 7
International Peace and Security in the Case of Climate Change

ABSTRACT

Climate change is one of the environmental problems that endanger life on Earth today; this phenomenon has caused negative consequences on the planet that affect the well-being and safety of living beings, which constitutes a great concern for humanity. For these reasons, the international community has drawn up various treaties and legal regulations, bearing in mind that to counteract this phenomenon requires all countries' cooperation and participation through effective international response, following their responsibilities regarding the damage caused to society. However, they have not managed to reverse most of the environmental problems humanity suffers as countries have not ratified international conventions, being among the most polluting worldwide. The general objective of this chapter is to establish, based on a theoretical and legislative study, the need to adhere to the international law on climate change with a view to mitigation and adaptation to its effects on an international scale.

INTRODUCTION

Environmental problems constitute a great problem in society; they are motivated by men, who tend to harm the environment. At present, many affect us; emphasizing our research on climate change as it is a recurring theme of great concern directly linked to human activity. Climate change may become the most complex and serious environmental problem of this century. In the international scientific and political community, it is recognized as one of the greatest problems that humanity must face, which can potentially significantly affect the physical conditions in which terrestrial and marine ecosystems exist; that is to say, every corner of the planet. One of the main causes that propitiate it is the high concentration of gases that cause the "greenhouse effect". Greenhouse gases (GHG) have increased considerably since the Industrial Revolution, showing the responsibility of the human being in this increase due to the indiscriminate burning of fossil fuels, livestock, deforestation, along with other economic activities responsible for the increase in the concentrations of carbon dioxide (CO_2) and methane, among others.

DOI: 10.4018/978-1-6684-7188-3.ch007

CO_2 concentrations in the atmosphere over the past 200 years have increased by almost a third. However, it is in 1960 when the figures begin to show an abrupt increase (Brown & McLeman, 2009).

Climate change has led to atmospheric warming; between 2002 and 2011 was the hottest decade since 1850 when these measurements began. Scientists believe that global average temperatures will continue to rise. Concern for this effect is not only the product of temperatures but also of the consequences of alterations, rainfall, and associated evaporation, which can lead to changes in the geographical distribution of crops, their reduction, and the levels of the seas. Some ecosystems and cultures are already at risk from climate change. Many systems with limited adaptive capacity, particularly those associated with Arctic sea ice and coral reefs, are subject to very high risks with an additional 2°C warming.

This subject has been dealt with in foreign kinds of literature by authors such as Anna Martínez Guallar in her work "Environmental Law. Climate Change in the United Kingdom "and Gustavo Alanis-Ortega in" The First Step of a Comprehensive Climate Policy in Mexico "who have supported their research in the study of legislative contributions on climate change, in countries such as Mexico and the United Kingdom. This issue has been approached from different regulatory and programmatic perspectives to create regulatory frameworks to face this environmental problem. In the national scenario, we must highlight authors such as Eduardo Orlando Planos Gutiérrez, Roger Rivero Vegas, Vladimir Guevara Velazco, and Aida Hernández Zanuy, those who developed the work "Impact of Climate Change and Adaptation Measures in Cuba," where they carry out an analysis scientific-statistical of the impact of climate change in Cuba. Still, they do not propose normative budgets for their mitigation; in addition, there is the work of Judith Cid Soto and the Leisy Pérez López, who dealt with the topic "Legal mechanisms to face climate change. Considerations about its regulation ", it studies different legal mechanisms to minimize climate change at the national and international level; Due to the importance of the protection of the environment and the treatment of climate change, this research is also directed towards the analysis of the legal norms that regulate these issues. Taking into account the topics previously discussed, international legal regulation against climate change is proposed as a new theoretical and normative vision, of which there is little treatment in international and national legal doctrine. The results that we propose to contribute with our research are the following (Penny, 2007):

- Propose a systematization through a doctrinal and legislative study about the current theoretical considerations on climate change and its impact on the environment as a legal basis.
- Carry out an analysis of international legal instruments and comparative law regulations to determine trends in international regulation in this regard.
- Provide a monographic material of interest for the study of climate change and mitigation and adaptation measures against its impacts through international legal norms that support permissive and prohibitive behaviors in this regard by the States.

The present article aims to systematize through a doctrinal and legislative study the problem of climate change so that its impact on the environment is known and carries out an exegetical-legal analysis of international legal instruments to show their legal regulation tendencies in this regard.

THEORETICAL DOCTRINAL ASPECTS OF CLIMATE CHANGE

To begin the doctrinal journey through the institution of climate change and before delving into its concept, the theoretical foundations of climate and atmosphere will first be enunciated, which are also subject to legal protection and are closely related to the subject under investigation. The word climate comes from the Greek Klima, which refers to the inclination of the Sun. It is the set of environmental conditions of a certain place and is characterized by being the average of the weather conditions, calculated through observations made over a long period (between 10 and 30 years). Climate describes the periodic succession of different types of weather that repeat themselves in a region in a characteristic way over a long period.

This exerts a great influence on our lives and the nature, the fauna and the flora of each place, the water, the crops, the way of being and the culture of each corner of the world, depend, among other factors, on the local climate. Its influence is easily perceptible in human activities based directly on ecosystems, on which the existence of humanity and the economy of each nation rests. Therefore it is understandable that climate change affects all areas of nature, including human life. Before conceptualizing climate change, it is necessary to describe the term that has been gaining more and more force and from which a series of considerations emerge. We are talking about Global Warming. Climate change should not be confused with global warming. The latter refers to the recent and continuous rise in global mean temperature near the earth's surface. Global warming is caused mostly by "increases in greenhouse gas concentrations in the atmosphere". Also, global warming is causing changes in weather patterns. This phenomenon itself is just one aspect of climate change. We can say that the main differences between global warming and climate change are (Scott, 2008):

Global warming has to do with the increase in temperature at a global level, while climate change refers to the change in climatic conditions, either globally or regionally (and not limited solely to temperature rise). Includes humidity, rain, wind, and other meteorological events. Global warming is mainly caused by the emission of greenhouse gases, while climate change, in addition to other natural and anthropic factors, is also caused by global warming. Global warming and climate change can be seen on a regional, local and global scale.

It has come to be considered that global warming and the greenhouse effect are not synonymous, considering their conceptualization. The greenhouse effect encompasses the phenomenon in which the earth's atmosphere retains the temperature on the planet, either naturally or through human intervention. One is seen as the cause and effect of the other, the greenhouse effect is the cause, and global warming is said to affect.

The greenhouse effect increased by pollution may be, according to some theories, the cause of the observed global warming. So we will cite the Anthropogenic Theory, which states that global warming is due to actions produced by human activity.

This theory predicts that global warming will continue if greenhouse gas (GHG) emissions, such as carbon dioxide or CO_2. The Intergovernmental Group of Experts on Climate Change (IPCC) indicates that most of the increases observed in the average temperatures of the terrestrial globe since the middle of the 20th century are probably due to the observed increase in the concentrations of anthropogenic greenhouse gases.

Global warming is often used inappropriately as a synonym for climate change to refer to climate changes that occur in the present, but this use is incorrect. Therefore, in addition to the reasons previously stated in our research that support their differences, it is necessary to define this concept to avoid such confusion.

Climate change is the modification of the climate with respect to the climatic history on a global or regional scale. Such changes occur at different time scales and on all meteorological parameters: temperature, atmospheric pressure, precipitation, cloudiness, etc. In theory, they originate from natural causes and due to human action, which is anthropogenic.

The Intergovernmental Panel on Climate Change (IPCC) defines climate change as "significant statistical variation in the mean state of the climate or in its variability, which persists for a prolonged period (usually decades or even longer)".

For its part, the United Nations Framework Convention on Climate Change, in its article 1, defines it as "a change in climate attributed directly or indirectly to human activity that alters the composition of the world atmosphere and adds to the natural variability of the climate observed during comparable periods." Therefore, it can be concluded that when we talk about climate change in the first sense, it refers to the phenomenon that manifests itself in an increase in the average temperature of the planet. But the one that has the most influence today, according to the United Nations Framework Convention on Climate Change of 9 May 1992, is the one that has its origin in anthropogenic causes, the latter being the object of our investigation (Davies & Riddell, 2017).

With a high degree of probability, the scientific community has estimated that climate change is due to the so-called greenhouse effect. It occurs because the atmosphere allows a part of the solar radiation to enter the biosphere. The seas react to this phenomenon by absorbing or returning heat in the form of infrared radiation. Greenhouse gases act like a mirror reflecting this radiation. This leads to the maintenance or increase of the planet's temperature.

This explains the constant change in temperatures that the earth has suffered, fluctuating continuously and through various events, between the ice age and times of global warming. However, the current problem of climate change lies in a super-acceleration of global warming to a point where it is not known for sure if it will become irreversible. This can be seen in the sustained increase in the earth's temperature from the 19th century on, and whose effects, at first merely theoretical, can now be seen as true.

For years there was uncertainty as to whether current global warming had natural or anthropogenic causes. The scientific area that denies the incidence of human beings in climate change has been called "Climate Change Denialist Current" and had its greatest apogee and acceptance between the years 1970 and 1990.

This current basis and its foundations on the fact that there is still no conclusive evidence on the human impact on global warming, and that consequently, the current phenomenon probably corresponds to the regular cooling-warming cycle that the earth undergoes periodically; however, this point of view is extremely narrow today, to the point where some scientists consider it pseudoscience.

This is due to the successive leak of documents that showed that the oil companies, the industry that generates most greenhouse gases, were responsible for fully financing the research carried out on this current for decades.

Scientists have estimated with almost no margin of error that the acceleration in global warming is due, among other causes, to anthropogenic greenhouse gases. When nature's processes produce the consequences of climate change, it tends to balance the environment by itself. However, the adverse effects on the climate that occur due to human action are irreversible, constituting a stigma that the next generations must bear(Warren, 2015).

Negative Impact of Climate Change on the Global Environment

In this section, the impacts of climate change on the environment will be addressed, one of the objectives set. These consequences are; rising sea levels, heatwaves, storms, drought, endangered species, disease, economic instability, and the destruction of ecosystems.

Regarding the sea level, it can be said that when the surface temperature increases, the ice of the glaciers, the sea ice, and the polar ice shelf melt. When this happens, the amount of water that flows into the world's oceans increases and leads to drastically rising sea levels, putting many cities below sea level at risk. Scientists have speculated that melting Antarctica and Greenland ice at this rate could raise sea levels by more than 20 meters by 2100.

It is also valid to note that severe heatwaves have become increasingly common, and this is because greenhouse gases are trapped inside the atmosphere. Studies indicate that they will continue to grow in the coming years and the future. This will lead to an increase in heat-related illnesses and consequently trigger countless fires.

The World Meteorological Organization has confirmed that the global mean surface temperature in 2015 beat all previous records by a surprisingly wide margin. For the first time, mean temperatures have been measured that are 1 ° C higher than those of the pre-industrial era (Nevitt, 2020).

When the temperature of the oceans gets warmer, the storms are more intense. Global warming will make storms extremely severe. Warm ocean water will fuel the force of storms, resulting in increased numbers of extremely devastating hurricanes. Effects like these are being felt today because, in the last 30 years, the severity and number of cyclones, hurricanes, and storms have increased and have almost doubled. This leads to flooding, loss of life, as well as property damage. Due to climate change, the frequency of the strongest hurricanes has increased by a ratio of 13 to 17, which amounts to an increase of 31%.

Droughts are another devastating effect of climate change, which is already wreaking havoc in various parts of our planet. This is heating up and, in turn, freshwater decreases, leading to poor conditions in agriculture. There is a great shortage of water that is causing difficulties in world food production, and hunger is becoming more and more widespread. There are historical lakes that are disappearing all over the world, such as the Poyang (the largest in China), the Poopó (Bolivia), or the Aral Sea.

The number of endangered species that currently exist is also an international concern. Desertification, rising ocean temperatures, and deforestation are contributing to the disastrous and irreversible changes that are taking place in the habitat and threaten to endanger several species. A crucial example is the case of the polar bear since we see that as the ice loss occurs in the Arctic regions, the numbers that were had for this species have dropped considerably.

A key feature for human success is biodiversity; the loss of flora and fauna due to mass extinction threatens our planet, such as sea turtles, penguins, orangutans, and whales, seriously endanger the continuity of humanity.

When there is a change in habitat, there is an immediate increase in disease worldwide. By combining warmer temperatures, floods, and droughts, the right conditions are created for rats, mosquitoes, and other disease-carrying pests to thrive, and diseases such as cholera, dengue fever, asthma, etc., are getting older and not limited to tropical climates (Tignino, 2010).

Due to climate change, there were approximately 30,000 deaths in Europe in the summer of 2003; for Spain, the National Epidemiology Center figures at 6,500. However, there are many more aspects that must be taken into account, among which are the increase in skin cancer, from 15 to 20% in the incidence of skin cancer in thin-skinned populations; cataracts and other eye injuries can increase from 0.6 to 0.8% for every 1% decrease in ozone and an increase in vulnerability in some infectious diseases as a result of the suppression of immunity caused by ultraviolet radiation; respiratory problems; poor nutrition; there is an increase of 5-10% in the number of undernourished people, especially in the tropics, among others.

The scientific community is also concerned about the increasing destruction of ecosystems. The increase in greenhouse gases is causing drastic changes in the atmosphere, but it is wreaking havoc on the entire planet, affecting the water supply, clean air, agriculture, and energy resources.

Plants and animals die or move to other "non-native" habitats when the ecosystems they depend on for survival; Like coral reefs, they are threatened by the influence of warming sea temperatures and acidic water. Due to the changes in the climate, the way of life of all living beings is considerably affected in a way that in some cases, humans will have to migrate, resulting in greater competition and war for the scarce quantity of natural resources to satisfy their socio-economic and cultural needs.

The coral reefs are being one of the great harmed; the increase in water temperature is causing their ecosystem to be destroyed. If not remedied, they could disappear by 2050, although today, many areas that are considered irrecoverable have already been lost.

We can also add that climate change is a social fact since its causes are largely due to human activities; Likewise, it is the global and specific societies, as well as the people that make up those societies, which will ultimately suffer its consequences either directly or indirectly through the change of the biogeophysical environment. It is a social fact also because its solution or resolution cannot be done by nature but by society.

For this reason, it is necessary to address the analysis of the social impact of climate change in a broad way, including the various spheres that make up societies.

Climate change also impacts social organization; social and political structure, conflicts, norms, and social values are already having an impact on various aspects of social organization. Specifically, social inequality is also increasing due to climate change, both in high-development and low-development countries. The risks that it brings with it greatly affect the most vulnerable sectors of all societies say the poor, the elderly, children, women, the weakest, among others, since they have fewer resources, which we must highlight not only the economic ones but also those of information, education and even those of the necessary encouragement and self-esteem to prevent and mitigate these effects (McClanahan & Brisman, 2015).

The economic base of society has also been severely damaged. When we talk about human settlements, be it rural and urban centers, housing, and infrastructure affected by climate change, it can be said that there is a direct risk of floods and earth movements in large parts of the world. These are aggravated by the expected increase in the intensity of the rains and coastal areas, the rise in sea level, and a greater number and intensity of temporary hurricanes. It can be said that the risk of flooding is greater for populations located on the slopes of rivers and the sea.

Urban flooding can become a problem in any area where the capacity of the sewerage, water supply, and waste management systems is insufficient, that is, the nuclei and societies with fewer infrastructure resources.

There are areas in which informal urban settlements with high population density are highly vulnerable, without having shelters for evacuation, with little access to resources such as drinking water and public sanitation services, and little capacity for adaptation. An example of this: in Europe, floods constitute 43% of all disasters in 1998-2002.

Human settlements that show little economic diversification and the main economic income from climate-sensitive sectors, such as agriculture, forestry, and fisheries, are more vulnerable to climate change than those with more economic economies.

Based on the above, it can be expressed that climate change is one of the most unfair challenges since it is generated by the consumption model of the richest countries, and instead, the effects have a greater impact on the most underdeveloped nations.

The route of money has prevented reaching the agreements required by the severity of climate change, nor do they have the financing that the poorest countries need to adapt to its current and future effects.

As Fidel put it: "It is the developed countries that have polluted the atmosphere, those that have saturated the atmosphere with carbon dioxide and the consequences are being paid by the poor. The solution cannot be to prevent development for those who need it most, the reality is that everything that today contributes to underdevelopment and poverty constitutes a flagrant violation of ecology".

Climate change has innumerable impacts worldwide, which can be observed in terms of its impact on sea-level rise, heatwaves, storms, drought, endangered species, diseases, the disappearance of glaciers, wars, economic instability, and the destruction of ecosystems, as well as having an impact on the social and political structure, conflicts, norms, and social values. Although it is necessary to clarify that many of these problems are cause-effect, that is to say, and sometimes deforestation, drought is causes caused by human activity and on other occasions effects of natural or anthropocentric climate change, denoting the dialectical nature of that causal relationship. For these reasons, it is necessary to study its international legal regulation (Barnett, 2003).

INTERNATIONAL LAW OF CLIMATE CHANGE

In the field of international environmental protection, the International Community has addressed issues related to climate change; As a result, important international actions have been carried out, and various highly relevant legal instruments have been promulgated, among which are: The Earth Summit, Agenda 21, the United Nations Framework Convention on Climate Change, the Conferences of the Parties to said convention and various international agreements.

The United Nations Conference on Environment and Development or Earth Summit was held between June 3-14, 1992, in Rio de Janeiro, Brazil. In this, the participating countries agreed to adopt a development approach that protects the environment. Heads of State or Government from all continents attended; At the same time, twenty thousand representatives of non-governmental organizations participated in one hundred events at the Global Forum in Rio, who brought the peoples' voice to the great ecological event. It was an important moment in the aspiration to achieve a fair balance between economic, social, and environmental needs while ensuring economic and social development.

The Rio 92 Declaration establishes a set of principles and actions to confront climate change. So we refer to two of them:

1. Strategies and action plans to reverse human-induced climate change should be created and implemented based on a precautionary principle; the lack of full consensus on scientific conclusions cannot be used to justify not acting.
2. All human beings should have equal access to a total amount of greenhouse gas emissions that the atmosphere can sustainably support.

As for the actions, they are indicated; promote awareness and mobilize society to identify and combat the causes and possible consequences of climate change and related problems and contribute to identifying the social and environmental impacts of climate change at the local and global level.

The celebration in Rio de Janeiro of the First Earth Summit was the most important ecological and nature protection event ever carried out by humanity. This event provided an undoubted impetus to the potential solution of some of the great environmental problems posed up to that moment.

Fidel Castro Ruz said in his speech delivered at this Summit (Buhaug et al., 2008):

An important biological species is at risk of disappearing due to the rapid and progressive liquidation of its natural conditions of life: man. Now we become aware of this problem when it is almost too late to prevent it. It is necessary to point out that consumer societies are fundamentally responsible for the atrocious destruction of the environment. They were born from the old colonial metropolises and imperial policies that, in turn, engendered the backwardness and poverty that today plague the vast majority of humanity.

With only 20 percent of the world's population, they consume two-thirds of the energy produced in the world. They have poisoned the air, weakened and perforated the ozone layer, and saturated the atmosphere with gases that alter climatic conditions with catastrophic effects that we are already beginning to suffer.

If humanity is to be saved from this self-destruction, the wealth and technologies available on the planet must be better distributed. Less luxury and less waste in a few countries so that there is less poverty and less hunger on much of the earth. No more transfers to the Third World of lifestyles and consumer habits that ruin the environment.

Cease selfishness, cease hegemony, cease insensitivity, irresponsibility, and deception. Tomorrow will be too late to do what we should have done a long time ago. "Without a doubt, our Commander foresaw from an early date how climate change would become one of the problems that would affect humanity in the 21st century. More than 20 years have passed since this Summit was held, and it can be seen that there have been no significant changes in the actions of most industrialized countries. Neoliberal globalization, together with trade liberalization and financial deregulation, has brought about a great increase in environmental deterioration, the elevation of the levels of inequality between the nations of the First World and those of the Third World, which has brought with it the viability to achieve sustainable development. This situation is enormously contradictory to the principles and agreements reached in Rio 1992. This Summit was the opportunity to adopt a program of action for the 21st century, called Agenda 21 (Agenda 21), which lists some of the 2,500 recommendations related to applying the principles of the Declaration. Agenda 21 is a strategic plan for this new century, a document that establishes the guidelines to move towards a more respectful world with the environment, and an action plan proposed by the United Nations. It was officially approved and signed by 173 governments, calling for local governments to start their Local Agenda 21 processes. It also covers issues related to climate change; in its article 18.84, it states that among its objectives are that all States carry out the following actions:

- Understand and quantify the threat of the impact of climate change on resources.
- Facilitate the adoption of effective national preventive measures for the effects of climate change, as long as the threat of the repercussions is considered sufficiently confirmed to justify such an initiative.
- Study the possible effects of climate change in areas prone to droughts and floods.

Its objectives are given in helping to carry out a sustainable development strategy in each region that achieves the well-being of the community. We can say that, although it is not binding, it has potential in terms of the interpretation of treaties and other instruments adopted following its provisions. United Nations Framework Convention on Climate Change (Ng, 2010):

The possibility of creating a normative body to cover the problems already discussed climate change and its repercussions had been installed since the end of the 1980s in the International Community. This is evidenced by Resolution 43/53 of the United Nations General Assembly on Global Climate Protection for Present and Future Generations of 1988, the International Meeting of Experts on Atmospheric Protection Policies held in Canada in 1989. and what was expressed by the Intergovernmental Group of Experts on Climate Change (IPCC) through the III Working Group in its II Session in 1989.

Thus, the United Nations in 1990, through Resolution 45/212, agreed on the Protection of the World Climate for Present and Future Generations. This agreement creates a commission in charge of presenting a convention project that manages to collect all that has been done on the matter.

We can say that the United Nations Framework Convention on Climate Change (UNFCCC) resulted from the well-known Brundtland Report of 1987 or the World Commission on Environment and Development, a direct antecedent of it. In this report, the concept of sustainable development is exposed, which denotes great importance for the international climate regime, since as proposed in it, the United Nations Conference on Environment and Development was convened in 1988, as well as Known as the "Rio Conference" or the "Earth Summit," held in 1992. Among the issues that would be analyzed in it was initially climate change. Said Summit resulted in the beginning of the signing by the States of the Convention. Finally, the convention was adopted on 9 May 1992 and entered into force after its fiftieth ratification on 21 March 1994.

As its name indicates, the convention functions as a framework, which implies a scant normative intensity, for the same reason it is that the States Parties are not coerced to comply with its objectives and regulations.

In other words, with its emergence, humanity would have a frame of reference within which governments would collaborate to apply new policies and programs, which would have a broad impact on how human beings and countries live and work. Parties to the Convention would have the primary purpose of working to achieve the fulfillment of said objective reflected in its article 2, which states:

The ultimate objective of this convention and any related legal instrument adopted by the Conference of the Parties is to achieve, following the relevant provisions of the convention, the stabilization of greenhouse gas concentrations in the atmosphere at a level that prevents dangerous anthropogenic interference in the climate system. That level should be achieved within a sufficient time frame to allow ecosystems to adapt naturally to climate change, ensure that food production is not threatened, and sustain economic development.

To carry out this purpose, the convention establishes universal duties for all parties and other specific ones for industrialized countries and economies in transition and only industrialized countries. It establishes a set of principles that function as a framework for action against climate change, which "fulfill the function of guiding the actions that the parties adopt to fulfill the final objective." Among the main axes of the United Nations Framework Convention on Climate Change are the principles of common but differentiated responsibility of the parties to the convention, intergenerational equity, sustainable development, and the precautionary principle. The convention also raises some guidelines regarding action against climate change, that is, measures on the climate problem must be as least costly as possible, the constant initiative of the countries classified as responsible for it, take into account the circumstances developing countries and a thorough review of the economic system. Finally, regarding the United Nations Framework Convention on Climate Change (UNFCCC), different multilateral funds have been developed (Maas & Scheffran, 2012):

- The Environment Fund (GEF). It is one of the most important funds for adaptation plans and implementation of projects in developing countries.
- The Fund for Least Developed Countries and the Special Fund for Climate Change aimed to adapt and activities that help developing countries diversify their economy.
- Fast-Start Finance and Long-Term Finance aimed at the immediate financing needs to combat climate change and meet developing countries' needs, respectively.
- The Green Climate Fund, which has as its final objective a paradigm shift in favor of development processes with low levels of carbon emission through support to developing countries. It also allows them to implement adaptation practices against climate change and mitigate its effects.

It could be said that from this Framework Convention, the idea that the protection of the environment is a common responsibility is consolidated worldwide. All states are involved, whether developed or developing, large or small, although not all in the same way.

This convention created the Conference of the Parties (COP), the supreme decision-making body that is in charge of examining, evaluating, and making recommendations regarding the obligations of the parties, promoting and facilitating cooperation and coordination of the parties; mobilize financial resources; among other matters necessary for the application of the convention.

Conferences of the Parties (COP):

In International Law, the Conferences of the Parties are going to be those bodies in charge of monitoring, carrying out, and applying an international convention. This Conference of the Parties to the United Nations Framework Convention on Climate Change is established by Article 7 of the treaty, which establishes: "The Conference of the Parties, in its capacity as the supreme body of this convention, shall examine regularly implement the convention and any related legal instrument adopted by the Conference of the Parties and, following its mandate, take the necessary decisions to promote the effective implementation of the convention (…)." The Convention, for its part, endows the Conference of the Parties with obligations and powers, which is why we wanted to mention some of them (Brzoska, 2009):

- Provide coordination between these parties to agree on measures that tend to reduce climate change.
- Because of the end of the convention, promote the development and improvement of comparable methodologies, whose objective is to carry out inventories of Greenhouse Gases, their sources, modes of absorption by sinks, and an evaluation of the measures that the parties adopt.
- Cooperate in preparations and develop plans for adaptation to the impacts of climate change, focusing on water resources and agriculture.
- Assess the considerations related to climate change in its policies and social, economic, and environmental measures, carrying out evaluations that impact the national level to minimize its adverse effects.

- Promote and support the full exchange of scientific, technological, technical, socio-economic, and legal information on climate change and its economic and social consequences.
- Promote and support education and public awareness on climate change, stimulating broader participation by all social sectors and non-governmental organizations.

International Agreements on climate change:

Among the most relevant international agreements that have been taken are the Kyoto Protocol, the Copenhagen Agreement, and the Paris Agreement.

After the Earth Summit, the Kyoto Protocol is carried out, considered the basis of international climate regulation. This arises from a negotiation process initiated by the so-called "Berlin Agreement" adopted at the First Conference of the Parties (COP) in 1995, which originated to reinforce the convention's commitments through a protocol or other instrument. To develop policies and establish quantified limitations and reduction targets within specified timeframes after the year 2000. The Protocol negotiations were complicated, arduous, and controversial and, despite reaching a final consensus, there was much work to be done to develop the subsidiary rules, guidelines, and methodologies necessary for its application.

It was signed in the city that gives its name on 11 December 1997, as part of the agreements reached at COP 3 of the Framework Agreement. It was signed by 193 parties, with the notable exclusion of the United States of America. This country signed the protocol, but the North American Congress rejected its ratification. Currently, the Kyoto Protocol has 192 member countries. This is due to Canada's controversial withdrawal in 2011, as it refused to meet its emission reduction targets.

The ultimate goal of this protocol is to concretely implement the Framework Convention on Climate Change approved at the Earth Summit. Article 2 states that:

The Parties included in Annex I shall endeavor to apply the policies and measures referred to in this article in such a way as to minimize adverse effects, including the adverse effects of climate change, effects on international trade and repercussions social, environmental and economic…taking into account the provisions of Article 3 of the Convention. (Cousins, 2013)

In the same way as the convention, it establishes as the governing body the Meeting of the Parties or Meeting of the Parties, which are held jointly with the Conference of the Parties since 2005 (in Montreal), and its objective is to ensure the realization of the agreements reached.

The main characteristic of the Kyoto Protocol is that it has mandatory objectives in which developed countries commit to the reduction and stabilization of atmospheric concentrations of greenhouse gases (GHG) to reduce the total emissions of these gases to a level no less than 5% lower than that of 1990, in the commitment period between 2008 and 2012. The protocol does not establish new obligations beyond those established in the convention for developing countries.

A. Obligations of all Parties:

The commitments that affect all the Parties constitute minimum obligations that are specified in the following actions indicated in Article 10 of the Protocol:

- The formulation of national and, where appropriate, regional programs which contain measures to mitigate climate change and to facilitate adequate adaptation to it; Such programs would relate, among other things, to the energy, transport, and industrial sectors as well as to agriculture, forestry, and waste management.

- Carrying out and periodically updating national inventories of anthropogenic emissions by sources and absorption by sinks of greenhouse gases.
- Cooperation in the promotion and transfer of technologies, specialized knowledge, practices, and ecologically sound processes related to climate change and the execution of programs for education and training in this area.
- Information on the programs and measures adopted to address climate change and its adverse impacts, to limit the increase in emissions, increase absorption by sinks, and build capacity and adaptation.

However, this protocol has received various criticisms. Industrialized countries, which are the biggest polluters and/or responsible for global warming, only made commitments with a minimal impact on reducing their emissions by using fossil fuels. In addition, some polluting countries try to avoid limiting their greenhouse gases, seeking to change it simply by planting trees in the territories of certain Third World countries. These actions are based on one of the Kyoto protocol mechanisms called the Clean Development Mechanism (CDM).

Unfortunately, it has not been able to meet its objectives of stopping or slowing down global warming, but instead, by opposing them, it is contributing to the intensification of the problem. Independent evaluations such as the Food and Agriculture Organization of the United Nations (FAO) (2002) have concluded that Clean Development Mechanism (CDM) projects often do not reduce emissions and thus are not met. With the objectives of sustainable development(Penny, 2018).

The Kyoto protocol has also been strongly criticized for having the Emissions Trading Regime mechanism. This focuses on the market, which tends to establish a trading system for the purchase and sale of carbon emissions, which allows especially those countries where there are more emissions, to negotiate them instead of reducing them.

In this way, one of the actions of the Kyoto Protocol, such as the "carbon market," is considered by some critics as a scheme to avoid real changes in industrialized countries. At first glance, it seems that the convention is more concerned with how much money can be saved by each country than with finding a real solution to the problem. Despite the above, we can say that, as Boisson de Chazournes pointed out, the entry into force of this instrument marks an important stage for the protection of the global environment since it represents a qualitative leap concerning previous environmental regimes and concerning techniques used for the development and application of international law, and has been one of the important advances in the fight against climate change at the international level.

After the signing of the Kyoto Protocol, international actions related to climate change were carried out. The Climate Conference in Marrakech, the Bali Conference, the Climate Summit in Doha, the Climate Summit in Warsaw.

The United Nations Framework Convention on Climate Change held in 2001 in Marrakech, Morocco, was the VII International Conference on Climate Change developed by the UN. Its primary objective was to verify the compliance regime of the Kyoto Protocol and the flexibility mechanisms. Among the decisions taken at this conference, we find the capacity building of developing countries and countries with economies in transition, the development of technology transfer, the implementation of forest management activities within the framework of the Kyoto Protocol, among others.

The Bali Conference or Conference of the Parties 13 was held in Bali, Indonesia, in 2007. Its objective was to establish a post-2012 regime at the XV Conference on Climate Change. It had an Action Plan, which contained four key elements: mitigation, adaptation, finance, and technology. The Plan also contains a non-exhaustive list of issues that should be considered for the treatment of "a shared vision for long-term cooperation."

The Climate Summit in Doha was held in Qatar, December 2012. Delegations attending the Summit agreed on an extension to 2020 for the Kyoto Protocol. These delegations agreed that the agreement did not meet the recommendations to avoid the dire consequences produced by climate change.

The Conference of the Parties 19th, or Climate Summit in Warsaw, was held between November 19 and 22, 2013, in Poland. Its objective was to create a binding global agreement to reduce greenhouse gases, applicable to industrialized countries and those not included in Annex 1 of the Kyoto Protocol.

Another relevant agreement on climate change is the Copenhagen Accord; it was taken at the XV International Conference on Climate Change that was held in Copenhagen, Denmark, from 7 December to 18, 2009, this being the most important meeting for the climate change regime since the "Kyoto Protocol" was approved, and the expectations in the international community were very high. In it, the different instances and groups that operate in the regime met in parallel, and 34,000 people were accredited among delegates from the 192 member countries of the United Nations Framework Convention on Climate Change (UNFCCC), climate experts, representatives of non-governmental organizations. government and press.

This Summit was the culmination of a preparatory process in Bali in 2007, which aimed to reach an ambitious international agreement to combat climate change. The objective of the conference, according to the organizers, was the conclusion of a legally binding climate agreement, valid worldwide, to apply from 2012 (Liberatore, 2013).

The final long-term objective sought was the global reduction of CO2 emissions by at least 50% in 2050 compared to 1990. To achieve this, countries had to set intermediate objectives. Thus, industrialized nations should reduce their greenhouse gas emissions between 25% and 40%, compared to 1990 in 2020, and should reduce between 80% and 95% by 2050.

The legal nature of the "Copenhagen Accord" is doubtful. It is not a binding instrument, and it is not a formal decision of the COP. It is, in short, a political document rather than a legal one, which represents a rather fragile consensus and where a large part of its content must be decided in future meetings through guidelines that the COP must agree on, implement and verify. We can express in a brief synthesis the most relevant aspects of the "Copenhagen Accord," such as:

a. **Joint Vision.** Although this expression is not used in the "Agreement," the introductory paragraphs are intended to establish a common and general vision, as a way to give context for future global action, highlighting the principles on which it should be developed and a series of objectives, including a global emission reduction goal. In this way, the signatory countries commit to keeping global temperature increases below an additional 2 ° Celsius above the current average temperature through deep emission reductions. Despite this claim, this type of language represents the lack of consensus to agree on specific goals that lead to the stabilization of greenhouse gas (GHG) emissions. In the same vein, although it is accepted as desirable that these reach their highest point as soon as possible, no particular date or deadline was specified.
b. **Mitigation Measures.** The mitigation measures to be agreed upon have been the most complex issue throughout the negotiation. With the preventions made in the previous section, the "Bali Action

Plan" had indicated that they could be different: reductions in greenhouse gas (GHG) emissions as an individual objective for each country, incentives to reduce emissions produced by deforestation and forest degradation, sector cooperation or the establishment of additional market instruments, among others.

One of the greatest contributions of the Copenhagen Accord is framed in the field of financing, with the commitment of developed countries to provide resources worth 30 billion dollars in the period 2010-2012 to finance actions for adaptation and mitigation of climate change in developing countries, as stated in Article 8: "Developed countries collectively commit to providing new and additional resources, including forestry and investment through international institutions, for a value of approximately US $ 30 billion for the period 2010-2012, with a balanced distribution between adaptation and mitigation". They also committed to jointly mobilize $ 100 billion annually by 2020 to meet the needs of developing countries, especially the least developed. To mobilize the promised funding, it was agreed to create a working group at the United Nations to study possible sources of income (Brown et al., 2007).

However, many participating governments described it as a failure and environmental groups since no binding agreements were reached. Most of the environmental movements and non-governmental organizations present at the Copenhagen conference expressed their disappointment, warning that the agreement reached was insufficient. They condemned that there are no targets for carbon cuts and that there has been no legally binding agreement. On creating a $ 100 billion global fund financed by rich countries to combat climate change in developing countries, some critics said that figure was still insufficient (McDonald, 2013).

Among the most important international instruments is the Paris Agreement, adopted on 12 December 2015, part of the United Nations Framework Convention on Climate Change approved at the Rio 92 Summit. It establishes measures to the reduction of Greenhouse Gas emissions through the mitigation, adaptation, and resilience of ecosystems for Global Warming. This would be applied for the year 2020 when the Kyoto Protocol ends. The agreement was negotiated during the XXI Conference on Climate Change (COP 21) by the 195 member countries and opened for signature on 22 April 2016 to celebrate Earth Day.

Until 3 November 2016, this international instrument had been signed by 96 countries individually, and the European Union ratified it on 5 October 2016. Said agreement expresses in its article 2:

This agreement, by improving the implementation of the convention, including the achievement of its objective, aims to strengthen the global response to the threat of climate change, in the context of sustainable development and efforts to eradicate poverty, and for it:

- *Maintain the global average temperature rise well below 2°C relative to pre-industrial levels, and continue efforts to limit such temperature rise to 1.5°C relative to pre-industrial levels, recognizing that This would significantly reduce the risks and effects of climate change;*

- *Increase the capacity to adapt to the adverse effects of climate change and promote climate resilience and development with low greenhouse gas emissions in a way that does not compromise food production;*

- *Raise financial flows to a level compatible with a path that leads to climate-resilient development with low greenhouse gas emissions.*

Achieving the ambitious goal of not exceeding 2°C, much more than 1.5 degrees. This is not possible with climate plans or national contributions (INDCs), which represent greenhouse gas reduction commitments, that nations have submitted to date. For this reason, they should be reviewed and be much more comprehensive. The contributions that each country can establish towards the overall goal are determined by all countries individually and are called Nationally determined contribution.

Article 3 states that:

In their nationally determined contributions to the global response to climate change, all Parties shall undertake and communicate the ambitious efforts defined in Articles 4, 7, 9, 10, 11, and 13 to achieve the purpose hereof. Agreement outlined in its Article 2. The efforts of all the Parties will represent a progression over time, taking into account the need to support the Parties that are developing countries to achieve the effective implementation of this agreement.

In other words, this article requires nationally determined contributions to be "ambitious," "represent progress over time," and be established "to achieve the purpose of this Agreement." Contributions should be reported every five years and recorded by the Secretariat of the United Nations Framework Convention on Climate Change. Each progress should be more ambitious than the previous one, known as the "progression" principle.

Regarding this agreement and due to the importance it connotes, the French Minister of Foreign Affairs, Laurent Fabius, expressed: "this" balanced "plan is a" historic turning point "in the objective of reducing global warming".

We can say that the international agreements set out above have not managed to reverse most of the environmental problems that humanity suffers. Many countries have refused to commit to these agreements, an example of this is the case of the United States, which emits a quarter of the world's greenhouse gases, and Australia, which refused to ratify the Kyoto Protocol, considering that it would hurt their economies by increasing energy prices, and it would encourage the loss of some five million American jobs. China and India are signatories to the protocol but, being emerging economies, they are not obliged, for the moment, to reduce their emissions. Saudi Arabia, Iraq, and Nigeria were absent from the signing of the Paris Agreement for taking a position contrary to its agreements (Ide & Scheffran, 2014).

In the negotiations on climate change, underdeveloped countries have struggled to preserve the concept of common but differentiated responsibilities for nations in different stages of development and accumulated emissions to date. Therefore, the International Community must establish a binding agreement that will most impact on confronting climate change.

As Fidel said: "Make human life more rational, apply for a just international economic order, use all the science necessary for sustained development without pollution, pay the ecological debt and not the eternal debt, hunger disappears and not man" (Garcia, 2018).

CONCLUSION

In the study of the legal regulation of climate change, we conclude that the approved international instruments have not established sufficient measures to mitigate the negative effects of this phenomenon; some countries have not adequately applied them, and others have not ratified them, the most developed being among them. This constitutes a challenge for International Environmental Law and humanity. However, countries that regulate the subject we are investigating in their regulations do not escape, so below, we will emphasize two countries that exhaustively study it in their legislation.

REFERENCES

Barnett, J. (2003). Security and climate change. *Global Environmental Change, 13*(1), 7–17. doi:10.1016/S0959-3780(02)00080-8

Brown, O., Hammill, A., & McLeman, R. (2007). Climate change as the 'new'security threat: Implications for Africa. *International Affairs, 83*(6), 1141–1154. doi:10.1111/j.1468-2346.2007.00678.x

Brown, O., & McLeman, R. (2009). A recurring anarchy? The emergence of climate change as a threat to international peace and security: Analysis. *Conflict Security and Development, 9*(3), 289–305. doi:10.1080/14678800903142680

Brzoska, M. (2009). The securitization of climate change and the power of conceptions of security. *S&F Sicherheit und Frieden, 27*(3), 137–145. doi:10.5771/0175-274x-2009-3-137

Buhaug, H., Gleditsch, N. P., & Theisen, O. M. (2008). *Implications of climate change for armed conflict.* World Bank.

Cousins, S. (2013). UN Security Council: Playing a role in the international climate change regime? *Global Change, Peace & Security, 25*(2), 191–210. doi:10.1080/14781158.2013.787058

Davies, K., & Riddell, T. (2017). The Warming War: How Climate Change is Creating Threats to International Peace and Security. *Geo. Envtl. L. Rev., 30*, 47.

Garcia, D. (2018). Lethal artificial intelligence and change: The future of international peace and security. *International Studies Review, 20*(2), 334–341. doi:10.1093/isr/viy029

Ide, T., & Scheffran, J. (2014). On climate, conflict and cumulation: Suggestions for integrative cumulation of knowledge in the research on climate change and violent conflict. *Global Change, Peace & Security, 26*(3), 263–279. doi:10.1080/14781158.2014.924917

Liberatore, A. (2013). Climate change, security and peace: The role of the European Union. *Review of European Studies, 5*(3), 83. doi:10.5539/res.v5n3p83

Maas, A., & Scheffran, J. (2012). Climate conflicts 2.0? Climate engineering as a challenge for international peace and security. *Sicherheit und Frieden (S+ F)/Security and Peace*, 193-200.

McClanahan, B., & Brisman, A. (2015). Climate change and peacemaking criminology: Ecophilosophy, peace and security in the "war on climate change". *Critical Criminology, 23*(4), 417–431. doi:10.100710612-015-9291-6

McDonald, M. (2013). Discourses of climate security. *Political Geography, 33*, 42–51. doi:10.1016/j.polgeo.2013.01.002

Nevitt, M. (2020). Is Climate Change a Threat to International Peace and Security? *Mich. J. Int'l L., 42*, 527. doi:10.2139srn.3689320

Ng, T. (2010). Safeguarding peace and security in our warming world: A role for the Security Council. *Journal of Conflict and Security Law*, *15*(2), 275–300. doi:10.1093/jcsl/krq010

Penny, C. K. (2007). Greening the security council: Climate change as an emerging "threat to international peace and security". *International Environmental Agreement: Politics, Law and Economics*, *7*(1), 35–71. doi:10.100710784-006-9029-8

Penny, C. K. (2018). Climate change as a 'threat to international peace and security. In *Climate change and the UN Security Council*. Edward Elgar Publishing. doi:10.4337/9781785364648.00009

Scott, S. V. (2008). Climate change and peak oil as threats to international peace and security: Is it time for the security council to legislate? *Melbourne Journal of International Law*, *9*(2), 495–514.

Tignino, M. (2010). Water, international peace, and security. *International Review of the Red Cross*, *92*(879), 647–674. doi:10.1017/S181638311000055X

Warren, P. D. (2015). *Climate Change and International Peace and Security: Possible Roles for the UN Security Council in Addressing Climate Change*. Academic Press.

ADDITIONAL READING

Liberatore, A. (2013). Climate change, security and peace: The role of the European Union. *Review of European Studies*, *5*(3), 83. doi:10.5539/res.v5n3p83

Maas, A., & Scheffran, J. (2012). Climate conflicts 2.0? Climate engineering as a challenge for international peace and security. *Sicherheit und Frieden (S+ F)/Security and Peace*, 193-200.

McClanahan, B., & Brisman, A. (2015). Climate change and peacemaking criminology: Ecophilosophy, peace and security in the "war on climate change". *Critical Criminology*, *23*(4), 417–431. doi:10.100710612-015-9291-6

McDonald, M. (2013). Discourses of climate security. *Political Geography*, *33*, 42–51. doi:10.1016/j.polgeo.2013.01.002

Nevitt, M. (2020). Is Climate Change a Threat to International Peace and Security? *Mich. J. Int'l L.*, *42*, 527. doi:10.2139srn.3689320

Ng, T. (2010). Safeguarding peace and security in our warming world: A role for the Security Council. *Journal of Conflict and Security Law*, *15*(2), 275–300. doi:10.1093/jcsl/krq010

Penny, C. K. (2007). Greening the security council: Climate change as an emerging "threat to international peace and security". *International Environmental Agreement: Politics, Law and Economics*, *7*(1), 35–71. doi:10.100710784-006-9029-8

Penny, C. K. (2018). Climate change as a 'threat to international peace and security. In *Climate change and the UN Security Council*. Edward Elgar Publishing. doi:10.4337/9781785364648.00009

Scott, S. V. (2008). Climate change and peak oil as threats to international peace and security: Is it time for the security council to legislate? *Melbourne Journal of International Law*, *9*(2), 495–514.

Tignino, M. (2010). Water, international peace, and security. *International Review of the Red Cross*, *92*(879), 647–674. doi:10.1017/S181638311000055X

KEY TERMS AND DEFINITIONS

Equity: Defined by UNEP to include intergenerational equity—"the right of future generations to enjoy a fair level of the common patrimony"—and intragenerational equity—"the right of all people within the current generation to fair access to the current generation's entitlement to the Earth's natural resources"—environmental equity considers the present generation under an obligation to account for long-term impacts of activities and to act to sustain the global environment and resource base for future generations.

Polluter Pays Principle: The polluter pays principle stands for the idea that "the environmental costs of economic activities, including the cost of preventing potential harm, should be internalized rather than imposed upon society at large."

Precautionary Principle: One of the most commonly encountered and controversial principles of environmental law, the Rio Declaration formulated the precautionary principle: To protect the environment, the precautionary approach shall be widely applied by States according to their capabilities.

Prevention: The concept of prevention can perhaps better be considered an overarching aim that gives rise to a multitude of legal mechanisms, including prior assessment of environmental harm, licensing or authorization that set out the conditions for operation and the consequences for violation of the conditions, as well as the adoption of strategies and policies.

Transboundary Responsibility: Defined in the international law context as an obligation to protect one's environment and prevent damage to neighboring environments, UNEP considers transboundary responsibility at the international level as a potential limitation on the sovereign state's rights.

Chapter 8
Waste Management Legislation

ABSTRACT

The chapter examines some of the most important jurisprudential and legal considerations arising from waste management with an analytical-descriptive method. Finally, it concludes that recycling should be done with the principle of precaution and prevention. But if it leads to damages, the fault-based system cannot compensate for the damages due to the difficulties it has in the proof process. Therefore, the use of a pure/absolute liability system and the promotion of specialized insurance in this regard are recommended.

INTRODUCTION

Due to severe quantitative and qualitative limitations of water resources and increasing the production and discharge of environmental pollutants, wastewater recycling is necessary for water resources management and environmental protection (Salgot & Folch, 2018). The use of raw wastewater or recycled wastewater has long been common in various countries around the world. The available data show that in the tenth century AH, sewage was used for agriculture in the suburbs of Isfahan (Rowe & Abdel-Magid, 2020). The use of effluents can be studied from different aspects that should be studied. In this regard, researchers have often addressed environmental, economic, and health issues (Rosiek, 2020). But in addition to the above, there are other requirements without which a comprehensive and effective result in the management and operation of wastewater will never be achieved. Issues such as legal, religious, social, and cultural considerations have direct and indirect effects on the management of wastewater supply and demand and complicate the presentation of these renewable resources' patterns. A group of researchers has studied and analyzed the possibility of using wastewater to find its compliance with legal and social requirements. According to these studies, the importance of considering legal requirements and social risks is such that if all aspects of reuse are anticipated, the effluent recycling system may also fail, and that is when the designers, legal and social considerations of wastewater reuse Have not been well studied. In water and wastewater management, the status of constitutions, civil, criminal, and penal laws is very important; As the necessary tools for encouragement and punishment should be determined following these laws, it should be based on the national policy of water resources and the environment

DOI: 10.4018/978-1-6684-7188-3.ch008

and emphasize the social considerations of meeting the needs and protection of the ecosystem. It also guarantees the right to use wastewater to allow for public investment and public and private participation in wastewater management.

In this research, an attempt is made to examine and analyze the legal considerations and responsibilities of wastewater recycling, especially from environmental law (Villarín & Merel, 2020). In fact, in wastewater recycling, what legal principles must be observed to comply with environmental law? Therefore, the most important legal principles and considerations are studied in this way. In addition, there will be a liability for wastewater recycling and environmental damage from effluents that may be harmful to the environment and society. The question is, what is such a responsibility based on? Therefore, it is necessary to address the principles on which the responsibilities arising from wastewater recycling are based. The present research is carried out with an analytical-descriptive method and based on the library process and using written and Internet sources, including technical and legal sources (laws and regulations, judicial procedure, jurisprudential sources, etc.). The results of this study will play a very important role in the management and planning of wastewater recycling and reduce the risks and dangers of reuse in the Iran.

This paper also develops the fundamental theoretical aspects of what should be the appropriate management policy for Urban Solid Waste (RSU) within a framework of sustainable development. The purpose of this article, in the first place, is to provide the reader with the necessary information to interpret the economic, social and environmental benefits that the implementation of this new model would provide in comparison with the shortcomings and limitations of the current policy model; so that after this analysis, as a secondary effect (or if you want to see it in another way, as an indirect effect), the importance that citizen participation would have within it is perceived. The form of policy that is being mentioned is a comprehensive management policy, whose structure and approach is completely different from what is currently being carried out. This new management approach is based on various international experiences that have been implemented for a few years in developed countries that have been the pioneers, as well as in developing countries where some of its components have already begun to be executed.

DEFINITION OF COMPREHENSIVE MSW MANAGEMENT POLICY

To begin with the subject, it is prudent to define what is meant by a "Comprehensive MSW Management Policy", so that in this way, the foundations of the concept to which we are referring are laid. A comprehensive MSW management policy is that policy that addresses the multilateral problem of waste management with a holistic and systemic approach. That is to say, it has a multi and interdisciplinary approach to the problem to be solved, which includes all the aspects involved in it, which are: political, legal, institutional, technical, economic, territorial ordering, and awareness, environmental education and citizen participation (Metson et al., 2015).

Central Elements of a Comprehensive MSW Management Policy

Within the line of comprehensive management, the Economic Commission for Latin America and the Caribbean (ECLAC) proposes in a study that a comprehensive MSW management policy should be made up of five central points: environmental education and community participation citizenship, legal and institutional aspects, the economy and the environment, the relationship between the spatial development of cities and waste management and finally, the technological solutions that can be implemented (Wingfield et al., 2021).

However, since I believe that the aspect of environmental education should be addressed separately from citizen participation; Due to the fact that they are two different and vitally important issues, I will dismember these two elements that were considered together, in such a way that there will now be six points or fundamental lines of action of a comprehensive management policy. These points are:

a. **Environmental Education.** This must be the central mechanism to consciously incorporate the population and all the agents involved.
b. **Citizen Participation.** Social groups and individuals must be provided with the means and channels to actively participate in the tasks that aim to solve the problems of MSW management.
c. **Legal and Institutional Aspects.** Modification and updating of the legal framework, as well as modification and restructuring of the institutional framework.
d. **Economy and Environment.** This refers to the commitment on the part of the authorities when allocating financial budgets to environmental items. As a complement to the financing of these items, economic instruments must be used that are adapted to the socio-economic conditions of the country.
e. **The Relationship Between Spatial Development of Cities and Waste Management.** This aspect has to do with urban development planning and land use planning policies to improve and facilitate the management of MSW.
f. **Technological Solutions That Can Be Implemented.** This aspect corresponds to all the technical issues that are intended to be implemented, such as, for example, the operating program of the waste collection, classification, treatment and disposal system; be in accordance with the economic, social and natural environment conditions of the environment.

Types of Measures

Following the same pattern of experiences of international institutions in Latin American countries with problems in the management of their RSU, in another ECLAC study project called "Policy for the environmentally sound management of waste", developed in some localities of Argentina, Brazil, Colombia, Costa Rica, Chile and Ecuador; It was observed that the strategies that have been used in these countries are not far from those applied in Mexico. In these, traditional instruments focused on corrective aspects prevail, and not, the measures linked to preventive environmental management of waste, an essential point to develop a comprehensive MSW management policy (Asano, 1998).

A proposal for a comprehensive MSW management policy must be proposed as far as possible, trying to respect the hierarchies of reduce, reuse, recycle, treat and dispose. These hierarchies come together from two approaches or perspectives: the preventive one that encompasses the hierarchies of reduce, reuse and recycle, and which aims to avoid or minimize the volumes of waste generation mainly through an assignment of responsibility to producers; and the corrective one, which includes the hierarchies of treatment and disposal, and which will refer to reducing the costs of final treatment and minimizing the risks associated with the waste already generated, by making an adequate collection, transportation, treatment and final disposal.

A vitally relevant skill for the development of an integrated MSW management policy is that the management measures must be able to adapt to variable conditions such as changes in the behavior of MSW generation for seasonal periods, and changes that may produced by future trends, that is, an increase in the generation of waste, either progressively due to a gradual and planned growth of the population, or untimely due to a wave of immigration. These possible changes should try to be identified or foreseen by the planning authority to anticipate the transition as much as possible. Among the most important factors to consider are (Russell, 2019):

- Changes in the amounts and composition of waste streams.
- Changes in specifications and markets for recyclable materials.
- Rapid technology changes.

Stages for the Adoption of a Comprehensive MSW Management Policy

For the implementation of a comprehensive RSU management policy, a series of stages must be followed. According to a proposal for a project carried out by a Research Institute in Spain, the process for adopting a comprehensive MSW management system in a territory can be divided into four parts (Lazarova & Bahri, 2004):

- **Decision to Implement a Comprehensive Management Policy for MSW.** This point is important and refers to the conviction and commitment of all the agents involved, that is, there must be awareness of the benefits that the comprehensive policy will bring to all; it is closely linked to consensus, education and participation.
- **Establishment of Policy Objectives.** These specific goals will preferably be by legal obligation to better encourage their achievement; but it should not be ruled out either, that in the initial stage of implementation, they were by public commitment or voluntary agreement. The foregoing, due to the inexperience and complexity involved in taking all the measures to be undertaken, giving all the agents involved an opportunity to adapt, regardless of whether the pilot tests have been successful. It is necessary not to rush into actions, since in this way a solid action of all the measures and agents involved will be ensured. Setting specific objectives or goals will serve to evaluate the operation of the policy and provide feedback. In this way, in the event that the result of the policy was not as expected; With the evaluation, malfunctions would be detected, which would help adapt or incorporate other measures to correct the malfunction. (Chu et al., 2004)
- **Appropriate Operating System Design.** This point focuses on the proposal of a technical executive project for the urban cleaning system that this policy will require for its proper functioning from the local scale.
- **Establish Comprehensive Policy Through Various Instruments.** The instruments that will globally shape the comprehensive waste management policy to try to achieve the objectives shown above; they are established by temporary priorities in such a way that they will begin to be implemented in different stages: short, medium and long term. At the same time, some will correspond to preventive aspects and others to corrective aspects, and some of them will have an influence on both aspects, either directly or indirectly. Each one may individually refer to an objective or have implications for several from the thematic point of view. The instruments are classified as: economic, legal, institutional, administrative, technological, social (focused on low-income popu-

lations), and educational. Regarding the jurisdictional level, there will be municipal, state and federal levels so that they can be achieved effectively and efficiently (Rhyner et al., 2017).

As this classification of stages can be interpreted as general for any comprehensive management system; For the hypothetical case of a policy proposal for a municipality such as Campeche, it must be taken into consideration that such a proposal must be adapted to the social, economic and environmental context of its specific location.

RULES ASSOCIATED WITH WASTEWATER RECYCLING MANAGEMENT

Despite the significant benefits of wastewater recycling, the improper use of recycled water and effluents can undoubtedly pollute water and soil resources. Due to the importance of the issue, legislators and governments of different countries have taken various measures to prevent pollution and negative changes in the quality of water resources. For example, the Chinese constitution states that the government is responsible for ensuring the "rational use" of all-natural resources, including water. The government also has full control over the effluent, and it can grant consumers the right to use the effluent or make the necessary regulations in this regard. The constitutions of Brazil, Mexico, and Vietnam specifically delegate responsibility for water and wastewater management to the government. According to Brazil, Kazakhstan, and Yemen, the government is responsible for the Iran's water and wastewater resources. The constitutions of China and Vietnam stipulate that the government has water resources in the people's name. In Iran, at the level of macro-national laws, the most important laws related to the issue of wastewater exploitation and management are as follows (Song et al., 2017):

- General "Environment" policies announced by the Supreme Leader
- Article 50 of the Constitution
- Law on Fair Distribution of Water
- Law on protection and improvement of the environment
- Islamic Penal Code
- Planning rules (third, fourth, fifth, and sixth development plans)
- Regulations for preventing pollution of water resources
- Executive instructions for water allocation
- Circulars and procedures for the use of recycled water

According to Article 8 of the General Environmental Policy, promulgated by the Supreme Leader in line with Article 110 of Article 1 of the Constitution, the development of a green economy using wastewater management using economic, social, natural, and environmental capacities and capabilities is emphasized. Is. Also, in Article 50 of the Iranian Constitution, environmental protection is considered a public duty. Therefore, economic activities other than those associated with environmental pollution or irreparable damage are prohibited. Therefore, if the discharge of sewage and various effluents into the environment is considered as a source of pollution, planning to use these sources is one of the issues related to Article 50 of the Constitution. In the Law on Fair Water Distribution, which references most laws related to water resources management, wastewater and returns water is listed as one of the water resources. Therefore, this law explicitly defines the organizational responsibility of the Ministry of Energy in managing these resources (Olsson et al., 2013).

The most important program rules include the creation and development of water pollution measurement networks and strengthening the principles of water quality management, issuing licenses for the use of water resources for large units subject to the implementation of wastewater collection facilities, wastewater treatment, and sanitation, and fines from polluting units, strengthening And the empowerment of structures related to the environment and natural resources, environmental and economic values of water resources, comprehensive management approach and both supply and demand throughout the water cycle with a sustainable development approach. The Executive Instruction on Water Allocation and the Directive on the Use of Recycled Water, citing the Law on Fair Distribution of Water, refer to the use of effluents to create a reasonable and stable balance between supply and demand with environmental considerations and simultaneous quantitative and qualitative water management. Therefore, it is considered that the use of wastewater is defined in the law, and in this regard, the principle of the issue is acceptable.

According to Article 688 of the Islamic Penal Code - Punishments and Deterrent Punishment adopted in 1392, any action that is considered a threat to public health, such as contaminating drinking water, unsanitary disposal of human waste, dumping poisonous substances in rivers, unauthorized use of raw sewage or wastewater treatment. It is prohibited for agricultural purposes, and the perpetrators will be sentenced to up to one year in prison if they are not subject to more severe punishment under special laws.

In Iran, the most important predicted and approved standards related to the use of wastewater include the standard of the Institute of Standards and Industrial Research, the standard of the Environmental Protection Organization, and the standard of the Iran's development plan law, as well as a study guide for justifying urban and rural wastewater treatment plans. Because of the above, the existing legal capacities have entrusted the Ministry of Energy with authority to allocate and exploit return water and effluents. Still, from a regulatory and health point of view, the two main authorities of the Environmental Protection Organization and the Ministry of Health and Medical Education have a decisive role. The Environmental Protection Organization has a more prominent and key responsibility and role (Chong, 2014).

In the discussion of wastewater recycling based on referendums made by several great authorities of imitation (Ayatollah Khamenei, Ayatollah Makarem Shirazi, Ayatollah Hashemi Shahroudi) by separating contaminants and microbes and others from sewage, Transformation does not take place. Unless the purification is done by evaporating water and converting steam to water again, the water (effluent from treatment plants) is impure; However, if it is connected to deaf or rainwater and then mixed, it will be cleaned. But it can be used for agriculture and drinking animals. It should be noted that the members of the Islamic Consultative Assembly, according to Article 2 of the Law on Development and Optimization of Urban Drinking Water and A villager in the Iran has obliged the Ministry of Energy to observe the religious aspects in terms of purity and impurity in recycling the effluent of treatment plants for agricultural purposes. Accordingly, the Ministry of Energy is obliged to recycle the effluent from sewage facilities to free the required water resources such as sanitary water supply, agriculture, and green space, observing the religious aspects of purity and impurity (Sun et al., 2016).

According to the above laws and rules, in Iran, managing wastewater recycling is the responsibility of the government and relevant agencies, which includes improving living conditions to provide society with a healthy environment and observe justice and intergenerational rights, prevent and prevent the spread of unauthorized pollution. And criminalization of environmental degradation and effective punishment and deterrent of polluters and environmental degraders and their obligation to compensate.

LEGAL CONSIDERATIONS OF WASTEWATER MANAGEMENT

To solve the problem of water shortage and to protect water resources and the environment, the reuse of treated wastewater has been considered in the programs of various countries, including Iran. In wastewater management, the position of legal and jurisprudential rules is very important; In such a way that while considering macro water policies, it is necessary to observe the considerations and rights of society and the environment. In this section, the main principles and legal considerations in wastewater recycling management are examined.

Principle of Precaution and Prevention

Undoubtedly, in today's world, wastewater recycling and reuse of treated wastewater have very beneficial effects in addressing water shortages and solving health and environmental problems. However, it may occur for various reasons, such as defects in the implementation of wastewater quality, non-compliance of wastewater with the type of consumption, the emergence of secondary effects, etc. Widespread hazards in society, water resources, and the environment. Some of these hazards exist potentially, but the truth is that misinterpretation of recycling principles, their incorrect application, etc., can turn wastewater into a detrimental factor for human life and the environment. Therefore, an attempt has been made to prevent losses by explaining the principles and regulations before the loss is raised. Therefore, the need for caution in using recycling and applying its principles is crucial. Therefore, in water management, the principle of precaution has been emphasized, and its purpose is to protect the environment and prevent its destruction and pollution. Some of these safeguards are to prevent actions that are destructive if they occur. For example, the raw use of some plants and aquatic animals that have been grown with trematode contaminated effluents can lead to various diseases, and some prevent cases for which there is insufficient knowledge about the consequences. According to the text of Article 15 of the Rio Convention, the precautionary principle is defined as follows: To protect the environment, governments must take extensive precautionary measures commensurate with their means. In the event of a risk of severe or irreparable damage, the lack of conclusive scientific evidence should not be used as an excuse to delay effective measures to prevent damage to the environment. However, governments are responsible for scientifically investigating and assessing the potential hazards of effluents, making them accountable to their neighbors and citizens.

According to the Rio (1992) Convention, the application of the precautionary principle is when there are reasonable grounds for concern; That is, if it is proven that substances can be harmful to human health and natural resources . For example, contamination of surface and groundwater sources with nitrate due to high concentrations of nutrients (nitrogen and phosphorus) in some effluents can be a risk factor. Therefore, the principle of legal precaution tries to prevent future losses due to danger, and problems die in the sperm before the damage occurs. On the other hand, some risks from wastewater may pose irreparable risks to future generations (Trianni et al., 2021). For example, incomplete treatment of municipal wastewater, which contains many infertility drug residues, can make the next generation infertile or have hormonal disorders. This means that meeting the current generation's needs, which is the same as demographic regulation, poses a risk to the future, undermining sustainable development. In this regard, governments are required to apply regulations and work based on fair action for the sake of public order so that the activities of the sectors under its jurisdiction are not harmful to the environment. Therefore, improper management or improper recycling of wastewater that does not comply with standard guidelines is contrary to the principle of prevention. In addition, today, with the spread of environmental effects of polluted effluents, it is possible that improper recycling of effluents or improper diversion of effluents to another Iran. International practice shows that countries must use their land harmlessly. This is also mentioned in the case of Lake Lano between Spain and France, and therefore the effects of misuse should not be extended to others (Guo et al., 2018).

Rule of "No Harm and Harmful Act"

The rule of no harm, known as the rule of no harm and no harm in Islam, is documented in verse 233 of Surah Al-Baqarah, numerous narrations, reason, and consensus. In fact, in society, no one has the right to inflict harm on another person, and the Islamic government can neither impose a harmful verdict nor can it commit an act with harmful effects. In wastewater recycling, the application of the no-harm rule leads to the prohibition of any act or omission and a law that has destructive effects on the environment, health, society, and so on. In this regard, attention is paid to the concept of environment in environmental law, which encompasses a wide range of the entire human environment. In addition, all its many dimensions (human, animal, plant, objects, landscape health, urban construction, etc.) must also be protected. For example, if the discharge of effluent from a treatment plant causes environmental pollution on the outskirts of a village, due to the wide scope of the no-harm rule, the above activity can be stopped in various ways be claimed (Spellman, 2008).

Rule of "Not to Waste"

Loss means the destruction of property. Whoever loses another's property will be the guarantor. This rule has Qur'anic principles ("I am the one who commits adultery against you, like us, I am the one who commits aggression against you...", so whoever transgresses against you, you should transgress against him as well [verse 194 of Sura Al-Baqarah]). In addition, early scholars such as Sheikh Tusi have also narrated this rule by quoting hadiths, and this rule has been mentioned in jurisprudential books. In the realization of loss, it is sufficient that there is a causal relationship between the person's action and the loss of property, even if there is no intention to result in it. In this theory, as soon as the people who recycle and, as a result, cause damage, they will be the guarantors. This theory is very useful for creating the responsibility of the people who recycle and purify, whether it creates responsibility without creating the intention of waste. In other words, if someone harms society and the environment by creating risks due to wastewater recycling that causes the destruction and loss of natural and public capital, he is obliged to compensate for the damage (Yong et al., 2019).

Rule of "Tasbib"

In this doctrine, whoever causes and causes damage to others must be able to cover the damage. Sahib Jawahir says about the guarantee resulting from Tasbib: "There is no difference between the jurists in this regard, and in addition, several news items indicate the existence of such a rule". Among them, we can refer to a correct hadith from Zararah, which states that I asked Imam Sadegh (AS) the ruling of a person who digs a well in his non-property and a passer-by falls in it while passing by? The Holy Prophet (PBUH) stated the guarantee of the one who digs the well; whoever digs a well in his non-property is a guarantor. Therefore, it is sufficient for the financial person who has not wasted directly but introduces causation and preparation to waste to be held responsible. Tasbib is a guarantee if it is considered aggression and aggression according to the custom, although it is not considered as a forbidden act in the Shari'a. Therefore, an action that is against the law and legal guidelines for wastewater recycling is guaranteed. As soon as the wastewater recycling is contrary to the guidelines and then causes the destruction of various plants or diseases, the responsibility arises. Although the theory of loss is more in line with environmental protection goals and can maximize this situation, it should be noted that competing theories in Imami jurisprudence refer to the well-known rule of Tasbi. This theory, which is accepted in the current law of many countries, is not a suitable theory for complete, rapid, and appropriate compensation for environmental damage. This is because, in most cases related to environmental damage, the documentation of the causal relationship faces many difficulties. For example, it is not easy to prove a problem in the refinery or that the farmer was not plowing his field due to plant pests and requires detailed expertise that may fail (Gallegos et al., 2015).

CIVIL LIABILITY FOR WASTEWATER USE

In our Iran, the basis of civil liability is based on fault. The fault, according to the laws of our Iran, is an exaggeration. In other words, the condition of liability resulting from a harmful act in a court is an excess or omission in the act or omission of the act, in a way that is out of the ordinary. The theory of fault is considered the most important basis for civil liability, and therefore, the only reason that can justify one's liability for damages is the existence of a causal relationship between his fault and loss. Article 1 of the Iranian Civil Liability Law states Anyone who, without legal permission, intentionally or as a result of negligence, inflicts damage to life or health or property or liberty or commercial reputation or any other right created by law for individuals that cause harm. Be material or spiritual, and he is responsible for compensating for the damage caused by his action. According to the theory of fault, the injured party must prove by citing why the other party's fault caused the damage. But this theory faces challenges in the modern and industrial world. Despite all the accurate calculations, the effluent may not be recycled properly and may be the source of many problems, in which case the innocent victim who has somehow used the treated wastewater will remain without compensation (Krenkel, 2012).

In most countries, there is a tendency towards absolute responsibility. That is, the perpetrators of environmental damage are given absolute responsibility. In other words, proving guilt has no role in creating liability, and according to this theory, the perpetrators are to blame. It has also been stated in this context that a person who has a dangerous activity in the field of the environment must bear the risk of compensating for its damages. Therefore, pure liability forecasting can be proposed to compensate for damages caused by wastewater recycling. This means that in the sole responsibility of the plaintiff, he must prove that the loss was caused by the act read. Therefore, only attributing the damage to the Cairo force can exempt the defendant from liability (such as the liability of owners of motor vehicles). Despite this sheer responsibility, it is an exceptional responsibility and should be interpreted in a limited and narrow way. In addition, the promotion of liability to absolute liability can lead to the perpetrator of the loss being liable, even if he can prove the innocence or power of Cairo (such as the usurper liability). Therefore, it is enough to prove that wastewater recycling has caused tangible and objective damage. This leads to the realization of responsibility (Copeland, 2008).

Regarding criminal liability and guarantee of criminal execution, it is necessary to explain that crimes have been dealt with as a result in some cases. This means that the occurrence of a crime is considered conditional on the achievement of a result of the act committed (restricted crime). For example, in paragraph (c) of Article 12 of the Law on Hunting, polluting the water of rivers and lakes is considered a crime if it destroys aquatic animals. This means that merely polluting water and destroying the environment is not a crime unless it leads to the extinction of aquatic animals. In the field of wastewater recycling and water and environment issues in general, it does not seem to be the result of correct policy crimes. Because some threats against water and the environment are not such that one or two acts, resulting in significant damage and loss, and if all these cases are to be ignored and the offender is not dealt with, the sum of these is enough to be in time. Certainly pose a serious risk to society and the environment. On the other hand, it is very difficult to prove guilt in such matters. Because such crimes pose dangers and it can hardly be proven that certain violations and activities against the environment have endangered human life and health or damaged natural resources and wildlife.

WAYS OF COMPENSATION

What causes the traditional rules of civil liability to fail in compensating for environmental damage can be due to various factors such as low motivation of victims to pursue, difficulty establishing a causal relationship, difficulty in assessing damages, and lack of conditions for civil liability. Therefore, mechanisms should be sought to compensate for the damage in a desirable way. In the meantime, restoring the previous situation is the first way to compensate. But can this be restored if treatment plants or wastewater recycling systems cause damage? In some cases, it is possible to return to the pre-damage position, and the refining process can be performed again, but in many cases, the arrow that has left the bow will not return to the bow and can not be restored. Some professors suggest providing equivalent and alternative to polluted natural resources with an equivalent source. For example, a standard treatment plant should be established and operated in a suitable environment, or a recycling guide should be modified to prevent plants from becoming pests by adding some materials. However, in many cases, compensation is tried to pay the equivalent amount of damages, but this method is not effective when the range of damages is very large and irreparable and even effective in the future.

However, to compensate for part of the damage, there is no choice but to resort to this method and receive money, because at least with this money, part of the damage can be compensated or used to improve and strengthen another part of the environment. In the Iranian legal system, the Civil Liability Law (adopted in 1960) has paid attention to the general principles of compensation as a result of fault or negligence, which can be used to oblige the offender to compensate the damage to the environment. In addition, liability insurance for wastewater recycling is a useful way to compensate, but insurers do not positively approach such cases. The effects of hazards usually occur over a long period, and insurers usually consider accidental damages with short-term effects such as fire, explosion, and burst water and sewer pipes, so a new model for insurers should be proposed cover such damages as well. It should be noted that due to the need to pay attention to the protection of natural resources and the environment of the National Environmental Fund in 2014 and to help to reduce environmental pollution, prevent degradation, protect natural resources, and sustainable recovery and exploitation of Basic resources, environment and biodiversity have been established. This fund is a public, non-governmental organization that, while equipping monetary resources from the place and income from turnover and activities of the fund and other incomes, uses these resources for environmental investments and pays facilities in line with the fund's goals. The fund provides financial facilities to factories and workshops to reduce pollutants or spend from the fund's funds for insurance of gardens and plants and animals harmed by wild animals (Article 5 of the fund's charter). However, there is no regulation regarding compensation for other cases. These funds can also be used to compensate for the damage caused by wastewater recycling.

CONCLUSION

What exists as effluent management laws in the world is not separate from the structure of the water law of countries. Most countries (especially countries with a similar approach to Iran in wastewater management) have not had the opportunity to develop comprehensive laws on wastewater management. This necessity is not institutionalized at the macro level of the Iran. Because of the above, it can be inferred that legal considerations and challenges in wastewater have more complex customary and infrastructural barriers than water. But effluent laws are part of water laws and are inseparable, especially in fundamental doctrines and analytical frameworks.

Based on the general environmental policies announced by the Supreme Leader, the development of a green economy using wastewater management using economic, social, natural, and environmental capacities and capabilities has been emphasized. In Iran, improving living conditions to provide society with a healthy environment, prevent and prevent the spread of unauthorized pollution and criminality of environmental degradation, and effective punishment and deterrent of polluters and environmental destroyers and their obligation to compensate the government and relevant agencies. Is. However, in considering the legal considerations and challenges of wastewater recycling management, it is very important to identify and define the damages and damages correctly and explain the types of civil, criminal, and related responsibilities. Although wastewater recycling plays an important role in conserving water resources and meeting basic human needs, the process of wastewater recycling and use faces legal risks that must be taken seriously when adopting development policies regarding recycling. In this regard, based on the principle of precaution and prevention, it is necessary to recycle, explain and inform the principles and regulations of recycling. While periodically updating and reviewing the results of treated wastewater to prevent possible harmful effects on people or the environment.

On the other hand, recycling, which is supposed to put wastewater in the path of public use and benefit, should not harm any creature, so the principles of precaution and no harm should be considered in the recycling stage. However, recycling damage is common, and losses may even be identified years after use of the effluent; Something that can threaten human and animal life. Therefore, a mechanism should be sought to compensate and identify the cause of the damage. In this regard, according to the rule of loss of responsibility is the responsibility of the person who is directly responsible for the loss; In this case, if it is proven that the damage is caused by the actions of the employees of the treatment plants or other employees in the recycling route, that person is the guarantor. However, such a burden of responsibility can hamper the courage and creativity of employees to do the job. In addition to wastage, according to the causal rule, if the cause of the damage can be proven, he is the guarantor. However, there are various factors involved in the recycling and using wastewater that makes it difficult to prove the main cause. In addition, the fault-based liability system requires proof of fault, which is not easy for the injured party, and the damages will remain irreparable. Therefore, a system should be sought that does not need to prove guilt and is easily accessible to victims of wastewater use. The system of pure or absolute responsibility is very useful and effective in this regard. But in the meantime, it will be important to strike a balance between staff and managers and the government that intends to use the treated wastewater to meet public needs and the community that needs to consume such products. Therefore, the establishment of specialized insurance in this field can compensate the victims and not impair the motivation of engineers and employees of the water and wastewater industry to do the job. In conclusion, the following is offered as a suggestion:

- Accurate training of wastewater recycling and incorrect recycling results to experts and staff intervening in the recycling process regarding the possible scope of their responsibility
- Explicit explanation of the system of pure or absolute civil liability in special laws in situations where there is an intention to abdicate liability based on fault
- Criminology in the field of non-compliance with the requirements of wastewater recycling and criminological investigation regarding the feasibility of formulating restricted or absolute crimes in the field
- Establishment and expansion of insurance fund services to compensate the victims of wastewater recycling

REFERENCES

Asano, T. (Ed.). (1998). *Wastewater reclamation and reuse: water quality management library* (Vol. 10). CRC Press.

Chong, J. (2014). Climate-readiness, competition and sustainability: An analysis of the legal and regulatory frameworks for providing water services in Sydney. *Water Policy, 16*(1), 1–18. doi:10.2166/wp.2013.058

Chu, J., Chen, J., Wang, C., & Fu, P. (2004). Wastewater reuse potential analysis: Implications for China's water resources management. *Water Research, 38*(11), 2746–2756. doi:10.1016/j.watres.2004.04.002 PMID:15207605

Copeland, C. (2008). *Cruise ship pollution: Background, laws and regulations, and key issues*. Congressional Research Service.

Gallegos, T. J., Varela, B. A., Haines, S. S., & Engle, M. A. (2015). Hydraulic fracturing water use variability in the U nited S tates and potential environmental implications. *Water Resources Research, 51*(7), 5839–5845. doi:10.1002/2015WR017278 PMID:26937056

Guo, J., Bao, Y., & Wang, M. (2018). Steel slag in China: Treatment, recycling, and management. *Waste Management, 78*, 318-330.

Krenkel, P. (2012). *Water quality management*. Elsevier.

Lazarova, V., & Bahri, A. (Eds.). (2004). *Water reuse for irrigation: agriculture, landscapes, and turf grass*. CRC Press. doi:10.1201/9780203499405

Metson, G. S., Iwaniec, D. M., Baker, L. A., Bennett, E. M., Childers, D. L., Cordell, D., Grimm, N. B., Grove, J. M., Nidzgorski, D. A., & White, S. (2015). Urban phosphorus sustainability: Systemically incorporating social, ecological, and technological factors into phosphorus flow analysis. *Environmental Science & Policy, 47*, 1–11. doi:10.1016/j.envsci.2014.10.005

Olsson, O., Weichgrebe, D., & Rosenwinkel, K. H. (2013). Hydraulic fracturing wastewater in Germany: Composition, treatment, concerns. *Environmental Earth Sciences*, *70*(8), 3895–3906. doi:10.100712665-013-2535-4

Rhyner, C. R., Schwartz, L. J., Wenger, R. B., & Kohrell, M. G. (2017). *Waste management and resource recovery*. CRC Press. doi:10.1201/9780203734278

Rosiek, K. (2020). Directions and challenges in the management of municipal sewage sludge in Poland in the context of the circular economy. *Sustainability*, *12*(9), 3686. doi:10.3390u12093686

Rowe, D. R., & Abdel-Magid, I. M. (2020). *Handbook of wastewater reclamation and reuse*. CRC press. doi:10.1201/9780138752514

Russell, D. L. (2019). *Practical wastewater treatment*. John Wiley & Sons.

Salgot, M., & Folch, M. (2018). Wastewater treatment and water reuse. *Current Opinion in Environmental Science & Health*, *2*, 64–74. doi:10.1016/j.coesh.2018.03.005

Song, J., Sun, Y., & Jin, L. (2017). PESTEL analysis of the development of the waste-to-energy incineration industry in China. *Renewable & Sustainable Energy Reviews*, *80*, 276–289. doi:10.1016/j.rser.2017.05.066

Spellman, F. R. (2008). *Handbook of water and wastewater treatment plant operations*. CRC press. doi:10.1201/9781420075311

Sun, Y., Chen, Z., Wu, G., Wu, Q., Zhang, F., Niu, Z., & Hu, H. Y. (2016). Characteristics of water quality of municipal wastewater treatment plants in China: Implications for resources utilization and management. *Journal of Cleaner Production*, *131*, 1–9. doi:10.1016/j.jclepro.2016.05.068

Trianni, A., Negri, M., & Cagno, E. (2021). What factors affect the selection of industrial wastewater treatment configuration? *Journal of Environmental Management*, *285*, 112099. doi:10.1016/j.jenvman.2021.112099 PMID:33588160

Villarín, M. C., & Merel, S. (2020). Paradigm shifts and current challenges in wastewater management. *Journal of Hazardous Materials*, *390*, 122139. doi:10.1016/j.jhazmat.2020.122139 PMID:32007860

Wingfield, S., Martínez-Moscoso, A., Quiroga, D., & Ochoa-Herrera, V. (2021). Challenges to water management in Ecuador: Legal authorization, quality parameters, and socio-political responses. *Water (Basel)*, *13*(8), 1017. doi:10.3390/w13081017

Yong, Y. S., Lim, Y. A., & Ilankoon, I. M. S. K. (2019). An analysis of electronic waste management strategies and recycling operations in Malaysia: Challenges and future prospects. *Journal of Cleaner Production*, *224*, 151–166. doi:10.1016/j.jclepro.2019.03.205

ADDITIONAL READING

Gallegos, T. J., Varela, B. A., Haines, S. S., & Engle, M. A. (2015). Hydraulic fracturing water use variability in the U nited S tates and potential environmental implications. *Water Resources Research, 51*(7), 5839–5845. doi:10.1002/2015WR017278 PMID:26937056

Guo, J., Bao, Y., & Wang, M. (2018). Steel slag in China: Treatment, recycling, and management. *Waste Management, 78*, 318-330.

Krenkel, P. (2012). *Water quality management*. Elsevier.

Rosiek, K. (2020). Directions and challenges in the management of municipal sewage sludge in Poland in the context of the circular economy. *Sustainability, 12*(9), 3686. doi:10.3390u12093686

Russell, D. L. (2019). *Practical wastewater treatment*. John Wiley & Sons.

Salgot, M., & Folch, M. (2018). Wastewater treatment and water reuse. *Current Opinion in Environmental Science & Health, 2*, 64–74. doi:10.1016/j.coesh.2018.03.005

Song, J., Sun, Y., & Jin, L. (2017). PESTEL analysis of the development of the waste-to-energy incineration industry in China. *Renewable & Sustainable Energy Reviews, 80*, 276–289. doi:10.1016/j.rser.2017.05.066

Spellman, F. R. (2008). *Handbook of water and wastewater treatment plant operations*. CRC press. doi:10.1201/9781420075311

Sun, Y., Chen, Z., Wu, G., Wu, Q., Zhang, F., Niu, Z., & Hu, H. Y. (2016). Characteristics of water quality of municipal wastewater treatment plants in China: Implications for resources utilization and management. *Journal of Cleaner Production, 131*, 1–9. doi:10.1016/j.jclepro.2016.05.068

Trianni, A., Negri, M., & Cagno, E. (2021). What factors affect the selection of industrial wastewater treatment configuration? *Journal of Environmental Management, 285*, 112099. doi:10.1016/j.jenvman.2021.112099 PMID:33588160

Villarín, M. C., & Merel, S. (2020). Paradigm shifts and current challenges in wastewater management. *Journal of Hazardous Materials, 390*, 122139. doi:10.1016/j.jhazmat.2020.122139 PMID:32007860

Wingfield, S., Martínez-Moscoso, A., Quiroga, D., & Ochoa-Herrera, V. (2021). Challenges to water management in Ecuador: Legal authorization, quality parameters, and socio-political responses. *Water (Basel), 13*(8), 1017. doi:10.3390/w13081017

Yong, Y. S., Lim, Y. A., & Ilankoon, I. M. S. K. (2019). An analysis of electronic waste management strategies and recycling operations in Malaysia: Challenges and future prospects. *Journal of Cleaner Production, 224*, 151–166. doi:10.1016/j.jclepro.2019.03.205

KEY TERMS AND DEFINITIONS

Equity: Defined by UNEP to include intergenerational equity—"the right of future generations to enjoy a fair level of the common patrimony"—and intragenerational equity—"the right of all people within the current generation to fair access to the current generation's entitlement to the Earth's natural resources"—environmental equity considers the present generation under an obligation to account for long-term impacts of activities and to act to sustain the global environment and resource base for future generations.

Polluter Pays Principle: The polluter pays principle stands for the idea that "the environmental costs of economic activities, including the cost of preventing potential harm, should be internalized rather than imposed upon society at large."

Precautionary Principle: One of the most commonly encountered and controversial principles of environmental law, the Rio Declaration formulated the precautionary principle: To protect the environment, the precautionary approach shall be widely applied by States according to their capabilities.

Prevention: The concept of prevention can perhaps better be considered an overarching aim that gives rise to a multitude of legal mechanisms, including prior assessment of environmental harm, licensing or authorization that set out the conditions for operation and the consequences for violation of the conditions, as well as the adoption of strategies and policies.

Transboundary Responsibility: Defined in the international law context as an obligation to protect one's environment and prevent damage to neighboring environments, UNEP considers transboundary responsibility at the international level as a potential limitation on the sovereign state's rights.

Chapter 9
Environmental Governance and Policy

ABSTRACT

Rio + 20 seeks to achieve goals such as participation in the principles of sustainable development, evaluating the implementation of the World Summit on Sustainable Development in the last 20 years, reviewing the results achieved, as well as paying attention to the implementation of national and international environmental laws and reviewing measures to achieve. The priorities are sustainability and social justice to achieve sustainable development. However, it should be noted that members of the international community should also think beyond Rio + 20 and seek to establish a strong framework to strengthen international environmental rules and provide effective environmental policy solutions for the world's future. In the chapter, this conference's formation process and objectives have been studied and analyzed.

INTRODUCTION

The Rio + 20 Summit kicked off on June 20, 2012, with clear goals for global sustainable development, with a focus on green economics, sustainable development, and international environmental governance. This great global gathering is based on the fundamental principles of justice, sovereignty and sustainable development, and looks forward to the help of global rulers and members of the United Nations to achieve new sustainable development in the 21st century and the rule of law. Rio + 20 seeks to achieve goals such as: participation in the principles of sustainable development, evaluating the implementation of the World Summit on Sustainable Development in the last twenty years, reviewing the results achieved, as well as paying attention to the implementation of national and international environmental laws and reviewing measures to achieve The priorities are sustainability and social justice in order to achieve sustainable development. However, it should be noted that members of the international community should also think beyond Rio + 20 and seek to establish a strong framework to strengthen international environmental rules and provide effective environmental policy solutions for the future of the world. In the present chapter, the formation process and objectives of this conference have been studied and analyzed.

DOI: 10.4018/978-1-6684-7188-3.ch009

Explaining and defining the concept of sustainable development is a development that meets the needs of the present without compromising the ability of future generations to meet their needs. In this definition, there are two key concepts: the concept of "needs", especially the basic needs of the poor world, to which important priority should be given; And the issue of "constraints" imposed by technological and social organization conditions and environmental capabilities to meet current and future needs (Björnsdóttir, 2013).

The spread of environmental pollution has led to measures in some developed and industrialized countries and then worldwide to prevent environmental pollution, prevent the destruction of resources and their sustainable use. Accordingly, the 1960s and 1970s should be considered a period of awakening and environmental awareness. In this period, human beings are aware of the signs threatening the yard and receive the convening of conferences and the ratification of numerous international documents to protect and align with sustainable development (Conway, 2010).

It is worth noting that international environmental law is a set of international legal rules aimed at protecting the environment. In fact, environmental issues and problems made the need to formulate environmental regulations at the international level, to be felt beforehand. Gradually it took on an international dimension, the environment emerged as a whole; At the domestic level, legislation to protect the environment increased dramatically, and governments established special administrative agencies to combat pollution; International organizations, in turn, reacted appropriately and strongly to the new issues raised, and a new chapter began (Conca et al., 2017).

Despite the holding of many international conferences and important developments in the field of environmental law, it can be acknowledged that the majority of these conferences have paid unilateral attention to the issue of sustainable development and less attention has been paid to other aspects of international environmental governance. . In fact, all global efforts in the field of environment are in line with the effective implementation of international environmental laws to improve the overall state of the global environment. This process requires the resolute determination of the governments of the world towards a new sustainable development centered on sustainable development, international environmental governance and a green economy, which means conscious participation in the implementation of international law in order to prevent threats to the global environment. Dissemination requires the protection of natural resources and sustainable development. However, the UN Commission on Sustainable Development has been organizing the Rio + 20 Summit on Green Economy and Poverty Alleviation for almost two years now, as well as the institutional framework for sustainable development (Conca et al., 2016). The overall goal of this global forum is to create a common vision and principles agreed upon by those who agree with the whole issue, and their ultimate goal is to lead a special focus on environmental law and achieve sustainable development. The role and internal relations of law, justice and governance as the basic and fundamental dimensions of environmental sustainability as well as environmental compatibility as an integral part of law, justice and governance for sustainable development are among the most important views of the participants (Edenhofer et al., 2011). Participants also expressed their views and opinions on the future vision of the global environment and new sustainable development, and shared their experiences regarding the effective implementation of environmental rules and policies in this area. In fact, the Rio + 20 Conference is the final point of approval for the decisions of the Preliminary Legal Sessions, the first Preliminary Session being held in Kuala Lumpur and the Second Session in Buenos Aires to finalize the structure of the Conference and the topics discussed. Rio +20 Summit to commemorate the 40th anniversary of the 1972 Stockholm Conference and the 20th anniversary of the 1992 Rio Conference at the level of heads of state and high-ranking representatives of countries, international organizations and non-governmental organizations. New and emerging environmental challenges were held. In this regard, the three main axes of the Green Economy Conference for Sustainable Development with emphasis on eradicating poverty, an institutional framework for sustainable development and international environmental governance were considered. It can be acknowledged that the Rio + 20 conference is a very important turning point and step towards achieving sustainable development and improving the global environmental situation.

INTERNATIONAL EFFORTS TO HOLD THE RIO +20 SUMMIT

Kuala Lumpur Summit

The first Preliminary Meeting on the World Conference on Justice, Sovereignty and the Law on Environmental Sustainability was held in Malaysia on 12 and 13 October 2011 with the aim of preparing the world for the legal issues in Rio. Statements were made at the meeting outlining their approach to the initial idea of the Rio Conference on Justice, Sovereignty and Law, Gaps in the implementation of environmental laws at the national, regional and global levels, and the formation of an international administration system. Environment, the National Environmental Agency, the Department of Environmental Affairs and the Green Economy, how to implement global environmental laws, have been key sustainable priorities for future generations and human rights and the environment. The most important views of the participants consisting of judges, legal officers and jurists in the Malaysian Preliminary Meeting are:

- The role of legal representatives of large communities in advancing national and international efforts towards sustainable environmental goals
- Use of international environmental laws for sustainable development
- Special focus on social justice and environmental impact assessment Principles related to access to information, public participation and some environmental considerations, development of judicial decisions and wider use of environmental auditing
- Recognize environmental justice to eradicate poverty and prevent environmental degradation and reduce climate change (Gupta, 2009)
- Use and strengthen existing capacities for environmental sustainability
- Strengthen the legal framework for the sustainable development of the environment and human rights
- Amend environmental agreements to help the environmental development process
- Strengthening international environmental governance for sustainability
- Meet the challenges of food security, energy, health, agriculture, water and poverty reduction
- The need for an integrated system for the consolidation of multilateral environmental agreements
- Optimal use of technology
- The unity of the world system in terms of environment, international peace and security
- Policy, evaluation, financing, technology in order to achieve sustainable development
- Capacity building at the national level to enhance national and regional cooperation through multilateral environmental agreements

- Establish a new structure called the World Environment Organization [4] and promote cooperation with the World Trade Organization and public support for the new structure based on the United Nations Environment Program.
- Encourage the gradual development of environmental law and the development of this category in the constitutions of countries
- Regional and international assistance and cooperation and sharing of environmental experiences and special attention to international environmental laws, customary rules and treaties
- Recognition of institutions at the national level and implementation of the mechanism and process of accountability for the implementation of environmental laws
- Establish mechanisms for combating environmental crimes and establish specialized courts related to environmental issues (Dodds & Sherman, 2009).

Buenos Aires Summit

The Second Preliminary Meeting on Justice, Sovereignty and Environmental Legislation Related to the Rio + 20 Summit was held on 23 and 24 April 2012 in Argentina with the aim of advancing the discussions of the first meeting and consensus on issues related to justice, governance and law to prepare the Rio Conference Document(Sicurelli, 2016; Tayebi & Zarabi, 2018). Statements on green economy, sustainable development, poverty eradication and an effective framework for sustainable development were presented at the meeting. The main points expressed in the second preliminary meeting of Rio +20 are (Hagen, 2016):

- Expressing concern about the destruction of the natural environment, ecosystems and vital natural resources
- Social justice, environmental governance and the development of environmental rights
- Recognition of the rule of law, use of appropriate standards in transparency, protection and promotion of human rights, commitment to the rights of generations, achieving sustainable development and environmental economy
- Emphasize the role of law as a valuable tool (Gupta, 2009)
- Access to information for sustainable development
- Concerned about the lack of implementation of important environmental achievements and lack of attention to environmental conventions and treaties
- Improving the effectiveness of environmental governance to achieve sustainable development and social justice and effective support of international environmental institutions (Kendall, 2012)
- Allocate adequate resources and commitment to raise awareness and strengthen educational capacity, especially in developing countries and the implementation of legal regimes
- Special attention to the main elements of the conference, including social justice and a sustainable environment, environmental challenges at the national, regional and global levels, and the future of environmental law.

Rio World Conference +20

Rio + 20 Summit at the level of Heads of State and High Representative with a view to achieving goals such as renewing political commitment to sustainable development, assessing major progress in sustainable development, addressing emerging environmental challenges centered on green economy and eradication Poverty, Sustainable Development and International Environmental Sovereignty was held on June 20, 2012 in Brazil.

The conference, which was designed on the basis of justice, sovereignty and law and its compatibility with the environment, pursued the following goals:

- Active participation in achieving sustainable development
- Assess the progress of national and international law enforcement
- Effective support for environmental governance at national and international levels
- Achieving social justice, human rights and dignity.

The summit has an international advisory committee with the participation of high-ranking representatives of the countries and an executive and strategic committee, and in this regard, the UNEP Secretariat technically supports the Sustainable Development Commission (Norouzi & Ataei, 2021).

It is worth noting that the Rio + 20 Summit paid special attention to the basis of environmental justice and human rights in the environment, as well as social justice and sustainable environmental governance. Also in relation to governance such as: the formation of an international governance framework for sustainable environment, the relationship between the environmental governance framework and the green economy, effective national environmental governance as the key to sustainable development, increasing environmental governance oversight for sustainable development, improvement National performance in environmental assessment and better protection of governance and management, law, gap in the implementation of environmental laws at the national, regional and global levels and creating a dichotomy between national and global environmental laws should be considered (Kelsen, 2000).

Intergovernmental Panel on Climate Change

The Climate Change Mission was established in 1988 by the United Nations Meteorological Agency and the United Nations Environment Program. Its main purpose was to study scientific, technical, social, and economic information related to climate change due to human activity, the potential effects of climate change, and options for mitigation and adaptation. The board has submitted four evaluation reports, guidelines for the developed method for the national greenhouse gas inventory, special reports, and technical chapters. The board has three working groups and a special executive force (Norouzi & Ataei, 2021):

- The first working group that studies the science of climate change.
- The second working group examines the adaptation, effects, and adaptation to climate change.
- The third working group examines the reduction of greenhouse gases.

The Special Task Force (TFI) checks the national greenhouse gas inventory.

The Executive Task Force was formed in 1998 by the Climate Change Board for its fourteenth meeting to oversee the GHG inventory program. This program has been carried out since 1991 by the first working group of the delegation, which worked closely with the Organization for Economic Cooperation and Development and the International Energy Agency. The objectives of this special force are to develop and improve the internationally agreed method and software for calculating and reporting the emissions of national greenhouses and eliminating them. Encourage the widespread use of this method by States Parties to the UN Security Council and the signatories of the United Nations Framework Convention on Climate Change in 2007 and implement the UN Joint Action Plan. Gore was awarded the former Vice President of the United States. In this regard, the head of the UN Environment Program stated that the award shows that climate change is very important for international peace and security. He went on to say in a statement that the Nobel Peace Prize Committee today made it clear that combating climate change is a fundamental policy for peace and security in the twenty-first century (Pachauri et al., 2014).

UNITED NATIONS CONFERENCE ON ENVIRONMENT AND DEVELOPMENT

History of Rio Summit +20

The United Nations Conference on Environment and Development (UNCED), also known as the Rio de Janeiro Earth Summit, the Rio Summit, the Rio Conference, and the Earth Summit (Portuguese: ECO92), this was a major United Nations conference held in Rio de Janeiro from June 3 to June 14, 1992.

Earth Summit was created as a response for member states to cooperate together internationally on development issues after the Cold War. Due to issues relating to sustainability being too big for individual member states to handle, Earth Summit was held as a platform for other member states to collaborate. Since the creation, many others in the field of sustainability show a similar development to the issues discussed in these conferences, including non-governmental organizations (NGOs).

Structural Framework at the Rio Summit +20

At the Rio + 20 Summit, the structural framework for sustainable development was designed as follows:

- Balanced integration of the three dimensions of sustainable development (environmental, economic and social)
- Paying attention to all cross-cutting issues in order to cooperate in the implementation of sustainable development, based on a pragmatic and result-oriented approach,
- Considering the importance of the links between key issues and challenges and the need for an organized approach to them at all levels of sustainable development;
- Improve cohesion, reduce fragmentation and overlap, and increase effectiveness, efficiency, and transparency in strengthening collaboration and coordination;
- Increase the effective and complete participation of all countries in the decision-making process;
- Interaction of high-ranking political leaders in providing political guidance and identifying specific measures to promote the effective implementation of sustainable development through the exchange of experience and knowledge (Sicurelli, 2016);
- Relationship between science and politics through transparent, comprehensive and evidence-based scientific evaluations as well as access to reliable, relevant and timely data in areas related to the three dimensions of sustainable development, in this regard, strengthen the participation of all countries in the international process Sustainable development and capacity building, especially for developing countries, is intended for monitoring and evaluation.

- Effective participation and interaction of civil society and other relevant stakeholders in international forums and, in general, the promotion of broad and transparent public participation in the implementation of the Sustainable Development Goals;
- Establish a monitoring plan to review and review the progress of implementation of all stated commitments for sustainable development (Saul, 2009).

Executive Framework at the Rio Summit +20

The implementation tools set out in Agenda 21, the Johannesburg Executive Plan, the Montreal Consensus, the International Conference on Development Financing, and the Doha Declaration on Financing for Development, to achieve the full and effective transfer of sustainable development commitments to practical results. It is considered a necessity (Sindico, 2007; Norouzi et al., 2021).

The Rio + 20 Sustainable Development Executive Framework was designed as follows:

- Financing, increase financial support for all countries to achieve sustainable development goals and developed countries' financial support of developing countries.
- Technology, technology transfer to developing countries as previously agreed in the Johannesburg Executive Plan.
- Capacity Building, increase capacity building for sustainable development and strengthen scientific and technical cooperation and human resource development including training, exchange of experiences and expertise, knowledge transfer and technical assistance to expand capacities.
- Business, consider international trade as an engine for sustainable economic development and growth, as well as the crucial role played by global multilateral trading systems based on non-discriminatory rules that can stimulate economic growth and development around the world and promote them. To play towards sustainable development, we must pay attention.
- Liabilities Registry, establish an office to record all the voluntary commitments raised at the United Nations Conference on Sustainable Development, expressed by all stakeholders and their networks to implement strong policies, plans, programs, projects and actions to promote sustainable development and eradicate poverty. And the information should be available to the public.

Rio + 20 Summit and Protecting the International Environment

At the Rio Summit, one of the most important and comprehensive priorities is to address the issue of international governance for sustainable development. In this regard, suggestions have been made to reform the framework of international governance:

- Transformation of the Sustainable Development Committee into a Sustainable Development Council under the auspices of the United Nations General Assembly
- Transformation of the UN Trusteeship Council into a Sustainable Development Council under the auspices of the Security Council, which addresses development challenges, including climate change. It seems that strengthening the international organizational framework for sustainable development is an evolutionary process and these arrangements should be maintained and the relevant gap should be examined and the economic, social and environmental dimensions of sustainable development should be considered (Schachter, 1951).

In fact, it should be noted that the foundation of the Rio World Summit is based on sustainable environmental governance, as Brazil stated in a letter to the UN Secretary-General to hold the Rio Conference, despite the international consensus and the need for increased cooperation. And coordination between international organizations and environmental agreements, there must be an integrated and broader movement towards sustainable development, and we must pursue the role of the Council for Sustainable Development (Norouzi, 2020).

Given the importance and fundamentality of this issue, the following points should be considered in order to continue it:

- Impact on security and effective action in the field of sustainable development
- Increase global knowledge on issues and the importance of sustainable development
- The existing method of governance, in other words, the attitude of governments towards sustainable development issues
- Execution process and financing of projects related to sustainable development
- Existence of international partnerships in the field of sustainable development (Saul, 2009)

Concerns about using sustainable development as an excellent framework for international environmental protection are largely due to the fact that today most environmental indicators are still far from global standards. One of the challenges in international environmental law is the congestion of treaties and the inefficiency of some of them. The international community needs to reconsider its approach to environmental issues with a focus on international environmental law (Norouzi, 2020).

The rapid growth of international environmental conventions since 1972, that is, since the Stockholm Conference, has led to the development of environmental activities, but it cannot be said that all these activities were necessarily aimed at protecting the global environment. One of these activities was the creation of numerous treaties, which raised concerns because these treaties were sometimes in conflict with each other and because the rest of the treaties lost their effectiveness in implementing some environmental treaties (Schachter, 1951).

The Rio + 20 World Summit is definitely different from previous conferences, as it has put sustainable development with a focus on the green economy and an emphasis on the protection and management of natural resources in line with economic and social development. In this regard, the goal is to protect and preserve the world environment by focusing on international environmental diplomacy and international environmental law.

Rio + 20 World Summit pursues specific goals such as the evolution of national and international environmental laws based on their fundamental principles and underpinnings, as well as its implementation, promoting effective national and international environmental governance, strengthening the link between social justice and environmental sustainability Environment and the presentation of a global framework for international cooperation and the institutionalization of global interactions, and this is in fact a look beyond the Rio + 20 summit, which strengthens environmental rights.

The Rio + 20 World Conference had two very important perspectives on its agenda, including the green economy in the field of sustainable development and eradication of poverty, as well as the organizational framework for sustainable development, and with this perspective, the future of environmental law and opportunities And emerging issues are social justice and environmental sustainability and the challenges of environmental governance at the national, regional and global levels with an emphasis on their improvement and effectiveness (Tayebi & Zarabi, 2018).

A point that should not be overlooked is a special look at the issue of international environmental sovereignty. This title may be a good strategy for organizing international environmental events, but it should not be easily overlooked, because it should be noted that the countries of the world are seeking international environmental custody and thus They seek to undermine the role of UNEP, and in fact these countries want to prove that developing and underdeveloped countries are procrastinating or unable to protect their environment(Euronews, 2014). The issue is being raised covertly, while the claiming countries themselves have been violating environmental regulations. It is noteworthy that in the context of this claim, they want to consider the Security Council competent in environmental matters and to advance this issue to the extent that they cover their goals, but it seems that in the face of these measures, a role should be played (Norouzi, 2021). It made Yunp even more colorful and controlled environmental events. One point that can be made is that environmental criminals can be prosecuted internationally. This is important from the point of view that environmental criminals are criminals against humanity and undermine international peace and security (Tayebi & Zarabi, 2018).

Paris Conference 2015

Climate change is a global emergency that goes beyond national borders. It is an issue that requires international cooperation and coordinated solutions at all levels. To tackle climate change and its negative impacts, world leaders at the UN Climate Change Conference (COP21) in Paris reached a breakthrough on 12 December 2015: the historic Paris Agreement. The Agreement sets long-term goals to guide all nations as substantially reduce global greenhouse gas emissions to limit the global temperature increase in this century to 2 degrees Celsius while pursuing efforts to limit the increase even further to 1.5 degrees; review countries' commitments every five years; provide financing to developing countries to mitigate climate change, strengthen resilience and enhance abilities to adapt to climate impacts. The Agreement is a legally binding international treaty. It entered into force on 4 November 2016. Today, 191 Parties (190 countries plus the European Union) have joined the Paris Agreement. The Agreement includes commitments from all countries to reduce their emissions and work together to adapt to the impacts of climate change and calls on countries to strengthen their commitments over time. The Agreement provides a pathway for developed nations to assist developing nations in their climate mitigation and adaptation efforts while creating a framework for the transparent monitoring and reporting of countries' climate goals. The Paris Agreement provides a durable framework guiding the global effort for decades to come. It marks the beginning of a shift towards a net-zero emissions world. Implementation of the Agreement is also essential for the achievement of the Sustainable Development Goals. The Paris Agreement works on a five-year cycle of increasingly ambitious climate action carried out by countries. Each country is expected to submit an updated national climate action plan every five years - known as Nationally Determined Contribution, or NDC. In their NDCs, countries communicate actions to reduce their greenhouse gas emissions to reach the Paris Agreement's goals. Countries also communicate in the NDC's actions to build resilience to adapt to the impacts of rising temperatures. To better frame the efforts towards the long-term goal, the Paris Agreement invites countries to formulate and submit long-term strategies. Unlike NDCs, they are not mandatory. The operational details for the practical implementation of the Paris Agreement were agreed on at the UN Climate Change Conference (COP24) in Katowice, Poland, in December 2018, in what is colloquially called the Paris Rulebook, with a few unresolved issues (Sicurelli, 2016).

Future and Prospects of Climate Change Prevention Measures

On 28 September 2012, the Presidents of St. Kitts and Nice announced that Pure Energy would be the also world's absolute power Emotions are reflected in the Sustainable Energy Session for the General Assembly's initiatives (Norouzi & Movahedian, 2021). To prevent the adverse effects of climate change from progressing, a basic global policy must be adopted and implemented quickly. For example, the International Climate Change Regime must ensure that greenhouse gas emissions, one of the main causes of these changes, are reduced by half by the middle of the 21st century to what is seen today. A change in the power structure must accompany this international policy in the world political order. An official note from the German government in 2011 outlined how the Security Council would deal with the assumption of new responsibilities related to the climate change crisis. The blueprint outlined several scenarios for dealing with the problems: These scenarios were: dealing with the effects of rising temperatures and rising sea levels, how to deal with displaced persons affected by the change, how to prevent conflicts in parts of Asia and the east; It will be obvious to identify new problems that the council has to face (S/2011/408). Other commentators have reflected these feelings; By making proposals such as proper planning, disaster preparedness, developing a public understanding of the particular dangers to nations, and understanding the international community as a whole. The Security Council should consider facilitating the creation of international agreements committing governments to reduce greenhouse gas emissions. Several other proposals have been made to the UN General Assembly in this regard. Another tool of the United Nations in dealing with this issue is the International Court of Justice, which can be a solution to this issue by exercising jurisdiction, and Chapter 7 of the Charter can also be a guarantee for the implementation of votes (Kendall, 2012). The UN Convention on Climate Change, as one of the most important instruments of international law, calls for the broad cooperation of countries and international assistance. However, in dealing with climate change, it is better not to make a military choice because war is inherently harmful; As the Rio Declaration acknowledges, war is inherently destructive and endangers peace, development, and environmental protection (Sindico, 2007). Therefore, military action to combat climate change is contrary to the principles of protection of international climate change treaties. Chapter 42 of the Charter can not be a good solution to combat this change because the absence of war and military conflict between countries guarantees peace and security. It is international, and the only restriction that the Security Council must adhere to is the scope of Chapter 39 of the Charter.

CONCLUSION

There is a strong consensus in scientific circles that climate change is taking place and that human activities play a key role in this process. According to the Fifth Interim Board's evaluation report, this role accounts for approximately 95%. While there is no difference of opinion among scientists as to the role of the human factor, international debate seeks to determine how the climate responds to rising greenhouse gas emissions over time and in different parts of the world. Will. If climate change is recognized internationally as a threat to international security, it can give hope to the integration of actions through the establishment and cooperation mechanisms to deal with its destructive effects. Otherwise, climate change will deepen conflicts and conflicts in the field of international relations and will lead to the beginning of international conflicts over the distribution of water resources, land and migration management, or the issue of compensation to the minorities that will be the main culprits of change in trends and countries affected by its destructive effects. The formation of international circles and regimes indicates an increase in public awareness and alarm bells of the effects of climate change worldwide and is gradually recognized by the international community. However, in addition to the Conference of the Parties to the Framework Convention on Climate Change, held annually, similar measures have been taken at the United Nations to bring about normalization and awareness-raising. The severity of the consequences of climate change has been raised on the Security Council agenda since 2007 as a threat to international peace and security. It seems that while the Security Council is discussing climate change and issues related to climate stability and security, the focus of future diplomats should be on the UN Framework Convention on Climate Change and the Kyoto Protocol.

Contrary to the fact that the countries participating in the Climate Change Negotiations of the Security Council have a consensus on the occurrence of climate change as a threat to international peace and security, but disagree on how to deal with the dangers of such changes. These differences are due to the views of these countries. Some see the issue as part of a conflict prevention policy, and others believe in the principle of climate change prevention. But the third group, which will receive the most impact in the future and understand the security of the issue, believes in immediate action in this regard. But in practice, these steps have always faced obstacles. For example, the United Nations Framework Convention on Climate Change and the Kyoto Protocol, a platform for global action in this area, is largely hampered by US opposition to the mandatory emission reductions set out in Annex I to the Kyoto Protocol have been exposed. The Security Council should facilitate the creation of international agreements that commit governments to reduce greenhouse gas emissions. In this regard, it can benefit from the tools of the Climate Change Framework Convention and even the International Court of Justice.

The first United Nations Conference on Sustainable Development, entitled “Earth Summit”, was held in Rio de Janeiro to examine the state of the environment and sustainable development. Among the achievements of this conference were memoranda of understanding such as climate change, combating expression and preserving biodiversity. The second conference, held in Johannesburg, once again united the nations of the world on the sustainability of these principles. Twenty years later, we witnessed the Rio World Summit, which focused on green economics, eradicating poverty, achieving an organizational framework for sustainable development and maintaining existing ecosystems, and working to renew countries’ political commitment to sustainable development and how to meet the challenges. New emergence has also been attempted. In a broad sense, the law plays a key role in shaping the behavioral changes needed to achieve environmental and social sustainability in the context of sustainable development. In the 21st century, paving the way for development is ensuring environmental and social sustainability. Legal and institutional developments over the past forty to fifty years show that environmental laws at the national and international levels can make a significant contribution to sustainable partnerships between the environment and sustainable development. The evolutionary process of sustainable development is inextricably linked to human survival, social justice, sustainable ethics and law, which are key tools for achieving the path of sustainable environmental and social development. The Rio + 20 Summit sought to unite the leaders of the United Nations to standardize international law and implement it at the macro level to achieve environmental justice. The full realization of sustainable development is an important goal for this summit, as it establishes a green economy and international environmental sovereignty.

ACKNOWLEDGMENT

This research received no specific grant from any funding agency in the public, commercial, or not-for-profit sectors.

REFERENCES

Björnsdóttir, A. L. (2013). *The UN Security Council and Climate Change: Rising Seas Levels, Shrinking Resources, and the 'Green Helmets'* [Master's thesis].

Conca, K., Thwaites, J., & Lee, G. (2016). Bully Pulpit or Bull in a China Shop? Climate change and the UN Security Council. *Annu. Meet. Acad. Counc. United Nations Syst, 1*.

Conca, K., Thwaites, J., & Lee, G. (2017). Climate change and the UN Security Council: Bully pulpit or bull in a china shop? *Global Environmental Politics*, *17*(2), 1–20. doi:10.1162/GLEP_a_00398

Conway, D. (2010). The United Nations Security Council and climate change: Challenges and opportunities. *Climate Law*, *1*(3), 375–407. doi:10.1163/CL-2010-018

Dodds, F., & Sherman, R. (2009). *Climate Change and Energy Insecurity:" The Challenge for Peace, Security and Development*. Routledge. doi:10.4324/9781849774406

Edenhofer, O., Pichs-Madruga, R., Sokona, Y., Seyboth, K., Kadner, S., Zwickel, T., & Matschoss, P. (Eds.). (2011). *Renewable energy sources and climate change mitigation: Special report of the intergovernmental panel on climate change*. Cambridge University Press. doi:10.1017/CBO9781139151153

Gupta, S. (2009). Environmental law and policy: Climate change as a threat to international peace and security. *Perspectives on Global Issues*, *4*(1), 7–17.

Hagen, J. J. (2016). Queering women, peace and security. *International Affairs*, *92*(2), 313–332. doi:10.1111/1468-2346.12551

Kelsen, H. (2000). *The law of the United Nations: a critical analysis of its fundamental problems: with supplement* (Vol. 11). The Lawbook Exchange, Ltd.

Kendall, R. (2012). Climate change as a security threat to the Pacific Islands. *New Zealand Journal of Environmental Law*, *16*, 83–116.

Norouzi, N. (2021). Post-COVID-19 and globalization of oil and natural gas trade: Challenges, opportunities, lessons, regulations, and strategies. *International Journal of Energy Research*, *45*(10), 14338–14356. doi:10.1002/er.6762 PMID:34219899

Norouzi, N., & Ataei, E. (2021). Covid-19 Crisis and Environmental law: Opportunities and challenges. *Hasanuddin Law Review*, *7*(1), 46–60. doi:10.20956/halrev.v7i1.2772

Norouzi, N., Khanmohammadi, H. U., & Ataei, E. (2021). The Law in the Face of the COVID-19 Pandemic: Early Lessons from Uruguay. *Hasanuddin Law Review*, *7*(2), 75–88. doi:10.20956/halrev.v7i2.2827

Norouzi, N., & Movahedian, H. (2021). Right to Education in Mother Language: In the Light of Judicial and Legal Structures. In Handbook of Research on Novel Practices and Current Successes in Achieving the Sustainable Development Goals (pp. 223-241). IGI Global.

Pachauri, R. K., Allen, M. R., Barros, V. R., Broome, J., Cramer, W., Christ, R., & van Ypserle, J. P. (2014). *Climate change 2014: synthesis report. Contribution of Working Groups I, II and III to the fifth assessment report of the Intergovernmental Panel on Climate Change*. IPCC.

Saul, B. (2009). Climate Change, Conflict and Security: International Law Challenges. *NZ Armed FL Rev.*, *9*, 1.

Schachter, O. (1951). *The Law of the United Nations*. Academic Press.

Sicurelli, D. (2016). *The European Union's Africa policies: norms, interests and impact*. Routledge. doi:10.4324/9781315239828

Sindico, F. (2007). Climate change: A security (council) issue. *Carbon & Climate L. Rev.,* 29.

Tayebi, S., & Zarabi, M. (2018). Environmental Diplomacy and Climate Change; Constructive strategic approach to reducer. *Human & Environment*, *16*(4), 159–170.

ADDITIONAL READING

Gupta, S. (2009). Environmental law and policy: Climate change as a threat to international peace and security. *Perspectives on Global Issues*, *4*(1), 7–17.

Hagen, J. J. (2016). Queering women, peace and security. *International Affairs*, *92*(2), 313–332. doi:10.1111/1468-2346.12551

Kendall, R. (2012). Climate change as a security threat to the Pacific Islands. *New Zealand Journal of Environmental Law*, *16*, 83–116.

Saul, B. (2009). Climate Change, Conflict and Security: International Law Challenges. *NZ Armed FL Rev.*, *9*, 1.

Sicurelli, D. (2016). *The European Union's Africa policies: norms, interests and impact*. Routledge. doi:10.4324/9781315239828

Tayebi, S., & Zarabi, M. (2018). Environmental Diplomacy and Climate Change; Constructive strategic approach to reducer. *Human & Environment*, *16*(4), 159–170.

KEY TERMS AND DEFINITIONS

Environmental Law: Environmental law is a collective term encompassing aspects of the law that protect the environment. A related but distinct set of regulatory regimes, now strongly influenced by environmental legal principles, focuses on managing specific natural resources, such as forests, minerals, or fisheries. Other areas, such as environmental impact assessment, may not fit neatly into either category but are nonetheless important components of environmental law. Previous research found that when environmental law reflects moral values for betterment, legal adoption is more likely to be successful, usually in well-developed regions. In less-developed states, changes in moral values are necessary for successful legal implementation when environmental law differs from moral values.

Equity: Defined by UNEP to include intergenerational equity—"the right of future generations to enjoy a fair level of the common patrimony"—and intragenerational equity—"the right of all people within the current generation to fair access to the current generation's entitlement to the Earth's natural resources"—environmental equity considers the present generation under an obligation to account for long-term impacts of activities and to act to sustain the global environment and resource base for future generations. Pollution control and resource management laws may be assessed against this principle.

Polluter Pays Principle: The polluter pays principle stands for the idea that "the environmental costs of economic activities, including the cost of preventing potential harm, should be internalized rather than imposed upon society at large." All issues related to responsibility for cost for environmental remediation and compliance with pollution control regulations involve this principle.

Precautionary Principle: One of the most commonly encountered and controversial principles of environmental law, the Rio Declaration formulated the precautionary principle as follows, to protect the environment, States shall widely apply the precautionary approach according to their capabilities. Where there are threats of serious or irreversible damage, lack of full scientific certainty shall not be used as a reason for postponing cost-effective measures to prevent environmental degradation. The principle may play a role in any debate over the need for environmental regulation.

Prevention: The concept of prevention etc. can perhaps better be considered an overarching aim that gives rise to a multitude of legal mechanisms, including prior assessment of environmental harm, licensing, or authorization that set out the conditions for operation and the consequences for violation of the conditions, as well as the adoption of strategies and policies. Emission limits and other product or process standards, the use of best available techniques, and similar techniques can all be seen as applications of the concept of prevention.

Public Participation and Transparency: Identified as essential conditions for "accountable governments...industrial concerns," and organizations generally, public participation and transparency are presented by UNEP as requiring "effective protection of the human right to hold and express opinions and to seek, receive and impart ideas, etc. a right of access to appropriate, comprehensible and timely information held by governments and industrial concerns on economic and social policies regarding the sustainable use of natural resources and the protection of the environment, without imposing undue financial burdens upon the applicants and with adequate protection of privacy and business confidentiality," and "effective judicial and administrative proceedings." These principles are present in environmental impact assessment, laws requiring publication and access to relevant environmental data, and administrative procedures.

Sustainable Development: Defined by the United Nations Environment Programme as "development that meets the needs of the present without compromising the ability of future generations to meet their own needs," sustainable development may be considered together with the concepts of "integration" (development cannot be considered in isolation from sustainability) and "interdependence" (social and economic development, and environmental protection, are interdependent). Laws mandating environmental impact assessment and requiring or encouraging development to minimize environmental impacts may be assessed against this principle. The modern concept of sustainable development was discussed at the

1972 United Nations Conference on the Human Environment (Stockholm Conference) and the driving force behind the 1983 World Commission on Environment and Development (WCED, or Bruntland Commission). In 1992, the first UN Earth Summit resulted in the Rio Declaration, Principle 3 of which reads: "The right to development must be fulfilled to equitably meet developmental and environmental needs of present and future generations." Sustainable development has been a core concept of international environmental discussion ever since, including at the World Summit on Sustainable Development (Earth Summit 2002) and the United Nations Conference on Sustainable Development (Earth Summit 2012, or Rio+20).

Transboundary Responsibility: Defined in the international law context as an obligation to protect one's environment. UNEP considers transboundary responsibility at the international level to prevent damage to neighboring environments at the international level as a potential limitation on the rights of the sovereign state. Laws that limit externalities imposed upon human health and the environment may be assessed against this principle.

Chapter 10
Environmental Court Procedure and Dispute

ABSTRACT

The need to protect and prevent the destruction and pollution of the environment is recognized by all governments and individuals. The possibility of litigation and litigation against harmful actions for the environment is one of the guarantees of effective implementation, which should be given more attention by the governments and regulatory mechanisms of the region and internationally, and the possibility of litigation based on benefits. Public courts or competent regional authorities, including the European Court of Human Rights, the American Court of Human Rights, and the African Court of Human and Peoples' Rights, appear to be an effective step in protecting the environment and respecting fundamental human rights. The chapter examines the procedure of the regional courts of human rights and the domestic courts in some countries regarding the possibility of public litigation.

INTRODUCTION

Since the adoption of the Stockholm Declaration in 1972, especially concerning the threat of ecological catastrophes, environmental awareness has increased to the degree that was unpredictable at the time. This progress can be seen in domestic courts, tribunals, and international authorities, which have faced many environmental lawsuits in recent years. The Rio Declaration of 1992 is also important in developing environmental norms, emphasizing greater access to justice in environmental matters. Thus, with the filing of environmental lawsuits in the International Court of Justice, the International Court of the Law of the Sea, and the dispute resolution bodies of the World Trade Organization, each of these institutions have been involved in some way in protecting the environment (Schall, 2008). The activity of human rights courts in the field of environmental protection has also increased significantly, and this seems to be the logical consequence of the first principle of the Stockholm Declaration, which relates human rights and the environment. Nevertheless, there are still doubts whether the three major regional human rights treaties, the European Convention on Human Rights, the American Convention on Human Rights, and the African Charter on Human and Peoples' Rights, provide effective protection of the environment.

DOI: 10.4018/978-1-6684-7188-3.ch010

Although the jurisdiction of the relevant courts allows them to hear environmental issues, the procedure of the European Court of Human Rights and the American Court of Human Rights seems to be a long way from ensuring effective protection of the environment (Hassan & Azfar, 2003).

In this chapter, the scope of "public interest litigation" with regional human rights institutions and national courts in some countries is examined as a tool to protect the environment further, although the jurisdiction of these authorities, especially regional institutions. It has conditions, including the realization of cases of human rights violations, the existence of a victim of such violations or violations, and the enjoyment of the legal status of the complaint.

CONCEPT OF PUBLIC INTEREST LITIGATION

In 1996, the United Nations Economic Commission for Europe presented its interpretation of the relationship between human rights and the environment. In this interpretation, the right to the environment is considered a means for humans to enjoy the fundamental right to life. This orientation was endorsed by the Aarhus Convention in 2001. Before the entry into force of this Convention, most national and international legal systems allowed only the aggrieved party to seek redress through the judiciary and the grievances of others. Individuals who did not personally suffer damages could not sue the victim or injured party. Thus, it was accepted as a general rule that unless a person has personally suffered damages, no one can claim compensation for certain actions, even if those actions violate the law (Goldston, 2006).

It should be noted that in this article, public interest litigation in human rights courts is only concerned with the assessment of acts and omissions performed by public authorities and does not consider its broader meaning. Also, anything that is considered to be in the public interest is not in the public interest.

In general, when there is a public interest in a case, several factors can be cited, some of which is where the public interest is gained from the outcome of the dispute; The outcome of the lawsuit has no personal, financial or material benefit, or if such a benefit arises, is not such as to justify the lawsuit economically; And litigation deals with issues that are more important than the immediate interests of the litigants.

In environmental claims, the European Commission has stated that the important nature of environmental law does not justify the exercise of those rights solely based on private interests. In addition, the environment is often referred to as a "common concern" or even a "shared heritage" of humanity. These factors in the public interest can be found in environmental litigation and should therefore be considered public interest litigation (Badwaza, 2005).

PUBLIC INTEREST LITIGATION IN REGIONAL HUMAN RIGHTS COURTS

Although the European Court of Human Rights, the American Court of Human Rights, and the African Court of Human Rights have all been established to uphold human rights, the formal and substantive rights guaranteed by each are somewhat different. It is very evident with the environment. Public interest litigation seems to be a formal rights issue in the first place. However, this formal aspect is closely related to the substantive rights guaranteed by human rights systems, and for this reason, an attempt has been made to examine both aspects of the case-law of the regional human rights courts (Alan, 2017).

European Court of Human Rights

The European Court of Human Rights and the Council of Europe have made great strides in defining and enforcing international environmental regulations that can be used in regional courts or domestic courts. However, it should be noted that the European Convention on Human Rights does not refer to the environment. However, the European Court of Human Rights has affirmed the right to a healthy environment with its broad interpretation of Articles 6, 2, and 8 of the Convention and some other articles (Leach, 2011).

Criteria for Accepting Lawsuits

According to Article 34 of the European Convention on Human Rights, any natural person and any non-governmental organization or group of individuals can file a lawsuit against the European Convention on Human Rights; However, the plaintiff must suffer a violation of the rights enshrined in the Convention. Thus, under the Convention, the complainant must be considered a victim of a violation of the rights outlined in the Convention or its protocols by one of the States Parties. This condition means that the claims of individuals whose right to guaranteed rights has not been violated in the Convention will not be accepted by the court (Mowbray, 2005).

A person is usually considered a victim of a violation when mandatory action is taken against them, and the court or the executive often accomplishes this action, and the other party will be considered a "direct victim." However, it should be noted that the concept of "victim" in some cases also includes potential victims if the European Commission of Human Rights places the rights of individuals who can show serious and immediate risk of human rights violations to enjoy legal status. The court has recognized the lawsuit. In addition, plaintiffs may be classified as direct victims and indirect victims directly affected by the violation of third-party rights.

However, filing a complaint on behalf of other individuals will not be accepted. The commission has stated that the plaintiff cannot sue on behalf of the people as a whole, as the Convention does not allow such "action on behalf of others," and the commission is only required to investigate a complaint in which the plaintiff himself is a victim. The European Court of Human Rights still follows the same approach. However, under certain circumstances, a non-governmental organization can represent a real victim. Nevertheless, it seems that the general interest-based proceedings, without being the victim of a violation of a right under the Convention, are hindered by the formal provisions of the European Convention on Human Rights (Greer & Wildhaber, 2012).

Of course, not only the formal laws but also the substantive rules are considered an obstacle to the admissibility of a public interest claim, where Article 34 of the Convention explicitly refers to the substantive rights enshrined in the European Convention on Human Rights: The complainant must identify the substance of the Convention, one of which has been violated. Concerning the codification of fundamental civil and political rights in the European Convention on Human Rights, the concept of sacrifice in Article 34 of the Convention is closely related to the fundamental rights enshrined in the Convention. This consideration can also be seen in the court's approach to qualifying victims in some cases.

Substantive Consideration and Some Issues

The European Convention on Human Rights does not include the right to protect the environment or explicitly address environmental issues specifically. However, several cases have been brought before the European Court of Human Rights that indirectly affect the protection of the environment. First, the right to life is enshrined in Article 2, which implies a positive commitment by States to take the necessary measures to protect human life. It is very important to pass laws that guarantee no harm to the environment. The prohibition of degrading or degrading treatment provided for in Article 3 of the Convention can also be invoked in the fight against environmental degradation, which leads to health problems. However, according to the court in Lo'pez Ostra, there must be a serious threat to human health to rely on these substances. In cases where environmental damage has affected individuals, the right to privacy, family, and home life guaranteed in Article 8 has often been invoked. While Article 8 does not deal with the relationship between the individual and the environment around him, it now encompasses the entire personal territory and therefore includes the broad right to make environmental decisions close to your home and place of residence. Other potentially relevant rights are the right to property and the right to a fair trial, as set out in Articles 6 and 13 (Christoffersen & Madsen, 2011).

American Court of Human Rights

The American Court of Human Rights is an important body in overseeing the implementation of States' obligations under the American Convention on Human Rights. The Protocol to the Convention recognizes the right to a healthy environment. It is noteworthy about the American human rights system that there is no admissibility stage in this system compared to European human rights arrangements. Upon receipt of the complaint and other preliminary information, the Commission's Secretariat shall consider whether the complaint is ostensibly admissible or not, and in the first instance, register the case and open it. Nevertheless, formal issues related to the admissibility of litigation and the case's merits have been examined separately (Pasqualucci, 2012).

Criteria for Accepting Lawsuits

Under the American Convention on Human Rights, individuals and non-governmental organizations may not sue directly with the American Court of Human Rights: Article 61 paragraph 1 of the Convention states that only States Parties and the Commission have the right to sue. However, Article 44 of the Convention provides for the right of individuals to sue the commission. According to this article, individuals, groups of individuals, and non-governmental organizations legally recognized in one of the member states can file a lawsuit (Buergenthal, 1982).

The plaintiff may be a spouse, a close friend, or even a stranger to the victim. According to this statement, according to the statements mentioned above, the person, even if he has not been directly affected by the violation of the law nor will he have a problem arguing with the commission, provided that his lawsuit is admissible in the light of other factors. Nevertheless, it is still important to the American Court of Human Rights that violations of a particular human right must be alleged against a particular individual. According to the commission's procedure, according to Article 44, the victims must be identified and be existed. Therefore, the plaintiff must claim that they are victims of a violation of the Convention or represent a known victim. As can be deduced from the advisory opinion of the American

Court of Human Rights, this court has been strongly influenced by the European Court of Human Rights in its interpretation and context of the term "victim." Thus, although the provisions of the American Convention are somewhat broader than the European Convention, the American system also does not allow litigation in the form of litigation on behalf of society, and thus violations that lead to the violation of human rights. Which is not possible and also is not subject to compensation. A cursory glance at the human rights enshrined in the American Convention on Human Rights reveals that only first-generation human rights are protected in this document, and as a result, economic, social, and cultural rights are not enforceable in the court. This basic idea, which considers only the first generation of human rights to be protected, is reflected in the conditions necessary for litigation, and according to which there is no enforcement right based on which a public interest claim can be made, and damages can be claimed (Lixinski, 2010).

Substantive Consideration and Some Issues

The American Court of Human Rights, like the European Court of Human Rights, has invoked rights such as the right to life, property, and privacy in lawsuits alleging environmental degradation. Although the 1989 Protocol to the US Convention includes the right to a healthy environment, this right cannot be the subject of individual complaints against government action (Antkowiak, 2007).

African Court of Human Rights and Peoples

On 25 July 2004, the Additional Protocol to the African Charter on Human and Peoples' Rights entered into force to establish the African Court of Human and Peoples' Rights. The court will fill the gaps in the African human rights system, governed by only one quasi-judicial body, the African Commission on Human Rights and Peoples. The role of the African Court of Justice in dealing with environmental issues is of great concern to jurists, as at the regional level, the African Charter on Human and Peoples' Rights is the first international document whose explicit right to a healthy environment is enshrined in Article 24 (Udombana, 2000).

Criteria for Accepting Lawsuits

According to Articles 55 and 56 of the African Charter on Human and Peoples' Rights, individuals and non-governmental organizations can take action before the African Commission on Human Rights and Peoples. Although human rights violations are not a requirement under the African Charter on Human and Peoples' Rights, these articles do not always guarantee access to justice through public interest litigation. For a claim to be admissible, the complaint must identify as many victims as possible and claim that a State party to the African Charter on Human and Peoples' Rights has violated at least one of the rights enshrined in the Charter. In addition, under the Protocol to the African Charter on Human and Peoples' Rights, the court has jurisdiction to receive complaints from individuals or non-governmental organizations based on a declaration by the State party of its full acceptance of the court. Given that, under the Protocol, States have reserved the right to hear individual complaints before the Court, Member States will settle any dispute in the public interest that is not in their best interests (Wachira, 2008).

It should be noted that the formal conditions for admissibility in the African Court of Human Rights are very much related to the substantive rights enshrined in the African Charter. However, the Charter covers not only human rights but also the rights of the people. The African Charter on Human and Peoples' Rights, in addition to civil and political rights, also upholds economic, social, and environmental rights, and this is one of the special features of the African Charter, which derives from the European and American conventions of law. It distinguishes human beings. Given that these rights in principle belong to a group of individuals or peoples, the condition that the plaintiff must have been personally affected by the alleged infringement does not seem to be appropriate. Another factor that leads to the lack of a victim condition is the many practical problems they face in claiming compensation in Africa: national or international compensation networks are not available to the victims themselves. It may be dangerous for these people to demand it (Bekker, 2007).

Substantive Consideration and Some Issues

As mentioned earlier, the African Court of Human and Peoples' Rights differs from the other two systems in that it enshrines not only civil and political rights but also environmental and social, and cultural rights. Of course, environmental issues can be raised under the right to life, health, family life, and the other two regional human rights systems. However, the African Charter is the first human rights treaty to enshrine the right to the environment, and this right is considered one of the rights of the third generation or the right of solidarity. The use of the latter term has been suggested in public interest litigation. However, as will be seen in the following case, the applicability of these rights is still in doubt (Eno, 2002).

EXAMINING PUBLIC INTEREST LITIGATION IN THE DOMESTIC COURTS

It is important to examine the practice and practice of domestic courts in dealing with environmental issues in public interest cases. However, these courts face several limitations in dealing with such cases, including First, given the transnational nature of some environmental issues and the need to resort to regional dispute resolution mechanisms; The international community considers it necessary to turn to recent authorities, and domestic courts often do not have jurisdiction to answer these questions; Second, even if the issue is purely domestic, domestic courts are reluctant to hear it, or in many cases, loopholes in domestic law and a reluctance to comply with international law. In addition, some domestic judges lack the necessary knowledge to enforce highly complex and often controversial environmental law. Despite such problems, domestic courts are a very important tool for the effective protection of the environment. These courts are well funded and can perform their duties even if international and regional organizations and authorities face financial difficulties (Schall, 2008). Although there are shortcomings regarding the impartiality, willingness, and ability of internal authorities, effective legal remedies have also been envisaged. Among other things, the international community can overcome some of these problems by emphasizing the expansion of international environmental treaties and the impact that these treaties have on domestic law. In this regard, the efforts made in some countries, especially in European countries, to address environmental claims based on public interest are undeniable, some of which are briefly reviewed. Considering their membership in the Aarhus Convention, the practice of European countries can be considered in litigation based on public interest.

Although the Aarhus Convention is a regional treaty, it is also important on a global scale, as Kofi Annan, the former Secretary-General of the United Nations, stated: "Although the territory of the Aarhus Convention is regional, its importance "This convention on" environmental democracy "is the most ambitious action ever taken under the auspices of the United Nations." Thus, the Convention can be considered a global framework for the extension of citizens' environmental rights (Bhagwati, 2008).

German Legal System

German law on access to justice and public interest litigation is fundamentally different from some countries, such as Britain and France. Clause 2 of Article 42 of the Administrative Courts Law applies the "Protective norm doctrine." Thus, access to justice in Germany is based on the protection of individual rights. The right to sue is granted only if the protection of the violated rights of the plaintiff is intended as a human being and not merely in the public's general interest. Therefore, NGOs and individuals must prove the existence of a personal interest in the case by demonstrating a violation of a protected substantive right. Thus, formal actions that generally significantly impact the environment cannot be challenged in the public interest. However, an exception to this rule has been made for non-governmental organizations: Article 61 paragraph 2 of the Federal Environmental Protection Act (2002) specifically provides for the protection of the environment by non-governmental organizations. Natural Biology to file a lawsuit. However, this regulation is relatively narrow and applies only to the subject matter of the Federal Environmental Protection Act; Thus, there is no jurisdiction to sue in case of other violations of the law (Ditzen, 1971).

The Aarhus Convention, in the case of Germany, now requires the exclusion of a small part of the German customary system. For example, paragraph 1 of Article 3 of the Environmental Information Act (2004), which governs the application of the first pillar of the Aarhus Convention, provides: Prove it. " The implementation of this unconditional access is worth considering, as this concept is completely new in German law. The third pillar of the Aarhus Convention has not yet been implemented in German law. Of course, the situation is not yet predictable, but the Germans have already expressed their unwillingness to ignore their customary system in the negotiations of the Aarhus Convention. Importantly, the German government has made efforts to show that Article 9 paragraph 2 of the Convention and its implementation in EU law now not only allows members to exercise jurisdiction based on "sufficient interest" but also Instead allows members to uphold the criterion of "violation of a right" in cases where the administrative procedures of a member state require it to be a prerequisite. In this case, the Member States have between the two options for access to individuals, the French concept of "benefit" or the condition of the existence of a substantive right in German legal practice (Freckmann & Wegerich, 2001). Therefore, it is not necessary to copy the German "protection rule theory" on the jurisdiction of individuals. However, there is still doubt as to whether the German government should extend its jurisdiction to NGOs. Of course, it seems that the Aarhus Convention allows members to impose more restrictive conditions for litigation if the German government can use its existing system, which deems it necessary to violate a substantive right, even in Also apply to non-governmental organizations (Jarass & DiMento, 1993). Of course, given the purpose of the Aarhus Convention, which is to expand access to justice in environmental matters, there have been criticisms that a narrow interpretation of the terms of the Convention is unreasonable and indefensible. Thus, even if the German government is not strongly obliged to change its legal system, it must align itself with the objectives of the Aarhus Convention and expand access to justice. For example, the concept of a substantive right may be extended, and in general, not only cases involving a violation of the rights of the plaintiff or other victims but any violation of the rule of law can be prosecuted in court. The Aarhus Convention helps extend jurisdiction in the courts and does not require establishing a specific legal system or the condition for the recognition of litigation on behalf of the collective, but only provides the basis and allows states. To exercise maximum flexibility to its standard. Concerning the German legal system, it can be said that with the passage of the Federal Law on the Protection of the Natural Environment, the first step was taken to expand the jurisdiction of litigation in this country, and further steps will be taken to implement the Aarhus Convention in this country (Wagner, 1999).

British Legal System

Under British law, individuals and non-governmental organizations must prove the necessary conditions for litigation to challenge executive decisions. According to Article 31, paragraph 3 of the Supreme Court Act (1981), the appellant must show that he has "sufficient interest" in the case relating to the complaint. The concept of "sufficient benefit" has been explained by the national courts. From a narrow interpretation in the early 1990s, the courts have moved steadily towards a broader approach, especially in the Rose Threat case. The court's decision in the Green Peace case indicated that the judges were greatly influenced by the verdict on the reactivation of a nuclear power plant. The court found that the Greenpeace Peace Group, a well-known and respected organization with a genuine interest in the issue at hand, had a sufficient interest condition. This extended approach has been continued and developed following the Pergau Dam and Dixon theorems . It should be noted that the Aras Convention has played a very important role in justifying recent cases and that if this trend continues, a broader approach will be strengthened (Vogel, 1983).

Therefore, it can be said that the British system has allowed all those directly and personally involved in the case to sue, and the acceptance of lawsuits based on the public interest is a matter for the court. In applying this discretion, the approach of the courts is considerably to accept the broad criterion for the jurisdiction of the litigation: the courts are willing to refuse to accept only an insulting complaint against the law, and otherwise, The way is open for law enforcement plaintiffs even if the plaintiff is not the victim of formal action. Therefore, the British courts have accepted a public interest lawsuit filed by a non-governmental organization or individual concerning environmental issues, and the proceedings are pending (Fitzgerald, 1996).

Dutch Legal System

The Netherlands is a party to the Aarhus Convention. In this country, organizations have played an important role in environmental litigation; for example, the European Commission's study on access to justice in environmental matters in eight European countries showed that in the period 1996-2001, environmental Biology has made the most claims in the Netherlands (Ostling, 1994). However, it has recently been argued in the Dutch parliament that access to the courts by environmental organizations should be restricted, and recently the country's highest administrative court has favored such restrictions.

The right of environmental organizations to review administrative decisions in the courts has been recognized since 1994 in paragraphs 1 and 3 of the second paragraph of Article 1 of the Law on Public Administrative Law. According to paragraph 1 of the said article, the interested party is: "a person whose interests have been directly affected by a decision." According to the jurisprudence, the said interests must belong to him, personal, specific, ordinary, and directly subject to restriction. According to this clause, natural persons can not protect the public interest, such as protecting the environment and natural protected areas or preserving historical monuments (Betlem, 1995). These are public interests that are outside the scope of each individual's interests. In this way, a person who seeks to protect a public interest will never have the necessary conditions; that is, the interests must belong to him, personally and definite. Paragraph 3 of this article makes it possible for legal persons to act as a beneficiary. This is achieved by demanding public interests (for example, in the field of the environment) and collective interests (for example, local residents) that legal entities can pursue in their own interests and by taking action based on their actions. The existence of two conditions is mandatory for an environmental organization that is allowed to access the court; in other words, a legal entity that seeks to protect the public interest must also meet its objectives. It should be relevant and practical action should be taken in this regard, so environmental organizations have the right to access the court if, following their statutes, their objectives are relevant to the issue and to take practical action accordingly. Since everyone can create a legal entity and set their own goals, it is obvious that individuals are free to form a group in the public interest and to be active in it. The stakeholder should take action against decisions that have directly affected the public interest in the matter relating to them. According to paragraph 3, the initiator is litigation on behalf of the collective that gives anyone access to judicial protection by the administrative courts (Verschuuren, 2010).

However, on 1 June 2005, a lawsuit was filed on behalf of the public concerning environmental permits, making access to the courts accessible to everyone in the urban design and environmental permitting process. It is not. The legislator wants the said permits to be placed under the legal system of the interested party stipulated in paragraph 1 of the General Administrative Law. Reasons for this change in the law have been cited in the Netherlands, including the fact that serious efforts are being made to reduce unnecessary complaints. However, it seems that a significant decrease in the number of complaints of environmental organizations should not be expected because, according to the existing practice, these organizations often file lawsuits under the title of stakeholder. But the most important reason given for the legal version of the lawsuit on behalf of the collective is the removal of heterogeneous patches of legal protection in the Dutch system. In the Netherlands, administrative law focuses primarily on protecting the rights of individuals, and it is argued that in a system based on the protection of individual rights, it is not appropriate for environmental organizations to question the legality of Administrative decisions that can enforce the law themselves. Although these organizations should participate in the preliminary decision-making phase, regardless of this stage, access to justice should only be available to those whose individual rights have been violated. In general, it can be said that until a few years ago, the access of environmental organizations to justice and the possibility of filing lawsuits based on public interest in this country could have been considered as a model for other countries, especially European countries that are members of the Aarhus Convention. However, recent developments in law and jurisprudence have led to restrictions on the access of environmental organizations to the courts, which some consider conflicting with the obligations of the Netherlands under the Aarhus Convention. Accordingly, some critics argue that the country should consider the need for public benefit measures in the light of international considerations (Betlem, 1995).

French Legal System

In France, the preconditions for litigation are similar to those in the United Kingdom: if there is an interest in litigation by the plaintiff, everyone, even environmental associations not recognized in French law, can file a lawsuit based on the public interest. Therefore, the plaintiff only has to prove his interest in the case by expressing grief, moral pain, and suffering. NGOs also have access to the courts if they can prove that there is a link between their goals and activities on the one hand and at-risk interests on the other. Thus, environmental associations may sue for damages in the event of a definite environmental crime (Bentata, 2014).

Pakistani Legal System

In Pakistan, the Environmental Protection Act sets out the principles, enforcement guarantees, and environmental protection methods (Sial et al., 2018). In this country, when a person files a lawsuit claiming to defend the public interest in the environment, the plaintiff's position concerning the complainant is examined. Although the right to sue has been widely considered in the country since 1990, no clear criteria of the concept of public interest have been presented in environmental litigation, and the lack of certainty in this area has caused many ambiguities. In general, it can be said that the recognition of this issue has been left to the courts, which, of course, have issued very contradictory decisions in this regard. Despite this, many rulings have been issued by Pakistani courts regarding civil and criminal damages. A review of Pakistan's practice shows that it is very difficult to distinguish between human rights and environmental claims. In many of the public interest cases, both environmental and human rights aspects have been addressed. In Pakistan, public interest claims cover all aspects of the environment, such as air or water pollution, waste management, and municipal pollution (Nadeem & Hameed, 2008).

Indian Legal System

The jurisdiction to hear complaints based on the public interest in India is free from the limitations of traditional judicial proceedings. Complaints based on the public interest in the Indian state are described in a collective approach that involves a variety of factors, including the flexibility of the procedure, the temporary appointments made by a higher judicial authority, and the compensation in advance. In their efforts, the judges obtained many findings from multiple litigants, formed oversight or review committees, and came up with effective and constructive solutions to the grievances raised by benevolent citizens. Judges have a great deal of power, especially concerning public interest claims: offering innovative solutions, direct strategic changes, proposing legislation, reprimanding government employees, and enforcing appointments. Judges also do not hesitate to exercise jurisdiction over what they consider to be in the public interest because there is a gap on the part of the government; Indian courts are rushing to compensate (Prasad, 2006). In the late 1970s and early 1980s, A judge extended public interest jurisdiction to individuals who were not economically and socially disadvantaged and therefore could not claim damages, the judge said. In recent years, the judiciary has focused on environmental protection. The right to life and liberty as fundamental rights protected by law, developed by judicial initiative, include vague and absolute rights, such as the right to a healthy environment and enjoy a climate free of pollution. Have taken. This right was recognized as part of the right to life in 1991, and over seven years, several environmental issues based on the public interest were raised in court. At this point, the impact of environmental quality on residents' lives became so apparent that it did not need to be emphasized or elaborated. Since then, the courts have sought to clarify this right and enrich India's jurisprudence, not only by enshrining established principles of international environmental law but also by its newly established principles. Some of these principles are (Sankar, 1998):

Principle of Equality of Human Generations (Principle 3 of the Rio Declaration), Principle of Sustainable Development (Principle 4 of the Rio Declaration), Principle of Precautions (Principle 15 of the

Rio Declaration), Principle of Acceptance of Costs by the Polluter (Principle 16 of the Rio Declaration) and Concept The government as the trustee of all-natural resources.

According to some well-wishers and officials, the court has made decisions regarding the protection of the Taj Mahal against erosive air pollution, the prevention of commercial diversions in the river, the "Canges of air pollution in Delhi and other major war-torn cities, " India's wildlife protection and municipal waste management . These cases, of course, are rare but important environmental cases based on the public interest that has been considered in Indian courts. The ability of well-wishers to go to court has been considered an effective criterion in the good governance of the Indian state, and judicial intervention has led to changes in the procedure, regulations, and improved environmental protection(Prasad, 2004).

Although this has been the practice of the courts in the past and present, the exercise of environmental jurisdiction based on the public interest has raised concerns about accessibility, participation, effectiveness, and sustainability that should be exercised in cases where public interest claims and actions Competence based on it should not be properly controlled and investigated should be reviewed and deepened.

Iranian Legal System

There are two types of lawsuits related to environmental degradation in Iran: lawsuits filed by public individuals and lawsuits filed by private individuals. The public entities that have the authority to complain about preventing environmental degradation are municipalities, the Forests and Rangelands Organization, and the Environmental Protection Agency. In some cases, private legal entities, whether natural or legal, can also complain to the competent public authority or authorities about one of the competent public institutions or that they have filed a lawsuit as a private plaintiff, in addition to prosecuting the perpetrator. And victim should also seek compensation for the damage caused by the crime. In some cases, people who see their health as endangered also complain to one of the competent public institutions about the pollution-causing industrial unit (Shohani et al., 2021). The peculiarity of these lawsuits is that they are the beneficiaries of the people who file lawsuits and are directly or indirectly damaged by the infection. For example, locals complained to the Kerman General Directorate of Environmental Protection about the discharge of factory wastewater in the surrounding area. After a visit, General Directorate issued a warning to the plant and asked it to discharge its sewage in the place designated by this department.

By referring to various laws and regulations, including the Islamic Penal Code and special environmental or public laws, as well as various and various existing approvals, instructions, and by-laws, it is determined that individuals, both natural and legal, Depending on the case, can act as a plaintiff, a private plaintiff, or both, due to the loss of life, property, or damage to the country's environment. Obviously, in this regard, the Environmental Protection Organization of Iran has a more important and inclusive role than other legal entities. In many of these cases, action, monitoring, follow-up, etc., are responsible for the legal rules and regulations (Heidari et al., 2017).

Thus, it seems that private individuals, whether individuals or non-governmental organizations, can pursue public interest litigation over the environment through other bodies, including the Environmental Protection Agency, and it is difficult for them to Independently have the authority to file a complaint or have their complaint accepted. However, the multiplicity and promotion of environmental NGOs in recent years, environmental protection, and the possibility of public litigation can be considered effective steps.

CONCLUSION

At the international level, the influence and impact of human rights on the environmental legal system are undeniable. Given the need to protect the rights of human beings, including future generations, in the enjoyment of a healthy environment, it seems The personal and thematic jurisdiction of regional authorities and institutions and domestic courts should be reconsidered to adjust the condition of damages and losses from harmful actions to the environment to file a lawsuit with the competent authority. All members of society or NGOs have the right to sue on behalf of the community and even future generations based on the public interest. Here we can point to the acceptance of action by the international community in litigation as a representation by the collective in the international legal system as a model. In some matters which are inherent in the interest of all States and all States have an interest in upholding these values, any State may take action in the event of a violation or violation by the international community. This is because these violations are, in fact, attacks on the whole of the international community and violate values to which all members of the international community share. Among these values is respect for fundamental human rights, which are among the universal obligations. Concerning the internal systems of countries in the field of environmental protection, the application of this model is considered possible concerning the individuals of a society because the protection of the environment and its care, including human rights and values, and Principles are considered to be fundamental to all members of society, and litigation based on the public interest and its consideration is considered necessary, even if the plaintiff has not personally suffered damages. The nature of the obligations of individuals and legal entities to protect the environment is one of the obligations that each individual has to all members of society and future generations, and therefore the right to sue in the public interest in environmental matters for each individual or organization. Non-governmental organizations that are concerned with some kind of concern and sense of responsibility seem necessary. However, it must be acknowledged that the gap between existing domestic, regional, and international mechanisms to protect the environment effectively is still very large. Therefore, creating an international environmental court has been suggested by some, which of course, has several opponents and supporters. It is noteworthy that, by reforming and strengthening countries' judicial and legal systems to enable public interest litigation on environmental issues and the expansion of active non-governmental environmental organizations, some believe that in the future, we can see changes in any level, be regional or global.

REFERENCES

Alan, B. (2017). *Human rights and the environment: where next?* Routledge.

Antkowiak, T. M. (2007). Remedial approaches to human rights violations: The Inter-american Court of human rights and beyond. *Colum. J. Transnat'l L.*, *46*, 351.

Badwaza, Y. M. (2005). *Public interest litigation as practised by South African human rights NGOs: any lessons for Ethiopia?* [Doctoral dissertation]. University of Pretoria.

Bekker, G. (2007). The African court on human and peoples' rights: Safeguarding the interests of African states. *Journal of African Law*, *51*(1), 151–172. doi:10.1017/S0021855306000210

Bentata, P. (2014). Liability as a complement to environmental regulation: An empirical study of the French legal system. *Environmental Economics and Policy Studies*, *16*(3), 201–228. doi:10.100710018-013-0073-7

Betlem, G. (1995). Dutch Soil Protection Act 1994. *The. Eur. Envtl. L. Rev.*, *4*, 232.

Betlem, G. (1995). Standing for Ecosystems—Going Dutch. *The Cambridge Law Journal*, *54*(1), 153–170. doi:10.1017/S0008197300083197

Bhagwati, P. N. (1984). Judicial activism and public interest litigation. *Colum. J. Transnat'l L.*, *23*, 561.

Buergenthal, T. (1982). The Inter-American Court of Human Rights. *The American Journal of International Law*, *76*(2), 231–245. doi:10.2307/2201452

Christoffersen, J., & Madsen, M. R. (Eds.). (2011). *The European court of human rights between law and politics*. Oxford University Press. doi:10.1093/acprof:oso/9780199694495.001.0001

Ditzen, U. (1971). Environmental Protection in West Germany. *Business Lawyer*, *27*, 833.

Eno, R. W. (2002). The Jurisdiction of the African Court on Human and Peoples'. *Rights. Afr. Hum. Rts. LJ*, *2*, 223.

Fitzgerald, E. A. (1996). The Constitutionality of Toxic Substances Regulation Under the Canadian Environmental Protection Act. *U. Brit. Colum. L. Rev.*, *30*, 55.

Freckmann, A., & Wegerich, T. (2001). *The German Legal System*. International Business Lawyer.

Goldston, J. A. (2006). Public interest litigation in Central and Eastern Europe: Roots, prospects, and challenges. *Hum. Rts. Q.*, *28*(2), 492–527. doi:10.1353/hrq.2006.0018

Greer, S., & Wildhaber, L. (2012). Revisiting the debate about 'constitutionalising'the European Court of Human Rights. *Human Rights Law Review*, *12*(4), 655–687. doi:10.1093/hrlr/ngs034

Hassan, P., & Azfar, A. (2003). Securing environmental rights through public interest litigation in South Asia. *Va. Envtl. LJ*, *22*, 215.

Heidari, F., Dabiri, F., & Heidari, M. (2017). Legal system governing on water pollution in Iran. *Journal of Geoscience and Environment Protection*, *5*(09), 36–59. doi:10.4236/gep.2017.59004

Jarass, H. D., & DiMento, J. (1993). Through Comparative Lawyers' Goggles: A Primer on German Environmental Law. *Geo. Int'l Envtl. L. Rev.*, *6*, 47.

Leach, P. (2011). *Taking a case to the European Court of Human Rights*. Oxford University Press.

Lixinski, L. (2010). Treaty interpretation by the inter-American court of human rights: Expansionism at the service of the unity of international law. *European Journal of International Law*, *21*(3), 585–604. doi:10.1093/ejil/chq047

Mowbray, A. (2005). The Creativity of the European Court of Human Rights. *Human Rights Law Review*, *5*(1), 57–79. doi:10.1093/hrlrev/ngi003

Nadeem, O., & Hameed, R. (2008). Evaluation of environmental impact assessment system in Pakistan. *Environmental Impact Assessment Review*, *28*(8), 562–571. doi:10.1016/j.eiar.2008.02.003

Ostling, M. (1994). Decision-Making in Dutch and Swedish Environmental Law. *Tilburg Foreign L. Rev.*, *4*(3), 209–234. doi:10.1163/221125995X00013

Pasqualucci, J. M. (2012). *The practice and procedure of the Inter-American Court of Human Rights*. Cambridge University Press. doi:10.1017/CBO9780511843884

Prasad, P. M. (2004). Environmental protection: The role of liability system in India. *Economic and Political Weekly*, 257–269.

Prasad, P. M. (2006). Environment protection: Role of regulatory system in India. *Economic and Political Weekly*, 1278–1288.

Sankar, U. (1998). *Laws and institutions relating to environmental protection in India. The role of law and legal institutions in Asian economic development*. Erasmus University.

Schall, C. (2008). Public interest litigation concerning environmental matters before human rights courts: A promising future concept? *Journal of Environmental Law*, *20*(3), 417–453. doi:10.1093/jel/eqn025

Shohani, A., Ataei, E., & Norouzi, N. (2021). Prevention and Suppression of Environmental Crimes in the Light of the Actions of Non-Governmental Organizations in the Iranian Legal System. *Research Journal of Ecology and Environmental Sciences*, *1*(1), 57–70.

Sial, S. A., Zaidi, S. M. A., & Taimour, S. (2018). *Review of existing environmental laws and regulations in Pakistan*. WWF-Pakistan.

Udombana, N. J. (2000). Toward the African Court on Human and Peoples'. *Rights: Better Late Than Never. Yale Hum. Rts. & Dev. LJ*, *3*, 45.

Verschuuren, J. (2010). The Dutch "Crisis and Recovery Act": Economic recovery and legal crisis? *Potchefstroom Electronic Law Journal/Potchefstroomse Elektroniese Regsblad, 13*(5).

Vogel, D. (1983). Cooperative regulation: Environmental protection in Great Britain. *The Public Interest*, *72*, 88.

Wachira, G. M. (2008). *African Court on Human and Peoples' Rights: Ten years on and still no justice*. Minority Rights Group International.

Wagner, S. (1999). Forest Legislation in a Constitutional State-The example of the Federal Republic of Germany. *Forstwissenschaftliche Beiträge der Professur Forstökonomie und Forstpolitik der ETH Zürich*, *21*, 41–48.

ADDITIONAL READING

Sankar, U. (1998). *Laws and institutions relating to environmental protection in India. The role of law and legal institutions in Asian economic development*. Erasmus University.

Schall, C. (2008). Public interest litigation concerning environmental matters before human rights courts: A promising future concept? *Journal of Environmental Law*, *20*(3), 417–453. doi:10.1093/jel/eqn025

Shohani, A., Ataei, E., & Norouzi, N. (2021). Prevention and Suppression of Environmental Crimes in the Light of the Actions of Non-Governmental Organizations in the Iranian Legal System. *Research Journal of Ecology and Environmental Sciences*, *1*(1), 57–70.

Sial, S. A., Zaidi, S. M. A., & Taimour, S. (2018). *Review of existing environmental laws and regulations in Pakistan*. WWF-Pakistan.

Udombana, N. J. (2000). Toward the African Court on Human and Peoples'. *Rights: Better Late Than Never. Yale Hum. Rts. & Dev. LJ*, *3*, 45.

Vogel, D. (1983). Cooperative regulation: Environmental protection in Great Britain. *The Public Interest*, *72*, 88.

Wachira, G. M. (2008). *African Court on Human and Peoples' Rights: Ten years on and still no justice*. Minority Rights Group International.

Wagner, S. (1999). Forest Legislation in a Constitutional State-The example of the Federal Republic of Germany. *Forstwissenschaftliche Beiträge der Professur Forstökonomie und Forstpolitik der ETH Zürich*, *21*, 41–48.

KEY TERMS AND DEFINITIONS

Environmental Law: Environmental law is a collective term encompassing aspects of the law that protect the environment. A related but distinct set of regulatory regimes, now strongly influenced by environmental legal principles, focuses on managing specific natural resources, such as forests, minerals, or fisheries. Other areas, such as environmental impact assessment, may not fit neatly into either category but are nonetheless important components of environmental law. Previous research found that when environmental law reflects moral values for betterment, legal adoption is more likely to be successful, usually in well-developed regions. In less-developed states, changes in moral values are necessary for successful legal implementation when environmental law differs from moral values.

Equity: Defined by UNEP to include intergenerational equity—"the right of future generations to enjoy a fair level of the common patrimony"—and intragenerational equity—"the right of all people within the current generation to fair access to the current generation's entitlement to the Earth's natural resources"—environmental equity considers the present generation under an obligation to account for long-term impacts of activities and to act to sustain the global environment and resource base for future generations. Pollution control and resource management laws may be assessed against this principle.

Polluter Pays Principle: The polluter pays principle stands for the idea that "the environmental costs of economic activities, including the cost of preventing potential harm, should be internalized rather than imposed upon society at large." All issues related to responsibility for cost for environmental remediation and compliance with pollution control regulations involve this principle.

Precautionary Principle: One of the most commonly encountered and controversial principles of environmental law, the Rio Declaration formulated the precautionary principle as follows, to protect the environment, States shall widely apply the precautionary approach according to their capabilities. Where there are threats of serious or irreversible damage, lack of full scientific certainty shall not be used as a reason for postponing cost-effective measures to prevent environmental degradation. The principle may play a role in any debate over the need for environmental regulation.

Prevention: The concept of prevention etc. can perhaps better be considered an overarching aim that gives rise to a multitude of legal mechanisms, including prior assessment of environmental harm, licensing, or authorization that set out the conditions for operation and the consequences for violation of the conditions, as well as the adoption of strategies and policies. Emission limits and other product or process standards, the use of best available techniques, and similar techniques can all be seen as applications of the concept of prevention.

Public Participation and Transparency: Identified as essential conditions for "accountable governments...industrial concerns," and organizations generally, public participation and transparency are presented by UNEP as requiring "effective protection of the human right to hold and express opinions and to seek, receive and impart ideas, etc. a right of access to appropriate, comprehensible and timely information held by governments and industrial concerns on economic and social policies regarding the sustainable use of natural resources and the protection of the environment, without imposing undue financial burdens upon the applicants and with adequate protection of privacy and business confidentiality," and "effective judicial and administrative proceedings." These principles are present in environmental impact assessment, laws requiring publication and access to relevant environmental data, and administrative procedures.

Sustainable Development: Defined by the United Nations Environment Programme as "development that meets the needs of the present without compromising the ability of future generations to meet their own needs," sustainable development may be considered together with the concepts of "integration" (development cannot be considered in isolation from sustainability) and "interdependence" (social and economic development, and environmental protection, are interdependent). Laws mandating environmental impact assessment and requiring or encouraging development to minimize environmental impacts may be assessed against this principle. The modern concept of sustainable development was discussed at the 1972 United Nations Conference on the Human Environment (Stockholm Conference) and the driving force behind the 1983 World Commission on Environment and Development (WCED, or Bruntland Commission). In 1992, the first UN Earth Summit resulted in the Rio Declaration, Principle 3 of which reads: "The right to development must be fulfilled to equitably meet developmental and environmental needs of present and future generations." Sustainable development has been a core concept of international environmental discussion ever since, including at the World Summit on Sustainable Development (Earth Summit 2002) and the United Nations Conference on Sustainable Development (Earth Summit 2012, or Rio+20).

Transboundary Responsibility: Defined in the international law context as an obligation to protect one's environment. UNEP considers transboundary responsibility at the international level to prevent damage to neighboring environments at the international level as a potential limitation on the rights of the sovereign state. Laws that limit externalities imposed upon human health and the environment may be assessed against this principle.

Chapter 11
Environmental Law and Gender

ABSTRACT

Legal challenges associated with environmental issues are among the biggest in the world, since it requires a comprehensive understanding of the environment without any political boundaries. Acknowledging new paradigms is only possible by integrating the quality of life for every human being to the law system as a supreme guardianship good. A common factor connecting the concept of environment and sustainable development is the fact that all three converge around the human being, the central protagonist responsible both individually and collectively for his own future and the future of the planet. The gender perspective must be integrated into the sustainable development management process in order to improve quality of life since the concept of equality it covers also implies gender equity. To analyze how gender is incorporated into international environmental protection and to make adequate conclusions about its current state, the most relevant juridical instruments have been considered.

INTRODUCTION

The environmental issue intertwines two basic concepts: environment and sustainable development. Its legal approach constitutes one of the greatest challenges that law has faced, since it requires an understanding of the environment as a whole, whose elements interact with each other with such a degree of interdependence that, not recognizing political borders, they reach a broader dimension by projecting at regional and global levels. This understanding can only be achieved by incorporating the quality of life of the human being into the legal system, as the supreme good of protection.

The law has been conceptualized, by a traditional thesis, as the fair social order: the set of norms adjusted to human nature that enables the best integral development of the human being, the achievement of intermediate or natural ends in the interrelation of life itself in society (Newman et al., 2004).

Not because it is classic, the outlined concept has lost validity or relevance, since its fundamental value is based on highlighting as a guiding guideline that the human being must be captured by the law in all its dimensions and that this subject is inserted and develops in an environment that it conditions it, at the same time that it is conditioned and modified by its action. This is its reality, formed by the natural

DOI: 10.4018/978-1-6684-7188-3.ch011

and the socio-cultural in mutual interrelation, constituting an intimate and inseparable environment-development equation, which the law must necessarily regulate to achieve its ultimate goal.

This is so and has been since the beginning of time, but not the capture of environmental problems, whose treatment cannot be outside the law.

Today confronts us with a world whose social and political structure becomes more complex day by day, whose productive and commercial phenomena continually change in terms of their protagonists and relationship of forces, which sees traditional borders disappear due to advances in communication and watches with astonishment the scientific and technological advances with the consequent impact on the environment. The law must elaborate a response so as not to fall into the request of principles of allowing the subject of its tutelage, the human being, to see his fundamental rights violated.

In the search for that response to the visualization and recognition of the great environmental problems—constantly evidenced by the daily reality of humanity—legal science has recognized the environment as a legal category (Buckingham & Le Masson, 2017).

Its conceptualization has presented epistemological difficulties, concealing other conceptual and even philosophical problems under the guise of semantic problems at the time of its definition.

We agree in considering the environment as the systematization of natural, social and cultural phenomena, processes and values that condition the life of man in society and the development of the rest of living organisms and the state of inert elements in a given time and space., in an integrating and dialectical synthesis of relations of exchange of man with the various natural resources, exhaustible and non-exhaustible in economic terms.

The environment, thus understood, is inextricably related to the concept of sustainable development and quality of life.

Sustainable development aims to achieve quality of life without compromising the ability of future generations to meet their own needs. Both concepts overlap around the central protagonist, addressee and individual and socially responsible for his destiny and the future of the planet: the human being.

From this approach clearly arises the need for a systemic and globalizing vision of reality, which allows the understanding of the whole and the valorization of the part as a bearer of aptitude for the generation and determination of changes. This necessarily implies the recognition of new paradigms.

In this sense, the law, as a social science, grants the individual his role as transformer of society, which leads to the hierarchization of the human being, the first and last subject and recipient of any legal system, while at the same time increasing his responsibilities. .

In its institutional face, there is a revaluation of local governments, based on the principle of immediacy that grants greater efficiency in the detection and action in the face of specific problems.

The legal response to the environmental question is substantially determined by the conclusions emanating from other scientific fields, given the interdisciplinary and globalizing nature of the subject, but it cannot be perceived as secondary, since it addresses fundamental questions about the orientation of a policy connected with the basic ideas of justice and equity.

The environmental issue crosses all legal disciplines and categories and requires an effort of synthesis and unity, which, beyond specializations, permeates the entire spectrum of law.

The protection of these categories and legal assets requires the establishment of a new value system and the change of socio-cultural guidelines that underlie the entire system, for which the law must contribute, through its exemplifying and behavior-forming function, leading to the strengthening a profile of sustainable and supportive development that advocates the rational use of natural resources to satisfy the current needs of man and at the same time ensure the improvement of the quality of life of the present generation and that of future generations.

From the foregoing it follows, then, that the new supreme legal good - quality of life - requires fairness and solidarity in this generation and between it and those to come.

That is why we believe it is necessary to incorporate the gender perspective in the management of sustainable development, since quality of life, as it has been understood, can only be achieved if the concept of equity that it presupposes includes gender equity (Maguire & Jessup, 2021).

As has been expressed, the environment does not recognize, as a system of interrelationships, the limits determined by the social systems that give rise to legal-political-administrative organizations. Both natural systems and subsystems, as well as the phenomena and processes that occur in them, do not strictly overlap with social systems. The environment does not recognize political borders and consequently, for the purposes of adequate environmental protection, the national and subnational organizations from which the particular legal systems arise come into conflict with the ultimate goal of environmental protection and the legal good of quality of life, making international cooperation necessary.

As stated, all branches of law must contribute to achieving these goals, therefore the issue intersects with International Law and that legal branch is also responsible for making its contribution.

Currently, there is a specific body of international standards for the protection of the environment, whose consolidated element is the recognition, as a protected legal right - from a substantive or material perspective - of the human vital environment, considered in a planetary and universal dimension.

In practice, it moves away from the incomprehensible totalizing approaches, to regulate more specific aspects of environmental problems. Therefore, although the concern is holistic, the action is sectoral.

Progressively, the evolution of its rules is reaching all sectors and is approaching a new totalizing objective: sustainable development and quality of life for human beings on a global scale (Galizzi, 2012).

Therefore, we propose to analyze the main legal instruments of international environmental protection, both those of a fundamental nature and those of sectorial regulation of environmental goods, in order to analyze the incorporation of the gender perspective in this matter and conclude regarding the approach, treatment and state of the question.

GENDER IN INTERNATIONAL LEGAL PROTECTION OF THE ENVIRONMENT

As has been expressed, the response to environmental problems from international law has been the provision of legal instruments to protect the environment. Some of them have been understood by the majority doctrine, with which we agree, as fundamental, since they have marked the evolution of the legal protection of the environment and allowed the establishment of essential principles, despite their declarative nature. Although this international regulation presents the characteristic profiles of what has come to be called "soft law", there is a hard core in this international protective order, which is made up of norms included in instruments, from a conventional source, of sectoral regulation of goods that, however, participate in the holistic vision -fundamental in the matter- and correspond to the protection of fundamental interests of humanity, thus accepted and recognized by the international community as a whole. Therefore, we have proceeded to analyze both types of instruments in the search for the inclusion of the gender dimension in environmental issues (Arora-Jonsson, 2014).

FUNDAMENTAL INTERNATIONAL LEGAL PROTECTION OF THE ENVIRONMENT

Declaration of the United Nations Conference on the Human Environment

The topic of environmental protection was included in the agenda of the XXIII General Assembly of the United Nations, which met in 1968. After the debate, it was unanimously decided that the UN Secretary General collect the maximum available data and propose a specific plan of measures to protect the environment. The official presented the report on him, "The man and his environment", on May 26, 1969.

The General Assembly decided that UNESCO would organize regional symposiums over the next two years, to be followed by a world conference on the theme of environmental protection.

The United Nations Conference on the Human Environment was held in Stockholm from June 5 to 16, 1972, under the chairmanship of the Swedish Minister of Agriculture, Ingemund Bengtsson, with the participation of 1,200 delegates representing 113 countries.

The greatest achievement of the Conference was that all the participants accepted an ecological vision of the world, in which it was recognized, among other things, that "Man is both the work and the creator of the environment that surrounds him...with an action on it that has increased thanks to the rapid acceleration of science and technology...to the point that the two aspects of the human environment, the natural and the artificial, are essential for its well-being."

The Instruments of a merely declaratory and recommendatory nature that emerged from the Conference are: a Declaration and the Action Plan for the Environment.

The established guidelines, notwithstanding the lack of a legally binding nature of the adopted instruments, were not a dead letter, since at the institutional level, through Resolution 2997 (XXVII) adopted by the UN General Assembly on December 15, 1972, The United Nations Environment Program (UNEP) was established, whose mission is "to lead and encourage participation in caring for the environment by inspiring, informing and empowering nations and peoples to improve the quality of life without endangering that of future generations" (Kurtz, 2007).

The Stockholm Declaration—consisting of a preamble and 26 principles—addresses the main environmental issues affecting the global environment. Set the applicable criteria for its treatment on an international and national scale.

Its principles refer in particular to: the fundamentals of the action to be carried out (principle 1), the objectives to be achieved (principles 2 to 7), the interconnection of environmental problems with other issues such as development disparities and effective protection of fundamental human rights (principles 8 to 17), the instruments of environmental policy and in particular planning and management at the national level (principles 18 to 20), the necessary international cooperation in the matter (principles 21 to 26).

Principle 1 marks the vision of the instrument: man, the human being without distinction of gender, is the subject since "he is both the work and the architect of the environment that surrounds him", it is man who is attributed the " fundamental right to freedom, equality and the enjoyment of adequate living conditions in a quality environment that allows them to lead a dignified life and enjoy well-being, and has the solemn obligation to protect and improve the environment for present and future generations. future".

However, that same principle, by expressing that "policies that promote or perpetuate apartheid, racial segregation, discrimination, colonial oppression and other forms of oppression and foreign domination are condemned and must be eliminated", must undoubtedly be considered as the seed of the condemnation of all discrimination based on gender issues as attacks on that fundamental human right that the Declaration proclaims. In the same sense, principle 16 referring to demographic policies whose application must necessarily contemplate respect for fundamental rights must be interpreted (Norgaard & York, 2005).

Rio Declaration on Environment and Development

The United Nations Conference on Environment and Development was held between June 3 and 14, 1992 in Rio de Janeiro. The instruments that emerged from the Conference are: the Rio Declaration on Environment and Development, the Declaration of Principles on Forests, the United Nations Framework Convention on Climate Change, the Convention on Biological Diversity, Agenda 21 and from the point of institutional view the creation of the Commission on Sustainable Development (CSD). The Declaration proclaims 27 principles that try to establish the criteria to reconcile the demands of development with those of environmental protection.

Principle 20 of the instrument expressly contemplates the gender perspective and its intersection with environmental matters. It clearly highlights the fundamental role of women in these issues and correlates the express recognition made with the principle of participation, since it postulates "their full participation to achieve sustainable development.

It should also be noted that the Declaration addresses in the subsequent Principles the participation of young people Principle 21-, of indigenous peoples Principle 22- and the necessary protection of the environment and natural resources of peoples subjected to oppression, domination and occupation Principle 23-, for which some doctrinaires consider that they "refer to the particular contribution of certain specific categories of people (MacGregor, 2017).

Agenda 21

Agenda 21, also called AGENDA XXI, also emerged from the Rio Conference, "outlines the measures related to human activities that affect the environment that must be taken between now and the next century by governments, United Nations agencies, development organizations, non-governmental organizations and independent groups." Its structure is as follows: a Preamble, Section I: Social and Economic Dimensions, Section II: Conservation and Management of Resources for Development, Section III: Strengthening the Role of Major Groups, Section IV: Means of Implementation. The gender perspective in the environment-development equation emerges from Chapter 24, Section III, entitled "Global measures in favor of women to achieve sustainable and equitable development." The instrument constitutes the fundamental milestone in the incorporation of the gender dimension in matters of environmental protection, since in the aforementioned chapter, when establishing the "Programs Area", it lays the foundations for action in them, indicating that there have been, at that time, "several action plans and conventions to achieve full, equitable and beneficial integration in all activities related to development", among which he mentions, in particular, the Nairobi Strategies, oriented towards the future for the advancement of women, because they "emphasize the participation of women in the national and international management of ecosystems and the fight against environmental degradation". Correlates the effective execution of programs such as those arising from the Convention on the elimination of all forms of discrimination against women, and ILO and UNESCO Conventions in order to eliminate discrimination based on sex and guarantee access to women to natural resources, education and secure employment under conditions of equality with "the active participation of women in decision-making of a political and economic nature", which is considered "decisive for the success of Agenda 21" (Kostruba, 2019).

Governments are proposed objectives that involve incorporating the gender perspective in the management of sustainable development, promoting the mainstreaming of gender criteria in environmental management and promoting the gender equity approach as a public policy within environmental processes, through its institutionalization (including equal opportunities in the environmental sector; elimination of all forms of discrimination against women, through actions that promote the well-being of women in harmony with a healthy social and natural environment, and under criteria that allow the current use and conservation of environmental resources for future generations).

In order to obtain the proposed objectives, and with the declarative and non-binding nature of the instrument, it indicates to the governments the activities that "they should actively dedicate themselves to putting into practice", at the internal level, and urges them, at the international level to "ratify all relevant conventions relating to women, if they have not already done so. Those who have ratified the conventions should enforce them and establish legal, constitutional and administrative procedures to transform agreed rights into national law and should take action to implement them, in order to strengthen the legal capacity of women and promote their full and equal participation in issues and decisions related to sustainable development.

The Agenda establishes as areas that require the adoption of urgent measures the prevention of "the rapid degradation that is occurring in the environment and the economy of developing countries and that, in general, affects the lives of women and children in rural areas suffering from the effects of drought, desertification and deforestation, armed hostilities, natural disasters, toxic waste and the consequences of the use of inappropriate agrochemicals".

At the level of international and regional cooperation and coordination, it indicates that: "The Secretary General of the United Nations should examine the adequacy of all the institutions of the Organization, including those that pay special attention to the role of women, in fulfilling environment and development objectives, and make recommendations to strengthen their capacity.... The review should consider how to strengthen the environment and development programs of each of the United Nations system bodies in implementation of Agenda 21, and to incorporate women into programs and decisions related to sustainable development" (Dankelman, 2010).

Malmö Declaration

Sponsored by UNEP, and hosted by the Government of Sweden, the "First Global Ministerial Environment Forum" was held from 29 to 31 May 2000, in Malmö, Sweden. More than 100 environment ministers from around the world, including nine ministers and deputy ministers from Latin America and the Caribbean, met to review important emerging environmental issues, and to help define the global agenda for the environment and sustainable development of the world. XXI century.

One of the most important results of the Forum is the Malmö Declaration 11, which contains information on several key issues addressed during the Forum, among which the identification of important environmental challenges for the 21st century stands out. In this sense, the Ministers highlighted the alarming discrepancy between the commitments and the actions of the international community with respect to sustainable development.

Among the challenges, and when addressing the issue that it calls "Civil society and the environment", the instrument expressly indicates: "19. Greater emphasis should be placed on the gender perspective within the decision-making process regarding the environmental management and natural resources", and in the Conclusion, highlights that "the emergence of a young generation with a clear sense of optimism, solidarity and values, women increasingly aware and with a strong and active role within society, all of this points to the birth of a new consciousness", which contributes to achieving sustainable development (Kurian, 2018).

Millennium Declaration

The General Assembly of the United Nations, meeting at the United Nations Headquarters in New York from September 6 to 8, 2000, at its 8th Plenary Meeting, approved, without reference to the Main Committee, what has been called the Declaration of the Millennium.

Under the heading "I. Values and principles," the Heads of State and Government, after reaffirming their faith in the Organization and its Charter, as essential foundations for a more peaceful, prosperous, and just world, acknowledge that "it is their collective responsibility to respect and defend the principles of human dignity, equality and equity at the global level", commit their "respect for human rights and fundamental freedoms; respect for the equal rights of all, without distinction based on race, sex, language or religion"

By listing the fundamental values that they consider to be essential for international relations in the 21st century, the wording of the instrument denotes the adoption of the gender perspective, as well as the environmental one. Thus, it is worth noting that when referring to the value of freedom, it does so in the following terms: "Men and women have the right to live their lives and raise their children with dignity and free from hunger and the fear of violence, oppression or injustice" and when doing so with respect to equality, it states: "No person or nation should be denied the possibility of benefiting from development. Equal rights and opportunities for men and women must be guaranteed". Regarding "respect for nature", it postulates that: "It is necessary to act with prudence in the management and planning of all living species and all natural resources, in accordance with the precepts of sustainable development. Only in this way can we conserve and transmit our descendants the immeasurable riches that nature provides us. The current unsustainable patterns of production and consumption must be modified in the interest of our future well-being and that of our descendants" (Ryan, 2014).

To translate these common values into actions, the instrument formulates a series of key objectives to which it attaches special importance. They are: Peace, security and disarmament; The development and eradication of poverty (within which the commitment is made to spare no effort to liberate men, women and children from the abject and dehumanizing conditions of extreme poverty, making the right to development a reality and to promote gender equality and women's empowerment as effective means of combating poverty, hunger and disease and of stimulating truly sustainable development); Protection of our common environment (with respect to which support for the principles of sustainable development is reaffirmed, including those set forth in Agenda 21, and the adoption of "a new ethics of conservation and protection in all our activities related to the environment is decided environment"); Human Rights, Democracy and Good Governance (in relation to which the commitment to "fight against all forms of violence against women and apply the Convention on the Elimination of All Forms of Discrimination against Women" is assumed); Protection of vulnerable people; Attention to the special needs of Africa; Strengthening of the United Nations.

From the highlights, it is clear that the instrument incorporates the gender perspective, with criteria of equality and equity, proposing a transversal integration of environment-development-gender issues (Mallory, 1999).

Johannesburg Declaration on Sustainable Development

The World Summit on Sustainable Development was held in Johannesburg from August 26 to September 4, 2002. The Secretary General of the United Nations, Kofi Annan, identified as central issues for the negotiations carried out during the Summit: Water and Sanitation, Energy, Agricultural Productivity, Biodiversity and Health. The instruments that emerged from the Summit are: the Johannesburg Declaration on Sustainable Development and the Implementation Plan for the decisions of the World Summit on Sustainable Development. In Point 20 of the Declaration, those attending the Summit assume the commitment to "ensure that the empowerment and emancipation of women and gender equality are integrated into all activities encompassed by Agenda 21, the Millennium Development Goals and the Plan of Implementation of the Decisions of the Summit" (Abels, 2011).

GENDER ISSUES IN INTERNATIONAL SECTOR REGULATION INSTRUMENTS

As stated, most environmental protection instruments at the international level recognize conventional sources and are made up of treaties, agreements and conventions for the sectoral regulation of environmental goods, without failing to pay tribute to the fundamental principles arising from the declarations that have been described and highlighted as instruments of environmental protection of general scope. Next, and after having proceeded to an exhaustive analysis, those that address the transversal integration of the environment-gender equation are highlighted (Liu, 2018).

Dublin Declaration on Water and Sustainable Development

The International Conference on Water and the Environment (CIAMA) held in Dublin, Ireland, from 26 to 31 January 1992, brought together 500 participants, including government-nominated experts from 100 countries and representatives of 80 organizations international, intergovernmental and non-governmental. The Conference adopted the Dublin Declaration and the Conference Report. The Declaration includes: an introduction, four guiding principles, a program of action and follow-up measures.

In the introduction, "The participants in CIAMA call for a radically new approach to the assessment, development and management of freshwater resources, and this can only be achieved through political commitment and participation ranging from the highest levels of government to the most elementary communities". In Principle No. 3 - Women play a fundamental role in the supply, management and protection of water -, it is established: "This fundamental role of women as providers and consumers of water and conservator of the living environment is rarely has been reflected in institutional arrangements for the development and management of water resources Acceptance and implementation of this principle requires effective policies that address the needs of women and prepare and empower them to participate, at all levels, in water resources programmes, including decision-making and execution, by such means as they may determine" (Wanjiru, 2012).

Convention on Biological Diversity

Within the framework of the Rio Conference on Environment and Development, the Convention on Biological Diversity was signed on June 5, 1992. In the preamble to the Convention, the Contracting

Parties expressly recognize "the decisive role played by women in the conservation and sustainable use of biological diversity and affirming the need for the full participation of women at all levels of the formulation and implementation of policies aimed at the conservation of biological diversity"

The objectives of the instrument are described in its article 1, in the following terms: "the conservation of biological diversity, the sustainable use of its components and the fair and equitable participation in the benefits derived from the use of genetic resources, through, inter alia, adequate access to those resources and appropriate transfer of relevant technologies, taking into account all rights to those resources and technologies, as well as through appropriate financing."

In such a way it is made clear that, although its provisions seek to achieve the conservation of biological diversity, they also respond to the need to ensure the economic development of the states, and as the ultimate goal, sustainable development and the quality of life of the being can be seen. and it is in this regard that the express recognition of the necessary participation of women at the levels of formulation and execution of policies aimed at fulfilling their primary objective acquires relevance. (Morrow, 2017)

United Nations Convention to Combat Desertification

United Nations Convention to Combat Desertification in Countries Experiencing Serious Drought or Desertification, Particularly in Africa is a Convention, signed in Paris in 1994, entered into force in 1996.

The preamble to the Convention highlights "the important role played by women in regions affected by desertification and drought, particularly in rural areas of developing countries, and the importance of ensuring at all levels the full participation of men and women in programs to combat desertification and mitigate the effects of drought."

The objective of the instrument is established in its article 2 and "is to combat desertification and mitigate the effects of drought, in countries affected by severe drought or desertification, particularly in Africa, through the adoption of effective measures in all levels, supported by international cooperation and association agreements, within the framework of an integrated approach in accordance with Agenda 21, to contribute to the achievement of sustainable development in the affected areas".

Article 5 of the instrument, when referring to the obligations of the affected country Parties, states: "d) promote awareness and facilitate the participation of local populations, especially women and youth, with the support of non-governmental organizations, in efforts to combat desertification and mitigate the effects of drought" (Großmann et al., 2017).

Countries affected by desertification implement the Convention by developing and carrying out action programs at the national, subregional and regional levels. The criteria for formulating these programs are detailed in the five "regional implementation annexes." Taking into account past experiences, the Convention stipulates that such programs must reflect a bottom-up, democratic approach.

When referring to national action programs, article 10 establishes that their objective is to determine the factors that contribute to desertification and the practical measures necessary to combat desertification and the effects of drought. ; specifying the respective functions of the government, local communities and land users, as well as determining the available and necessary resources. These national action programs must "ensure the effective participation at the local, national and regional levels of non-governmental organizations and local populations, both women and men, especially resource users, including farmers and herders and their representative organizations, in policy planning, decision-making, implementation and review of national action programmes".

For its part, in Article 19, included in Section 3: Support Measures, when contemplating the Promotion of capacities, education and public awareness, the parties commit to promote these capacities, as appropriate, through "(a) the full participation of the population at all levels, especially at the local level, in particular of women and youth, with the cooperation of non-governmental and local organizations".

From the summary made, it is clear that the convention is dependent on the principles and lines of action contained in Agenda 21.

The fundamental principle that marks its normative content is that of full and equitable popular participation, committing governments to create a "facilitating environment" so that in an environment of absolute respect for the characteristics of the populations, they are integrated into programs, national, regional and subregional.

This participation expressly contemplates the participation of women in the terms in which it has been conceived in the already analyzed Program 21.

From April 29 to May 1, 1998, the Fourth Regional Meeting of the Countries of Latin America and the Caribbean within the Framework of the United Nations Convention to Combat Desertification was held in St John's, Antigua and Bearded.

Decision 18 of the final report of the meeting contemplates the explicit incorporation of the gender perspective in the implementation of the Programs to combat Desertification.

In this regard, it considers that women play a fundamental role in the process that links development with the environment, that currently both men and women must be recognized as agents of development and that they must, therefore, have the same possibilities of access to decision-making, resources and development benefits and point out that when gender and development considerations referring to arid regions are introduced in the formulation of policy, the planning process and the implementation of programs and projects, the challenge is to define strategies for development that explicitly recognize the role of women in social and economic life, without ignoring the other concepts that make development sustainable.

It recognizes that the participation of women in the process of political planning and decision-making is justified, not only from an ethical point of view but also because economic development requires the full participation of both sexes (Swanston, 1993).

It considers that the conditions already exist to definitively promote the adoption of a gender perspective both at the level of national strategies and in the field of new initiatives and specific projects in an advanced stage of formulation and execution.

For all these reasons, it was decided to welcome the initiatives of the focal points of Argentina and Chile and the Regional Office of the Non-Governmental Organization EarthAction Network to submit to the Meeting the draft entitled "Incorporation of gender perspectives in the implementation of programs to combat desertification, particularly with regard to its emphasis on subregional and binational projects, as evidenced in the binational project between Argentina and Chile, incorporated as an annex to the document". Therefore, it requires the Secretariat to submit the document for the consideration of the governmental and non-governmental focal points of the Region to obtain their observations in order to enable the preparation of its final version, which may be presented by the Regional Executive Committee to international agencies. for its consolidation.

From the foregoing, the importance of the meeting arises, both because of the progress that evidences the reception of the gender perspective in its intersection with environmental and development issues, as well as because of its regional character and the significance that the initiative that emerged from the Argentina-Chile binational relationship.

The meeting and the project correspond to the concept of regional, subregional and national programs whose construction and implementation constitute one of the most outstanding characteristics of the Agreement under treatment.

That is why it seems illustrative to us to briefly address the aforementioned Binational Project between Argentina and Chile - Explicit Incorporation of the Gender Perspective in the Implementation of the Programs to Combat Desertification-.

On April 7, 1998, delegates from Argentina and Chile met to analyze and propose action strategies that included the gender perspective in the National Programs to Combat Desertification in Argentina and Chile, within the framework of the Macro binational regions.

The cooperation between Chile and Argentina has multiple historical antecedents and in this framework, it has been considered that the initiative to apply a strategy of integration of the gender perspective in all the levels of application of the UNCCD jointly in both countries, will give greater consistency and vigor to this process, consequently benefiting both countries (Joyner & Little, 1996).

The project has taken into account that the Convention exalts the importance of the role of women in the fight against desertification, especially in rural areas, as well as the need to incorporate them at all levels of action programs.

Likewise, it has taken into account that the need to integrate the gender perspective has also been explicitly expressed by the women potentially affected by desertification. The 1st Latin American Assembly of Rural Women, held in Brasilia on November 3, 1997, states in the conclusions of one of its work commissions: "we women from rural sectors ask governments to guarantee gender equality, equally, must be contemplated in the processes of literacy and education".

That is why it has been concluded that a binational initiative is justified by several concepts. Undoubtedly, among the main ones we can point out, on the one hand, the fact that the geographical space comprised by the macro-region corresponds, in turn, to the same macro-ecosystem where a similar behavior occurs in the processes of environmental transformation, both of anthropogenic and natural origin and, on the other hand, the fact of the homogeneous character of the population on both sides of the border with social, commercial and consanguinity relations, with a great transnational mobility(Detraz, 2017).

Among the activities planned for Phase I is the incorporation of the gender dimension in the initiatives derived from the convention, the awareness and training of relevant actors and the identification of project profiles oriented by the gender perspective and the vindication of the woman and family.

Stockholm Convention on Persistent Organic Pollutants (POPs)

Signed in Stockholm in 2001, entered into force on May 17, 2004. Its objective is established by its article 1: "Bearing in mind the precautionary principle enshrined in principle 15 of the Rio Declaration on Environment and Development, the objective of this Agreement is to protect human health and the environment against to persistent organic pollutants. The persistent organic pollutants reached, known as the first "dirty dozen" are: 8 Pesticides (Aldrin, chlordane, DDT, dieldrin, endrin, heptachlor, mirex and toxaphene), polychlorinated biphenyls PCBs, dioxins, furans and hexachlorobenzene. The express recognition of being tributary to the Principles contained in the Rio Declaration removes any strangeness about the express inclusion of the gender perspective in a matter as specific as the one regulated by the sectoral treaty. That is why its article 7, referring to the Implementation Plans, establishes that "the Parties, when appropriate, shall cooperate directly or through global, regional or subregional organizations, and shall consult the national stakeholders, including women's groups and groups dealing with children's health, in order to facilitate the development, implementation and updating of their implementation plans." And article 10, dedicated to Information, public awareness and training, establishes that each party, within its capacities, will promote and facilitate "the development and application of public training and awareness programs, especially for women, children and the less educated, on persistent organic pollutants, as well as on their effects on health and the environment and on their alternatives" (Lindquist & Lindquist, 1997; Nnoko-Mewanu et al., 2021).

CONCLUSION

Before presenting the conclusions to which the analysis carried out has led us, we have found it convenient to present a table in which all the international instruments analyzed are listed, in chronological order. We believe that this will make it possible to more clearly visualize the result of the research carried out and for the reader to arrive at their own considerations or use the material for their own work. We conclude that the turning point in the introduction of the gender dimension in the instruments of international protection of the environment occurs from those that emerged from the Rio Conference on Environment and Development: in the Rio Declaration, but with precision and forcefulness in Agenda 21. Therefore, even those later instruments that do not address in their text the interrelation of the gender perspective in matters related to the legal protection of the environment, are imbued with the spirit of those instruments, of the which are tributaries. It would have been desirable that they contemplated it expressly, since as we have pointed out in the introduction of this work, the concern of the law for environmental issues has implied the recognition of a supreme legal good that is the quality of life and this can only be achieved with equity, which includes gender equity and until this final objective is fully achieved, there will never be little reiteration of fundamental principles such as those described. Both the Malmö Declaration and the Millennium Declaration advance along this line. Special mention deserves, within the sectoral protection instruments, the United Nations Convention to Combat Desertification in Countries Affected by Serious Drought or Desertification, in particular Africa, and the regional and subregional instruments that have arisen within its framework. The terminology used is exact and denotes the complete understanding of the role of women, by postulating the necessary participation of women in the process of political planning and decision-making, based on ethics and on the requirements of sustainable development, that requires the full participation of both sexes at all decision-making levels in the formulation of policies under equal conditions to enable the improvement of the quality of life of the present and future generations. We hope that the principles established in the fundamental instruments highlighted continue to guide the course and the agreement becomes a model to follow.

REFERENCES

Abels, G. (2011). Gender equality policy. In Policies within the EU Multi-Level System (pp. 325-348). Nomos Verlagsgesellschaft mbH & Co. KG. doi:10.5771/9783845228266-325

Arora-Jonsson, S. (2014, November). Forty years of gender research and environmental policy: Where do we stand? In *Women's Studies International Forum* (Vol. 47, pp. 295-308). Pergamon.

Buckingham, S., & Le Masson, V. (Eds.). (2017). *Understanding climate change through gender relations*. Taylor & Francis. doi:10.4324/9781315661605

Dankelman, I. (Ed.). (2010). *Gender and climate change: An introduction*. Routledge.

Detraz, N. (2017). *Gender and the Environment*. John Wiley & Sons.

Galizzi, P. (2012). *Missing in action: Gender in international environmental law*. Fordham Law Legal Studies Research Paper, (2779320).

Großmann, K., Padmanabhan, M., & Afiff, S. (2017). Gender, ethnicity, and environmental transformations in Indonesia and beyond. *ASEAS-Austrian Journal of South-East Asian Studies*, *10*(1), 1–10.

Joyner, C. C., & Little, G. E. (1996). It's Not Nice to Fool Mother Nature-The Mystique of Feminist Approaches to International Environmental Law. *BU Int'l LJ*, *14*, 223.

Kostruba, A. (2019). The rule of law and its impact on socio-economic, environmental, gender and cultural issues. Kostruba AV, The Rule of Law and its Impact on Socio-Economic, Environmental, Gender and Cultural Issues. *Space and Culture, India*, *7*(2), 1–2.

Kurian, P. A. (2018). *Engendering the environment? Gender in the World Bank's environmental policies*. Routledge. doi:10.4324/9781315185101

Kurtz, H. E. (2007). Gender and environmental justice in Louisiana: Blurring the boundaries of public and private spheres. *Gender, Place and Culture*, *14*(4), 409–426. doi:10.1080/09663690701439710

Lindquist, C. H., & Lindquist, C. A. (1997). Gender differences in distress: Mental health consequences of environmental stress among jail inmates. *Behavioral Sciences & the Law*, *15*(4), 503–523. doi:10.1002/(SICI)1099-0798(199723/09)15:4<503::AID-BSL281>3.0.CO;2-H PMID:9433751

Liu, C. (2018). Are women greener? Corporate gender diversity and environmental violations. *Journal of Corporate Finance*, *52*, 118–142. doi:10.1016/j.jcorpfin.2018.08.004

MacGregor, S. (Ed.). (2017). *Routledge handbook of gender and environment*. Taylor & Francis. doi:10.4324/9781315886572

Maguire, R., & Jessup, B. (2021). Gender, race and environmental law: A feminist critique. In *International Women's Rights Law and Gender Equality* (pp. 107–127). Routledge.

Mallory, C. L. (1999). *Toward an ecofeminist environmental jurisprudence: Nature, law, and gender*. University of North Texas.

Morrow, K. (2017). Integrating gender issues into the global climate change regime. *Understanding climate change through gender relations*, 31-44.

Newman, M. K., Lucas, A., LaDuke, W., Berila, B., Di Chiro, G., Gaard, G., & Sturgeon, N. (2004). *New perspectives on environmental justice: Gender, sexuality, and activism*. Rutgers University Press.

Nnoko-Mewanu, J., Téllez-Chávez, L., & Rall, K. (2021). Protect rights and advance gender equality to mitigate climate change. *Nature Climate Change*, *11*(5), 368–370. doi:10.103841558-021-01043-4

Norgaard, K., & York, R. (2005). Gender equality and state environmentalism. *Gender & Society*, *19*(4), 506–522. doi:10.1177/0891243204273612

Ryan, S. E. (2014). Rethinking gender and identity in energy studies. *Energy Research & Social Science*, *1*, 96–105. doi:10.1016/j.erss.2014.02.008

Swanston, S. F. (1993). Race, gender, age, and disproportionate impact: What can we do about the failure to protect the most vulnerable. *The Fordham Urban Law Journal*, *21*, 577.

Wanjiru, L. (2012). Gender, climate change and sustainable development. *Fordham Environmental Law Review*, 1-6.

ADDITIONAL READING

MacGregor, S. (Ed.). (2017). *Routledge handbook of gender and environment*. Taylor & Francis. doi:10.4324/9781315886572

Maguire, R., & Jessup, B. (2021). Gender, race and environmental law: A feminist critique. In *International Women's Rights Law and Gender Equality* (pp. 107–127). Routledge.

Mallory, C. L. (1999). *Toward an ecofeminist environmental jurisprudence: Nature, law, and gender*. University of North Texas.

Morrow, K. (2017). Integrating gender issues into the global climate change regime. *Understanding climate change through gender relations*, 31-44.

Newman, M. K., Lucas, A., LaDuke, W., Berila, B., Di Chiro, G., Gaard, G., & Sturgeon, N. (2004). *New perspectives on environmental justice: Gender, sexuality, and activism*. Rutgers University Press.

Nnoko-Mewanu, J., Téllez-Chávez, L., & Rall, K. (2021). Protect rights and advance gender equality to mitigate climate change. *Nature Climate Change*, *11*(5), 368–370. doi:10.103841558-021-01043-4

Norgaard, K., & York, R. (2005). Gender equality and state environmentalism. *Gender & Society*, *19*(4), 506–522. doi:10.1177/0891243204273612

Ryan, S. E. (2014). Rethinking gender and identity in energy studies. *Energy Research & Social Science*, *1*, 96–105. doi:10.1016/j.erss.2014.02.008

Swanston, S. F. (1993). Race, gender, age, and disproportionate impact: What can we do about the failure to protect the most vulnerable. *The Fordham Urban Law Journal*, *21*, 577.

Wanjiru, L. (2012). Gender, climate change and sustainable development. *Fordham Environmental Law Review*, 1-6.

KEY TERMS AND DEFINITIONS

Environmental: Law: Environmental law is a collective term encompassing aspects of the law that protect the environment. A related but distinct set of regulatory regimes, now strongly influenced by environmental legal principles, focuses on managing specific natural resources, such as forests, minerals, or fisheries. Other areas, such as environmental impact assessment, may not fit neatly into either category but are nonetheless important components of environmental law. Previous research found that when environmental law reflects moral values for betterment, legal adoption is more likely to be successful, usually in well-developed regions. In less-developed states, changes in moral values are necessary for successful legal implementation when environmental law differs from moral values.

Equity: Defined by UNEP to include intergenerational equity—"the right of future generations to enjoy a fair level of the common patrimony"—and intragenerational equity—"the right of all people within the current generation to fair access to the current generation's entitlement to the Earth's natural resources"—environmental equity considers the present generation under an obligation to account for long-term impacts of activities and to act to sustain the global environment and resource base for future generations. Pollution control and resource management laws may be assessed against this principle.

Polluter Pays Principle: The polluter pays principle stands for the idea that "the environmental costs of economic activities, including the cost of preventing potential harm, should be internalized rather than imposed upon society at large." All issues related to responsibility for cost for environmental remediation and compliance with pollution control regulations involve this principle.

Precautionary Principle: One of the most commonly encountered and controversial principles of environmental law, the Rio Declaration formulated the precautionary principle as follows, to protect the environment, States shall widely apply the precautionary approach according to their capabilities. Where there are threats of serious or irreversible damage, lack of full scientific certainty shall not be used as a reason for postponing cost-effective measures to prevent environmental degradation. The principle may play a role in any debate over the need for environmental regulation.

Prevention: The concept of prevention etc. can perhaps better be considered an overarching aim that gives rise to a multitude of legal mechanisms, including prior assessment of environmental harm, licensing, or authorization that set out the conditions for operation and the consequences for violation of the conditions, as well as the adoption of strategies and policies. Emission limits and other product or process standards, the use of best available techniques, and similar techniques can all be seen as applications of the concept of prevention.

Public Participation and Transparency: Identified as essential conditions for "accountable governments...industrial concerns," and organizations generally, public participation and transparency are presented by UNEP as requiring "effective protection of the human right to hold and express opinions and to seek, receive and impart ideas, etc. a right of access to appropriate, comprehensible and timely information held by governments and industrial concerns on economic and social policies regarding the sustainable use of natural resources and the protection of the environment, without imposing undue financial burdens upon the applicants and with adequate protection of privacy and business confidentiality," and "effective judicial and administrative proceedings." These principles are present in environmental impact assessment, laws requiring publication and access to relevant environmental data, and administrative procedures.

Sustainable Development: Defined by the United Nations Environment Programme as "development that meets the needs of the present without compromising the ability of future generations to meet their own needs," sustainable development may be considered together with the concepts of "integration" (development cannot be considered in isolation from sustainability) and "interdependence" (social and economic development, and environmental protection, are interdependent). Laws mandating environmental impact assessment and requiring or encouraging development to minimize environmental impacts may be assessed against this principle. The modern concept of sustainable development was discussed at the 1972 United Nations Conference on the Human Environment (Stockholm Conference) and the driving force behind the 1983 World Commission on Environment and Development (WCED, or Bruntland Commission). In 1992, the first UN Earth Summit resulted in the Rio Declaration, Principle 3 of which reads: "The right to development must be fulfilled to equitably meet developmental and environmental needs of present and future generations." Sustainable development has been a core concept of international environmental discussion ever since, including at the World Summit on Sustainable Development (Earth Summit 2002) and the United Nations Conference on Sustainable Development (Earth Summit 2012, or Rio+20).

Transboundary Responsibility: Defined in the international law context as an obligation to protect one's environment. UNEP considers transboundary responsibility at the international level to prevent damage to neighboring environments at the international level as a potential limitation on the rights of the sovereign state. Laws that limit externalities imposed upon human health and the environment may be assessed against this principle.

Chapter 12
Environmental Law and Non-Governmental Organizations

ABSTRACT

Non-governmental actors play a crucial role both internationally and nationally. Since these new actors are geographically and functionally diverse, it is impossible for government actors to ignore them. The non-government sector is at once a rival and a partner for the government. When it comes to human rights, these organizations are more of an opponent of governments; however, when it comes to health, development, and the environment, they are more of a partner. Although these organizations play an active role in Iran's legal system when it comes to environmental protection, there are still many gaps in the reaction phase. Appropriate conditions must be provided to ensure their participation in environmental litigation so that they can attain their desired position in criminal proceedings as quasi-prosecutors.

INTRODUCTION

Unbridled industrial development has transformed nature and caused irreparable damage to the human environment. Therefore, taking measures at the national and international levels as one of the important measures is on the agenda of many countries; Many conferences have been held in this regard so far. One of the major and decisive mechanisms to achieve this important goal is identifying and involving NGOs in environmental issues. The prevention principle is one of the most well-known principles of international environmental law and is considered the "golden rule" in this field. This principle is explicitly enshrined in the Rio 1992 Declaration on Environment and Development; Achieving this principle in line with the implementation of Chapter 27 of Agenda 21 requires the role of non-governmental organizations. This agenda focuses on the role of social actors; In general, these actors are classified into two types of individuals, associations, and groups. Women, youth, indigenous peoples, farmers, scientists are among the real people who have been considered in environmental protection; While local councils, trade unions, and NGOs are among the group actors; There are roles for them. Reflection on these actors reveals several facts; First, these actors are divided into two large groups of individuals and legal

DOI: 10.4018/978-1-6684-7188-3.ch012

entities; second, they share a non-governmental nature; and finally, they share a direct or indirect goal in environmental protection (Tarlock, 1992).

Today, non-governmental organizations contribute to the criminal justice system in achieving its goals in two ways. On the one hand, by performing their duties in reporting and reporting crimes, they help the criminal justice system detect and prosecute crimes, and on the other hand, they help specific victims by playing a role in the criminal process. Sensitizing public opinion and timely advertising and information can play an important role in protecting the environment (Li et al., 2018).

Achieving high environmental goals is difficult regardless of the role of NGOs in the various stages of the formation of environmental norms and their monitoring and implementation. The social nature of the crime phenomenon requires that all social capacities be used as much as possible to prevent and suppress crime. The participation of non-governmental organizations in the criminal process is considered one of these capacities; Something that can be effective in the success of prevention programs in the community. Therefore, a participatory approach and preventive measures can prevent many environmental crimes.

In addition to prevention, it is very important to take a reactive approach to criminalize acts of environmental degradation and pollution; By identifying and expanding the competence of those in charge, from public and governmental to private and non-governmental, the protection of privacy will become a better and easier process. Therefore, the interaction of non-governmental organizations with the government and the competence of these organizations in environmental criminal cases has an important role in protecting the environment in the control and enforcement phase (Spiro, 2007).

The question is whether Iranian law has a role for NGOs in preventing environmental crimes. If so, what is this role in Iranian law, and if not, what action should be taken in the field of legislation? The answers to these questions reveal the type of attitude of the Iranian legislature towards the environment in environmental policies and explain the criteria and mechanisms adopted in the form of participatory prevention for environmental protection and sustainable use of the country's natural resources.

Depending on the interpretation of the word prevention, its mechanisms also vary. In providing a broad interpretation of prevention in environmental crimes, the purpose is various criminal and non-criminal measures and mechanisms. In the case of providing a narrow interpretation of this concept, only criminal mechanisms are considered. In other words, prevention in its broadest sense includes a wide range of actions and mechanisms; While in its narrow and specific sense, it includes only non-criminal acts (Wu et al., 2017).

Accordingly, if we refer to the specific meaning of prevention, we have only action measures in front of us, and if we refer to its general and broad meaning, in addition to active measures, we have also considered reactionary measures. With this interpretation, in this article, each of the non-criminal and criminal measures and mechanisms in the field of environment is examined in the light of the actions of non-governmental organizations.

NGOs AND ENVIRONMENTAL CRIME PREVENTION

So far, several definitions of NGOs have been provided nationally and interna-tionally, but the diversity of these organizations is such that it is difficult to provide a single definition of them; However, these organizations have something in common in terms of their private and non-commercial nature. The definition of environmental NGOs reads as follows:

Environmental non-governmental organizations refer to non-governmental, non-profit and non-political organizations that are organized from the gathering of natural persons voluntarily and in such a way that these persons by drafting a statute, from the date of registration in the official authorities of the country as a Legal entity to achieve common goals and ideals in the field of protection and protection of the en-vironment to work in the cities and villages of the country and if possible at the inter-national level. (Edele, 2005)

These organizations in Iran are proof that these organizations have existed in practice for a long time in the form of charitable organizations, whether religious or welfare, such as boards and groups of breastfeeding and chain and managing the res-ervoirs in Iran. In other words, these institutions, except for registration, had all the components of an NGO. In Iranian law, perhaps the Commercial Code and the Regu-lations for Registration of Non-Commercial Organizations and Institutions approved in 1958 can be considered the first regulations defined by non-governmental organiza-tions.

Some countries, such as France and Switzerland, have pioneered in this area; For more than a century, the status of these organizations has been regulated by the legis-lature. France, for example, sought to regulate the formation of these organizations bypassing the Association Law on 1 July 1901. Article 1 of this law states: "Association is a contract according to which two or more persons permanently share their knowledge and activities to achieve a goal other than profit" (Loi relative au contrat d'association). This article has some structural elements of non-governmental organi-zations such as private nature, non-profit nature and continuity, and continuity of ac-tivities. Has been postponed to registration, while in Swiss civil law, associations have legal personality as soon as they are formed and do not need to be registered (Nomura, 2007).

Therefore, this law is considered the cornerstone of the formation and establish-ment of non-governmental organizations in the French legal system and has provided a suitable platform for the participation of these organizations. In France, many NGOs are active in various fields, including the environment; This indicates the importance of the environment and an appropriate and legally defined context for forming and establishing these organizations. In addition to Law 1901, other laws, such as the French Environmental Act of 2005, directly address the role of environmental associa-tions and prescribe their participation in the actions of public institutions and bodies related to the environment and administrative and judicial authorities.

The Constitution of the Islamic Republic of Iran envisages two important princi-ples in the field of environmental protection as well as the legitimacy of the establish-ment and activity of non-governmental organizations. In principle 50 of this law, envi-ronmental protection is considered a public duty. In addition, activities that cause pol-lution or destruction of the environment are prohibited. According to this principle, all Iranians are obliged to protect and protect their environment; Accordingly, the establishment and participation in environmental non-governmental organizations and membership in them can be considered in line with Article 50 of the Constitution and the second principle according to which the creation and activity of non-governmental organizations are inferred is Article 26 of the Constitution. Adopting the principle of freedom of association as enshrined in Article 20 of the Universal Declaration of Human Rights, the legislation stipulates that "Parties, associations, political and trade unions and Islamic associations or recognized religious minorities are free, provided that the principles of independence Do not violate freedom, national unity, Islamic norms and the basis of the Islamic Republic. "No one can be barred from participating in them or forced to participate in any of them." In principle, the goals and ideals out-lined in the constitutions are quite general; It is on this basis that they refrain from go-ing into details and refer its examination to ordinary laws. In this regard, the principle deals with two issues related to the formation of non-governmental organizations: the principle of freedom of association and the principle of free participation of individu-als in these communities. The legislator has used terms such as populations and associ-ations to include popular gatherings (non-governmental organizations); In the last part of this principle, the issue of the right of individuals to participate or not to partic-ipate in any of these gatherings is explicitly raised. A noteworthy point in this principle is the same two features of non-opposition to the general principles of the constitution and the non-obligation to participate in them, which is also mentioned in paragraph A of Article 1 of the Executive Regulations on the Establishment and Activities of Non-Governmental Organizations approved in 2005. Given the above, the role of non-governmental organizations in the participatory prevention of environmental crimes is examined (Doh & Guay, 2004).

PARTICIPATION CONCERNING ENVIRONMENTAL CRIMES PREVENTION

In general, the relationship between government and non-governmental organi-zations can be considered from different aspects. This relationship varies according to the study area and is very diverse. In general, the presence and participation of non-governmental organizations in the two processes of formation of international norms and their monitoring and implementation is realized. This diversity gives rise to different roles for these organizations; As in a field as a complement to government and international organizations; While in another area, it manifests itself as a rival to the government. Therefore, the participation of these organizations is not equal in all levels and fields; In some areas, such as humanitarian law, environmental law, and development, they have an effective and prominent presence; While in some areas, such as human rights, these organizations stand up to governments and are seen as their fiercest rivals (Esty, 1998).

As mentioned above, the fifty-sixth principles of the Constitution of the Islamic Republic of Iran are defined as two basic principles on the importance of the environ-ment and its protection and the freedom of association. The combination of these prin-ciples, on the one hand, and the existence of other legislative and sub-legislative provi-sions related to environmental protection and the establishment and establishment of non-governmental organizations, on the other hand, raises some fundamental ques-tions about participatory prevention.

The diversity of non-governmental organizations and their different goals give rise to different functions for these organizations. Given the diverse functions of these organizations, the important question is whether environmental crime prevention is on their agenda? Answering this question requires the following points: First, a reflec-tion on the prevailing regulations indicates that in the Iranian legal system, only a handful of different materials are allocated to the participation of non-governmental organizations; There is no comprehensive and independent law on participatory pre-vention; Because crime prevention programs are generally formed in a completely formal and governmental context and do not consider a suitable position for non-governmental organizations, that the issue of environmental crime prevention less attention has been paid.

More than two decades after the adoption of the constitution, the Iranian judici-ary has recently, through the proposal of the "Crime Prevention Bill," considered the implementation of part of its duties in the field of prevention, which is the subject of the fifth paragraph of Article One Hundred and Fifty-six of the Constitution. Article 1 of the bill defines crime prevention as "predicting, identifying and evaluating the oc-currence of crime and taking the necessary measures and measures to prevent its oc-currence." This article has prescribed three types of social, judicial, and disciplinary crime prevention and has established general and specific elements for them (Hendry, 2006).

Social prevention is very important. This type of prevention allows the presence of non-governmental organizations. According to the bill, social prevention is "educa-tional, cultural, economic and social measures and methods of the government, non-governmental organizations and non-governmental organizations in the field of im-proving the social environment and the physical environment to eliminate or reduce the social factors of crime." To implement the goals of this bill and to achieve social prevention, Article 7 of the bill envisages a pillar called "Social Commission for the Prevention of Crime." Paragraph "f" of this article makes one of the duties of the commission "assisting in the establishment and strengthening of non-governmental organizations, associations, and non-governmental organizations in the field of crime prevention." Facilitating the establishment of these institutions plays a unique role in preventing environmental crimes and protecting the environment (Lindblom, 2005).

Despite the judiciary taking a big step towards implementing the constitutional duty of prevention, the crime prevention bill seems to be disregarding environmental crimes. However, the bill does not seem to have the capacity to cover such crimes, de-spite its implicit nature. In particular, there is a strong belief that access to a healthy environment is a prerequisite for realizing other fundamental rights protected by criminal law and intergenerational environmental justice. Today, crime prevention programs in the official and governmental form alone are not very popular; The suc-cess of these programs depends on the participation of all civil society actors. In this regard, important international environmental instruments, such as the Rio Declaration and Order 21, contain elements that emphasize the participation of all non-governmental actors in the protection of the environment. Chapter 23 of Order 21 is dedicated to strengthening the role of major groups and emphasizes the serious com-mitment and participation of all social groups in achieving their goals, policies, and mechanisms (Shohani et al., 2021).

Chapter 27 of the order is dedicated to strengthening non-governmental organiza-tions; This document introduces partners for sustainable development. The chapter be-gins with the phrase: "NGOs play a vital role in the process of democracy." These or-ganizations are very diverse in activity (local, regional, national and global) and the subject of activity (human rights, humanitarian law, development, environment, etc.). To further explain, the issue of participatory prevention is examined in the light of the activities of non-governmental organizations.

Participation in the Field of Environment

Non-governmental organizations move within the framework of the goals set out in their statutes and make decisions on various issues following the established pur-pose and subject matter. Compared to government organizations and institutions, these organizations generally face many financial and technical limitations. Therefore, these organizations use tools and methods to achieve their goals more or less used by other organizations. Non-governmental is also used. Applying this according to para-graph 5 of Article 3 of the recent law is one of the cases included in the law of count-ing, and to properly implement the provisions of the law, supervision of all matters mentioned in Article 3 and inspection in these areas is entrusted to the legal unit "Food from radiation contamination, including the protection of people and future genera-tions in general, is the oversight role of the Atomic Energy Organization." Applying the prevention principle and using non-criminal applications is considered the first and easiest way to protect natural elements. Experience and scientific expertise prove that prevention should be the "golden rule" of the environment for ecological and eco-nomic reasons (Moghimi, 2007).

In some cases, it is impossible to compensate for the damage to the environment: the extinction of animal or plant species, the erosion and discharge of persistent pollu-tants into the sea, create uncontrollable and irreversible conditions. Even when the damage is compensable, the costs of environmental remediation are very high. In many cases, it is impossible to prevent all possible risks of such injuries. In such cases, to protect the environment and the rights of others, measures may be taken to allow risky activities to be minimized.

The material realm of the environment is vast and can be violated in various forms. The close relationship between the elements of the environment is a matter of great importance; Because this relationship distinguishes environmental crimes from other crimes; In fact, it brings crimes in this area to a new level that should be given special attention. Overall, the breadth of the environmental issue leads to a variety of crimes committed against it. Therefore, the diversity of criminals is the logical result of the diversity of crimes; In practice, determining the level of involvement of each per-petrator of environmental crimes will not be easy. According to the above explana-tions, the perpetrators of environmental crimes can be divided into natural and legal persons. However, the exact determination of the perpetrator or perpetrators of these crimes and the degree of participation of each of these persons in these crimes is not the same[58]. For example, in air pollution, determining the exact amount of gases emitted by each of the air pollutants (industrial plants and car owners) on the one hand and the contribution of each manufacturer (non-standard car manufacturers and producers of unsuitable and non-standard fuel) in causing pollution from, On the other hand, it isn't easy. At the same time, transboundary pollution must be added to it; This is due to the close relationship between environmental elements and the impact and impact of each of them on each other, and as a result, it is difficult to identify the pol-luters and the source of pollution (Hashemi et al., 2019).

Identifying the culprit against environmental crimes faces several major prob-lems; If the environment is considered as insane, this insane has no real or legal per-sonality; The value of the environment is also inherent; Its value, in essence, should not be weighed against the benefit it owes to humankind; First of all, respect for it should depend on its intrinsic value, not the importance it brings to human beings.

Given the perceived characteristics of environmental crimes, is the prevention of these crimes different from other crimes? It is certain that the prevention of environ-mental crimes is very important and requires an integrated and comprehensive pro-gram; NGOs will play a pivotal role in this.

Functionally, the nature of the actions of non-governmental organizations is more in favor of non-criminal prevention; the tools and methods used by these organizations are not police and repressive. These methods are usually non-violent; In this respect, they can not be compared with government institutions' coercive and reactionary mechanisms. For example, these organizations prevent environmental degradation by publishing various scientific articles on environmental problems in their publications or publishing statistics in their reports(Moosavi et al., 2015). Therefore, environmental NGOs try to take the following preventive measures by informing, educating, and mobilizing public opinion and announcing a threat to the state of the environment.

Awareness and Information on the Environmental Status

Information is one of the common methods used by these organizations. In gen-eral, public access to and participation in information is not limited to environmental regulation but is a pre-eminent right of the government and its citizens in all democra-cies. Here the issue of information and information and citizen participation is very important; Because with their participation in decision-making, a suitable platform for democracy is provided. On the other hand, the observance of this right for citizens shows the government's attention to pluralism and the people's efforts in managing the country's affairs and their right to self-determination. Citizens' access to infor-mation and their participation in environmental issues is very important; Because ac-cess to environmental data and their participation in the decision-making process will directly affect citizens.

Conceptually, the right to access environmental information is conceived in two ways: First, the narrow meaning of this right is the freedom to seek information. Ac-cording to this interpretation, the competent authorities provide the requested infor-mation to them at the request of natural or legal persons. This type of information has a passive aspect; Because competent public authorities provide information according to the request submitted by natural or legal persons; In other words, these authorities do not provide any information to individuals in the first place unless there is a prior request. For example, an environmental NGO may request information from the rele-vant authorities regarding implementing a development project, such as a dam or a highway. In this case, the public authorities are obliged to inform and respond to the relevant organization; Unless the requested information is one of the exceptions to the right to information, the information is refused (Asadi, 2021).

Second, the right of access to environmental information in the broadest sense of the word, which is, in fact, the right of access to information or the right to receive in-formation. Here we are talking about active information; Which means that the com-petent administrative authorities provide information on their initiative and do not need to request natural or legal persons. In other words, every citizen has the right to receive information about their environment, without the need for a prior request; Such as informing citizens about air pollution in metropolitan areas.

It should be noted that some non-governmental organizations active in the field of the environment have accurate specialized information; This information can help de-cision-makers and implementers of large industrial projects in terms of environmental issues. For example, these organizations can collect data and analyze it, thus being ef-fective as experts in the decision-making process. A clear example of this participation can be seen in specialized matters such as the study of the environmental effects of large-scale development projects; Thanks to the methods and tools and the expert role of these organizations, public rights in the environment can be defended (Li et al., 2018).

Public Acceptance and Interest

One of the powerful tools of these organizations at the national and international levels is their external demonstrations, which turn them into a powerful pressure group. The actions of these organizations are very diverse; It ranges from the harsh protest of an environmental NGO against a country's nuclear tests at sea to the for-mation of a coalition to fight anti-personnel mines and the establishment of an inter-national criminal court. Therefore, with the help of the formal and informal status that has been considered for them in the international arena, these organizations can be in-fluential in international relations. It should be noted that in terms of effectiveness and credibility, the actions of these organizations are not the same; Because they are very diverse both in terms of the type of activity and geographical distance. At the na-tional level, these organizations, to achieve their goals to protect the environment, take measures such as preparing a petition to support an environmental issue, such as the relocation of the road connecting the north of the country to Khorasan province or opposing unauthorized destruction of natural resources. have given. Sometimes the actions of these organizations are such that they can be considered in the formation of minimum criteria, for example, companies producing products and food products or even health and pharmaceuticals and clothing and services to maintain their credibility in the market and Also, their convenience from the buyer point to the logo of the li-censing of these institutions as a proof of the high quality of their products on their goods. Despite the many efforts of non-governmental organizations in Iran, these or-ganizations have not yet been able to find their true place properly and act in an inte-grated manner in mobilizing public opinion to protect the environment and prevent environmental crimes (Hashemi et al., 2019).

Environmental Education

Education plays an important role in preventing environmental degradation and pollution; Therefore, non-governmental organizations, by forming environmental networks, have started training to acquaint people with nature and how to interact with it. The establishment of specialized workshops to educate various people, includ-ing women, students, and children, has been in this direction. The main purpose of these pieces of training is to become more familiar with the environment and exchange information and experiences. Researching various environmental issues to collect ac-curate information and statistics on the state of the environment in different areas and research on animal and plant species are among the things that have been done within the framework of the actions of these organizations (Hashemi et al., 2017).

RIGHT TO SUE NGOs IN ENVIRONMENTAL LITIGATION

Explaining the role of non-governmental organizations in preventing environ-mental crimes requires that the position of the Iranian legal system in recognizing the right to sue for these organizations be examined and pathologically examined. Recog-nition of this right is one of the criteria for assessing the status and importance of the principle of participation; This plays an important role in implementing environmen-tal regulations. In this regard, the existing laws and then some new orientations are examined.

Comparative Study

Stakeholders in environmental criminal cases are one of the most con-troversial theoretical and practical issues; Because environmental crimes are inherent-ly special and cannot be compared to other crimes. In fact, in such crimes, the victim is the victim of silence who cannot protest and defend himself. In this situation, two as-sumptions are conceivable.

The first assumption is when other natural or legal persons are harmed by a crime against the environment; In this case, the legal system can also protect the environ-ment due to the damaged private complaint.

The second assumption is when the crime is committed solely against the envi-ronment, without directly harming other people. Pollution of open rivers, seas, over-fishing, poaching, and deforestation are some of the issues that result in the environ-ment being a victim of environmental crimes; Something that does not seem to cause direct harm to individuals (Ormazdi, 2013).

Regarding the latter premise, the crime should be considered an example of crime against the public interest, in which only the prosecutor or the public prosecutor has the necessary legal position to declare a crime and file a criminal lawsuit in a compe-tent court. However, in many countries, including Iran, crimes against the environ-ment are of secondary importance compared to other public crimes; prosecutors do not pay much attention to cases suspected of committing environmental crimes. So the question remains, what should be done about such crimes, and what can be done?

Undoubtedly, strengthening the role of non-governmental organizations in envi-ronmental litigation will be one of the important solutions to this problem. The success of these organizations and the recognition of the right to sue is not limited to domestic law but is also considered in some international instruments, such as the Convention on the Protection of the Environment through Criminal Law. Article 11 of the Conven-tion designates environmental groups, including non-governmental organizations, as beneficiaries of environmental claims. In this regard, the 21st agenda calls on govern-ments and legislators to establish judicial and administrative procedures to ensure le-gal redress and to compensate for environmental actions that are illegal or violate le-gal rights (Pourhoseyni et al., 2019).

A comparative study of other countries' laws reveals that countries have tried to solve this problem by involving non-governmental organizations in the criminal pro-cess to answer this problem. For example, Articles 2-1 to 2-21 of the French Code of Criminal Procedure set out detailed provisions for the protection of victims' aid asso-ciations; According to these articles, if there are some conditions, these associations can play the role of plaintiff and even file a lawsuit. These associations can greatly help prosecutors and victims to achieve the goals of the criminal justice system.

In this legal system, in addition to the Code of Criminal Procedure, in other laws such as the Law on Consumer Protection and Environmental Law, environmental as-sociations have been granted legal positions so that they can appear before adminis-trative and judicial authorities under the conditions outlined in these laws. Thus, asso-ciations can exercise the rights conferred on the private plaintiff regarding acts that harm the public interest subject to the protection of these associations (Bromideh, 2011).

Lawmaking is one of the most important and main steps in environmental protec-tion. Legislation alone is not enough to protect natural resources; Because the regula-tions must be implemented by the executive bodies. Achievement and goals outlined in the laws and regulations will be tangible if there is an executive guarantee for their non-implementation and the competent judicial authorities guarantee this perfor-mance guarantee. Accordingly, the legal system in the field of environment can be hoped for if each of the three powers performs its duties well.

In France, "the Administrative Court ruled that an association for the promotion of tourism and nature conservation may, following the objectives outlined in its arti-cles of association, object to the issuance of a permit for the construction of a recycling plant following its objectives." Showed that trade unions, especially those related to the chemical industry interested in retaining the license, will have the right to sue. Contract. However, due to the inability of the association to prove the occurrence of material damages, it could not obtain a ruling on compensation".

Therefore, in most countries, efforts are made to provide a suitable basis for im-plementing environmental regulations based on the principles of prevention by admin-istrative law, counteraction by criminal law, and compensation by civil law. The reali-zation of this importance depends on the role of these organizations and their partici-pation in the prevention and combating of environmental crimes. As mentioned, the establishment and participation of these organizations in French law has long been considered and has been able to play a significant role in the implementation of envi-ronmental regulations in this country (Khan et al., 2020).

Iranian Legal System

In Iranian law, no law explicitly recognizes the right of non-governmental organ-izations to access the judiciary and restricts such a right to government officials ac-cording to the subject matter of the lawsuit. This legal vacuum creates many prac-tical problems for environmental organizations seeking to respond to environmental damage. However, in the case-law of Iran, the right of access to judicial authorities can be inferred based on some scattered regulations and their extensive presentation and interpretation.

Constitutional Law

Article 34 of the Constitution states: Litigation is the inalienable right of every person, and everyone can refer to the competent courts for litigation. All members of the nation have the right to access such courts, and no one can. The principle of this constitutional principle must be found in Article 8 of the Universal Declaration of Rights (Rabani et al., 2020).

Man searched. Paying attention to terms such as "every person," "everyone," "all members of the nation," and "no one" in the constitution suggests that there is a legal basis for the presence of non-governmental organizations in the proceedings; At the same time, this has been identified in other principles of this law and under different titles. The question is whether the non-specification of the right of non-governmental organizations to sue can prevent the presence of these organizations?

Given the current facts and the nature of environmental crimes, depriving these organizations of access to judicial authority does not seem logical. In the international arena, these organizations play a role in various ways; One of the most obvious exam-ples of this participation is their presence in the presence of some international judicial authorities, who are referred to as "friends of the court." This institution is inspired by the rights of the Anglo-Saxon countries and has gradually found its way into interna-tional law.

Does the fundamental question remain whether these organizations can be con-sidered genuine in environmental litigation and be directly involved in litigation? Examining some general principles of the constitution, such as this law's fortieth and forty-fifth principles, can be helpful as a response.

The latter principle, while emphasizing "public interests" and "public interests," enumerates some environmental elements and refers to them as "Anfal and public wealth," which is at the disposal of the Islamic government to consume following pub-lic interests.

Also, according to Article 50 of the Constitution: "In the Islamic Republic, the protection of the environment, in which today's generation and future generations should have a growing social life, is considered a public duty ...". Can file a lawsuit to protect the environment be considered a "public duty"? A positive answer to this question is reinforced by invoking the eighth and ninetieth principles of this law. Ac-cording to the eighth principle, "In the Islamic Republic of Iran, calling for good, en-joining the good and forbidding the evil is a public and reciprocal duty of the people towards each other, the government towards the people and the people towards the government." "Its conditions, limits, and quality are determined by law ...". According to the ninetieth principle, "Anyone who has a complaint about the work of the parlia-ment or the executive or the judiciary can submit his complaint in writing to the Is-lamic Consultative Assembly ..." The term "anyone" mentioned in this principle can apply to all persons. Include real and legal. Therefore, it seems that non-governmental organizations can not be deprived of the right of access to judicial authorities to pro-tect the environment (Moghaddam et al., 2008).

By-Laws on the Establishment and Activities of Non-Governmental Organizations

As mentioned, the right of NGOs to sue for environmental crimes under certain constitutional principles; In addition, among other governing regulations, we can mention the executive regulations for the establishment and activity of non-governmental organizations approved by the Cabinet on 20 June 2005, which di-rectly addresses this issue.

This regulation allows non-governmental organizations to act following the sub-ject of their activity, comply with this regulation and other related laws and regula-tions, and exercise the right to sue in judicial and quasi-judicial authorities. According to Article 16 of this regulation, "the organization [non-governmental organizations] have the right to file lawsuits against natural and legal persons in the courts in the field of their activities and to protect the public interest."

This is the only article that allows non-governmental organizations to sue in the name of the public interest. Therefore, the right to sue is generally accepted for these organizations; These organizations can file lawsuits in line with their activities' subject matter and protect the public interest. The reference to "public interest" in this article is in line with Article 50 of the Constitution; Because environmental protection is con-sidered a "public duty" in this principle. Accordingly, litigation by environmental NGOs to protect the environment is a "public duty" performed by these organizations during the proceedings.

Despite what has been said, the legal framework for such a right is a matter for re-flection; Because, formally, by-laws are rules, not laws; Therefore, there is no guarantee of legal implementation, and it is based on the fact that according to Article One Hun-dred and Seventy, "judges of courts are obliged to refrain from enforcing government decrees and regulations that are contrary to Islamic laws and regulations or outside the powers of the executive branch." Anyone can request the annulment of such regulations from the Court of Administrative Justice". The provision of the right to sue for non-governmental organizations in the said regulation alone is not sufficient; To fulfill Article 50 of the Constitution, the legislature must prepare the ground for the active participation of these organizations in the proceedings.

In response to this legal vacuum, the judiciary in 2008, following the submission of some demands by non-governmental organizations on the one hand and the confron-tation of judges with numerous lawsuits by non-governmental environmental net-works, on the other hand, seeks to identify the position for this. According to Article 66 of the new draft Code of Criminal Procedure: "It is a matter of protecting citizens' rights. They can file charges for crimes committed in the above areas and protest at all stages of the proceedings to provide evidence of participation and against the rulings of the judicial authorities (Pourbafrani & Hemati, 2016)."

To strengthen the role of non-governmental organizations in the criminal process and to intend to support the victims of specific crimes, this article confirms their effec-tive and useful presence in the criminal proceedings. It should be noted that the way the article is written can involve two conflicting interpretations; The first interpreta-tion is that the article only grants the right to "plead guilty" to non-governmental or-ganizations active in the areas in question and refuses to grant them the position of the litigant. The second interpretation, which is more compatible with the objectives of civil rights in general and environmental law in particular, is that according to the last part of the article, which refers to the organizations mentioned above, "in all stages of the proceedings Judiciary "grants the right to object, the legislature has granted posi-tions to non-governmental organizations in lawsuits subject to the article. In other words, it does not make much sense for the legislature to grant these organizations the right to object or appeal the rulings on those disputes without granting them the right to sue.

In addition, granting the right to participate in all stages of litigation and the right to object to the opinions of judicial authorities is a significant consideration of the ex-istence of "legal interest" for these organizations as the beneficiary of these lawsuits. Furthermore, this right is enshrined in Chapter 3 of the bill, entitled "Duties and Powers of the Prosecutor," which implies granting a quasi-judicial function to non-governmental organizations; This is an unparalleled innovation in Iranian law (Tahbaz, 2016).

CONCLUSION

In extrajudicial trials, granting legal status to non-governmental organizations for "declaring a crime," filing a lawsuit, presenting evidence and proof of a crime, participating in all stages of criminal proceedings, and finally "challenging court rulings" are major developments. Which, on the one hand, eliminates one of the gaps in the Iranian legal system and, on the other hand, ensures the effective participation of these organizations in the implementation of environmental norms; In addition, the realization of the effective participation of such institutions in the criminal process will lead to the expansion of environmental protection authorities from governmental to non-governmental authorities.

In a general conclusion, the main obstacles facing these organizations in preventing and suppressing environmental crimes caused by the following factors; Some of these factors are related to these organizations and others are beyond their authority: A) Excessive diversity of non-governmental organizations in the field of environment and lack of coordination in determining the goals and executive mechanisms between them; B) weak relations of these organizations with the government and government institutions; C) the dispersion of environmental duties and competencies between government organizations and insti tutions and the extensive ownership and domination of the government and government institutions over the environment; D) Lack of comprehensive and independent legislation regarding the establishment and activities of non-governmental organizations; E) Lack of effective legal mechanisms for the active participation of these organizations in the criminal process; F) The emptiness of the Ministry of Environment and the lack of accountability of the head of the Environmental Protection Organization before the Islamic Consultative Assembly.

However, what can be most effective in preventing and suppressing environmental crimes is creating a favorable environment for public participation in environmental protection in the form of non-governmental organizations; Achieving this partnership requires free access to environmental information, public participation in the decision-making process; decision-making, litigation, and environmental justice.

REFERENCES

Asadi, R. (2021). Analysis of Challenges of Environmental Non-Governmental Organizations in Iran with Riggs Prismatic Society Theory. *Research Political Geography Quarterly, 6*(2).

Bromideh, A. A. (2011). The widespread challenges of NGOs in developing countries: Case studies from Iran. *International NGO Journal*, *6*(9), 197–202.

Doh, J. P., & Guay, T. R. (2004). Globalization and corporate social responsibility: How non-governmental organizations influence labor and environmental codes of conduct. In *Management and international review* (pp. 7–29). Gabler Verlag. doi:10.1007/978-3-322-90997-8_2

Edele, A. (2005). *Non-governmental organizations in China*. Geneva, Switzerland: The Programme on NGOs and Civil Society, Centre for Applied Studies in International Negotiation, CASIN.

Esty, D. C. (1998). Non-governmental organizations at the World Trade Organization: Cooperation, competition, or exclusion. *J. Int'l Econ. L.*, *1*(1), 123–148. doi:10.1093/jiel/1.1.123

Hashemi, F., Sadighi, H., Chizari, M., & Abbasi, E. (2017). Influencing Factors on Emerging Capabilities of Environmental Non-Governmental Organizations (ENGOs): Using Grounded Theory. *Journal of Applied Environmental Biological Sciences*, *7*(3), 173–184.

Hashemi, F., Sadighi, H., Chizari, M., & Abbasi, E. (2019). The relationship between ENGOs and Government in Iran. *Heliyon*, *5*(12), e02844. doi:10.1016/j.heliyon.2019.e02844 PMID:31890931

Hashemi, F., Sadighi, H., Chizari, M., & Abbasi, E. (2019). Influencing Factors on Emerging Capabilities of Environmental Non-Governmental Organizations (ENGOs): Using Grounded Theory. *OIDA International Journal of Sustainable Development*, *12*(03), 39–54.

Hendry, J. R. (2006). Taking aim at business: What factors lead environmental non-governmental organizations to target particular firms? *Business & Society*, *45*(1), 47–86. doi:10.1177/0007650305281849

Khan, M., Chaudhry, M. N., Ahmad, S. R., & Saif, S. (2020). The role of and challenges facing non-governmental organizations in the environmental impact assessment process in Punjab, Pakistan. *Impact Assessment and Project Appraisal*, *38*(1), 57–70. doi:10.1080/14615517.2019.1684096

Li, G., He, Q., Shao, S., & Cao, J. (2018). Environmental non-governmental organizations and urban environmental governance: Evidence from China. *Journal of Environmental Management*, *206*, 1296–1307. doi:10.1016/j.jenvman.2017.09.076 PMID:28993017

Lindblom, A. K. (2005). *Non-governmental organisations in international law*. Cambridge University Press.

Moghaddam, M. R. A., Maknoun, R., & Tahershamsi, A. (2008). Environmental engineering education in Iran: Needs, problems and solutions. *Environmental Engineering and Management Journal*, *7*(6), 775–779. doi:10.30638/eemj.2008.103

Moghimi, S. M. (2007). The relationship between environmental factors and organizational entrepreneurship in non-governmental organizations (NGOs) in Iran. *Iranian Journal of Management Studies*, *1*(1), 39–55.

Nomura, K. (2007). Democratisation and environmental non-governmental organisations in Indonesia. *Journal of Contemporary Asia*, *37*(4), 495–517. doi:10.1080/00472330701546566

Ormazdi, M., Pourfikouhi, A., & Amar, T. (2013). The necessities of verifying the policies of Non Governmental Organizations development in planning and management of rural tourism of Iran. *Life Science Journal*, *10*(3s).

Pourbafrani, H., & Hemati, M. (2016). *A Critique on Iranian Criminal Policy towards Environmental Crimes*. Academic Press.

Pourhoseyni, B., Masoud, G., & Shekarchizadeh, M. (2019). The Role of Non-Governmental Organizations in Iran's Criminal Policy towards Environmental Protection. *Iranian Journal of Medical Law, 13*, 337–351.

Rabani, H., Jalalian, A., & Pournouri, M. (2020). Typology of Environmental Crimes in Iran (Case Study: Crimes Related to Environmental Pollution). *Anthropogenic Pollution, 4*(2), 78–83.

Shohani, A., Ataei, E., & Norouzi, N. (2021). Prevention and Suppression of Environmental Crimes in the Light of the Actions of Non-Governmental Organizations in the Iranian Legal System. *Research Journal of Ecology and Environmental Sciences, 1*(1), 57–70.

Spiro, P. J. (2007). Non-governmental organizations and civil society. The Oxford Handbook of International Environmental Law, 770-790.

Tahbaz, M. (2016). Environmental challenges in today's Iran. *Iranian Studies, 49*(6), 943–961. doi:10.1080/00210862.2016.1241624

Tarlock, A. D. (1992). The role of non-governmental organizations in the development of international environmental law. *Chi.- Kent L. Rev., 68*, 61.

Wu, J., Chang, I. S., Yilihamu, Q., & Zhou, Y. (2017). Study on the practice of public participation in environmental impact assessment by environmental non-governmental organizations in China. *Renewable & Sustainable Energy Reviews, 74*, 186–200. doi:10.1016/j.rser.2017.01.178

ADDITIONAL READING

Moghaddam, M. R. A., Maknoun, R., & Tahershamsi, A. (2008). Environmental engineering education in Iran: Needs, problems and solutions. *Environmental Engineering and Management Journal, 7*(6), 775–779. doi:10.30638/eemj.2008.103

Nomura, K. (2007). Democratisation and environmental non-governmental organisations in Indonesia. *Journal of Contemporary Asia, 37*(4), 495–517. doi:10.1080/00472330701546566

Ormazdi, M., Pourfikouhi, A., & Amar, T. (2013). The necessities of verifying the policies of Non Governmental Organizations development in planning and management of rural tourism of Iran. *Life Science Journal, 10*(3s).

Pourhoseyni, B., Masoud, G., & Shekarchizadeh, M. (2019). The Role of Non-Governmental Organizations in Iran's Criminal Policy towards Environmental Protection. *Iranian Journal of Medical Law, 13*, 337–351.

Rabani, H., Jalalian, A., & Pournouri, M. (2020). Typology of Environmental Crimes in Iran (Case Study: Crimes Related to Environmental Pollution). *Anthropogenic Pollution, 4*(2), 78–83.

Shohani, A., Ataei, E., & Norouzi, N. (2021). Prevention and Suppression of Environmental Crimes in the Light of the Actions of Non-Governmental Organizations in the Iranian Legal System. *Research Journal of Ecology and Environmental Sciences, 1*(1), 57–70.

Tahbaz, M. (2016). Environmental challenges in today's Iran. *Iranian Studies*, *49*(6), 943–961. doi:10.1080/00210862.2016.1241624

Tarlock, A. D. (1992). The role of non-governmental organizations in the development of international environmental law. Chi.-. *Kent L. Rev.*, *68*, 61.

Wu, J., Chang, I. S., Yilihamu, Q., & Zhou, Y. (2017). Study on the practice of public participation in environmental impact assessment by environmental non-governmental organizations in China. *Renewable & Sustainable Energy Reviews*, *74*, 186–200. doi:10.1016/j.rser.2017.01.178

KEY TERMS AND DEFINITIONS

Environmental Law: Environmental law is a collective term encompassing aspects of the law that protect the environment. A related but distinct set of regulatory regimes, now strongly influenced by environmental legal principles, focuses on managing specific natural resources, such as forests, minerals, or fisheries. Other areas, such as environmental impact assessment, may not fit neatly into either category but are nonetheless important components of environmental law. Previous research found that when environmental law reflects moral values for betterment, legal adoption is more likely to be successful, usually in well-developed regions. In less-developed states, changes in moral values are necessary for successful legal implementation when environmental law differs from moral values.

Equity: Defined by UNEP to include intergenerational equity—"the right of future generations to enjoy a fair level of the common patrimony"—and intragenerational equity—"the right of all people within the current generation to fair access to the current generation's entitlement to the Earth's natural resources"—environmental equity considers the present generation under an obligation to account for long-term impacts of activities and to act to sustain the global environment and resource base for future generations. Pollution control and resource management laws may be assessed against this principle.

Polluter Pays Principle: The polluter pays principle stands for the idea that "the environmental costs of economic activities, including the cost of preventing potential harm, should be internalized rather than imposed upon society at large." All issues related to responsibility for cost for environmental remediation and compliance with pollution control regulations involve this principle.

Precautionary Principle: One of the most commonly encountered and controversial principles of environmental law, the Rio Declaration formulated the precautionary principle as follows, to protect the environment, States shall widely apply the precautionary approach according to their capabilities. Where there are threats of serious or irreversible damage, lack of full scientific certainty shall not be used as a reason for postponing cost-effective measures to prevent environmental degradation. The principle may play a role in any debate over the need for environmental regulation.

Prevention: The concept of prevention etc. can perhaps better be considered an overarching aim that gives rise to a multitude of legal mechanisms, including prior assessment of environmental harm, licensing, or authorization that set out the conditions for operation and the consequences for violation of the conditions, as well as the adoption of strategies and policies. Emission limits and other product or process standards, the use of best available techniques, and similar techniques can all be seen as applications of the concept of prevention.

Public Participation and Transparency: Identified as essential conditions for "accountable governments...industrial concerns," and organizations generally, public participation and transparency are presented by UNEP as requiring "effective protection of the human right to hold and express opinions and to seek, receive and impart ideas, etc. a right of access to appropriate, comprehensible and timely information held by governments and industrial concerns on economic and social policies regarding the sustainable use of natural resources and the protection of the environment, without imposing undue financial burdens upon the applicants and with adequate protection of privacy and business confidentiality," and "effective judicial and administrative proceedings." These principles are present in environmental impact assessment, laws requiring publication and access to relevant environmental data, and administrative procedures.

Sustainable Development: Defined by the United Nations Environment Programme as "development that meets the needs of the present without compromising the ability of future generations to meet their own needs," sustainable development may be considered together with the concepts of "integration" (development cannot be considered in isolation from sustainability) and "interdependence" (social and economic development, and environmental protection, are interdependent). Laws mandating environmental impact assessment and requiring or encouraging development to minimize environmental impacts may be assessed against this principle. The modern concept of sustainable development was discussed at the 1972 United Nations Conference on the Human Environment (Stockholm Conference) and the driving force behind the 1983 World Commission on Environment and Development (WCED, or Bruntland Commission). In 1992, the first UN Earth Summit resulted in the Rio Declaration, Principle 3 of which reads: "The right to development must be fulfilled to equitably meet developmental and environmental needs of present and future generations." Sustainable development has been a core concept of international environmental discussion ever since, including at the World Summit on Sustainable Development (Earth Summit 2002) and the United Nations Conference on Sustainable Development (Earth Summit 2012, or Rio+20).

Transboundary Responsibility: Defined in the international law context as an obligation to protect one's environment. UNEP considers transboundary responsibility at the international level to prevent damage to neighboring environments at the international level as a potential limitation on the rights of the sovereign state. Laws that limit externalities imposed upon human health and the environment may be assessed against this principle.

Chapter 13
Environmental Law and Green Constitutions

ABSTRACT

The general theory of environmental orthodoxy is based on the idea that by introducing the issue of the environment and the need to protect and protect it in the constitution as a reference norm, we will see the establishment of the legal order in this field. In this chapter, an attempt has been made to analyze the content of the constitutions of Latin America in the context of environmental issues and the conservation of natural resources with a comparative approach and as an example. The basic hypothesis of this chapter is based on the fact that today, paying attention to the issue of the environment at the level of constitutions means following sub-norms and the legal system from the norm of reference and the effects that this right can have in the legal system. This is well reflected in the constitutions of Latin America.

INTRODUCTION

Orientalism and fundamentalism in the field of environment is a relatively new topic in the field. Studies in public law (Deville, 2014; Randrianandrasana, 2016). For example, in France after 2005 and 2008, with the adoption of the Basic Charter on the environment and constitutional amendments, it has shown itself at the level of reference norms (Mashhadi, 2013; Georgian, 2009). Although constitutional law has traditionally been the science of examining political institutions and initial perceptions of this discipline are also based on the study of the state-country structure and organizational relations between institutions. (Laferriére, 1974) Now constitutional rights are not merely the formulation of political institutions or political power. The most important and prominent content and function of constitutional rights, along with the framework of political institutions and phenomena, is identifying and guaranteeing citizens' rights. In other words, fundamental rights are a branch of domestic public law that deals with the political relations between rulers and rulers (Foster & Grundmann, 2001). For this reason, in the institutional approach, the links between this field and environmental rights can not be relied upon. In other words, whenever we define fundamental rights in terms of the study of the regulation of powers and the guarantee of rights; Guaranteeing the right to a healthy environment is included in the second

DOI: 10.4018/978-1-6684-7188-3.ch013

part of this definition. With this description, after the formation of the environmental crisis, especially after the seventies, many world constitutions have addressed this issue. In this article, while expressing the general theory of environmental protection of the environment, we have tried to introduce and analyze this issue in the constitutions of Latin American countries.

Environmental orthodoxy has been thematically introduced into the constitutions with some delay (formal fundamentalism) and has played its role as the reference environmental norm. (Fundamentalism of Mahavi) (Taghizadeh, 2004, Mashhadi, 2014; Ghazi Shariat Panahi, 1995). There is no doubt that the content of the issues raised in the constitutions is a function of the fundamental issues of the political society, which can be placed in the framework of the general theory of the observance of the issues of the constitution from the fundamental issues of the political society.

In the early stages of nation-state formation, the issue of the environment was not given much attention. It is not surprising that most of the first constitutions, such as the 1958 Constitution and previous French laws, and the Iranian constitution, do not explicitly mention the need to protect the environment.

Attention to the environment in the constitutions results from the evolution of the environmental problem from a simple and unimportant issue to a macro, important, and sovereign matter. We are concerned that a modern constitution has dealt with it or will deal with it in the future. Thus, the historical and generational evolution of constitutional issues was considered inevitable.

After the emergence of environmental issues and the subsequent introduction of the right to the environment as a human right and even the duty of the government and political society to protect it, a new generation of basic laws emerged, most of which dealt with environmental issues. Some countries, including Iran, tried to explicitly and extensively mention the importance of this issue (Ramazani Ghavamabadi: 2009). Others, in general, and indirectly, paid attention to the environment at the level of the reference norm. In Iranian law, the process of environmental constitutionalism can be found clearly in the two principles 50 and 45 of the constitution (Mashhadi, 2013). Article 50, as the most specific principle, was approved by a consensus of constitutional experts without any objection. The last principle approved in the 54th session of the final review of the constitution was Article 50 of the constitution. This principle was approved without any dissenting or abstaining vote and with the affirmative vote of all members present at the meeting (64 people) (Detailed Negotiation Form, 1982). According to this principle, "in the Islamic Republic, the protection of the environment, in which the present and future generations should have a growing social life, is considered a public duty. Therefore, economic activities other than those caused by environmental pollution or irreparable damage "It is forbidden to find it."

Those constitutions that were passed before the 1970s (traditional constitutions) did not reference the protection of the environment, and this naturally did not mean that the drafters of this part of the constitution were neglected. Because in the same period, we can refer to some constitutions that have dealt with natural resources. In some opinions and texts, efforts have been made to use some expressions indirectly to protect the environment. But it means neglecting the protectionist approach in expressing the principles of the constitution (PINI, 1997). For example, in French law before environmental constitutionalism, we see this crystallization at the level of legislative norms (Inserguet-Brisset, 2005).

The second is those constitutions that were passed from the 1970s to the 1990s (Green Constitutions). In terms of the date of enactment of constitutions in the second half of the twentieth century due to the emergence of new governments, this category indicated the birth and multiplicity of new constitutions. The sixties and seventies of the twentieth century, in terms of the emergence of the issue of the environment and its extension to the political level and public opinion, also coincided with this historical multiplicity of constitutions. Thus, it can be seen that those constitutions that have been adopted at this historical moment; The issue of the environment, have been considered and addressed (Laferrière, 1947).

In the third stage, we can refer to those basic constitutions adopted since the nineties or have been gradually amended; In addition to the environmental issue, they have also addressed sustainable development (Cans, 2005; Morand-Deviller, 2014). Attention to sustainable environmental development was also a function of developments in international environmental law and the holding of the Rio de Janeiro Conference and had a tangible effect on the constitutions. Most of the constitutions of Latin America are among the laws that have been enacted since the second half of the twentieth century. Therefore, attention to the environment is strong in the constitutions of these countries. In the following, an attempt has been made to study and analyze the position of this issue in the constitutions of Latin America, for example, with a comparative approach.

CASE STUDY IN LATIN AMERICA

In Latin America, the environment issue has always been the focus of fundamental norms due to the ecological importance of biological resources. However, the constitutions of this field do not pay attention to the uniformity and extent; some in this field directly and some indirectly, some countries have moved earlier, and some others have moved towards constitutionalism in the field of environment. Eco-Marxist approaches also influence several countries (Foster, 2001, Mashhadi, 2013). In the following, the content of environmental grammar is briefly reviewed in the constitutions of this region.

Chile: Right to the Environment and the Duties of the Government

The 1980 Constitution of Chile, with subsequent amendments to Article 19, paragraph 8, denies the right to live in a pollution-free environment. The law also introduces oversight of the protection of this right and protection of nature from the duties of the state. In addition, the constitution allows the government to enact laws to establish restrictions on rights and freedoms to protect the environment (Carruthers, 2001). In addition, the government will be allowed to adopt methods for acquiring property and using and exploiting them to protect the environmental heritage. Another important point emphasized in Article 20 of the Constitution is that living in a healthy environment violates the right to sue (de Gliniasty & Morand-Deviller, 2018).

Brazil, an Example of a Green Constitution

The Brazilian constitution addresses the issue of the environment in detail. Article 225 of the 1998 Brazilian Constitution defines the outline of the environment in several paragraphs (Drummond & Barros-Platiau, 2006). According to this article, "Everyone has the right to a balanced environment where the public interest belongs to the people and is necessary for a healthy life. The government and society are obliged to protect and protect the environment for present and future generations." In the continuation of the article and the first paragraph regarding the duties of the government, the following items have been established (Drummond & Barros-Platiau, 2006).

- Conservation and restoration of vital ecological resources and ecological management of species and ecosystems.
- Protecting the diversity and evolution of the country's genetic heritage and supervising institutions actively researching and reproducing genetic material.
- Determining all parts of the country and those places of territorial territory and cases that must be protected specially and any change or encroachment will be allowed only by law. Prohibiting any exploitation that endangers their integrity and privacy must be justified.
- Prioritize environmental impact assessments recognized by public authorities for facilities and activities that may lead to major environmental degradation.
- Production control, economic methods, methods, and materials harmful to life, quality of life, and the environment.
- Promoting environmental education at all levels of education and public awareness of the need to protect the environment.
- The protection of plants and animals and the prohibition of all measures endanger their biological functions and lead to species extinction or animal cruelty.

Another very important issue addressed in the Brazilian constitution; The existence of the Amazon rainforest as the lifeblood of the planet and the need to protect it. The constitution states: "The Amazon rainforests of Brazil, the Atlantic forests, etc. shall be part of the national wealth and shall be exploited following the law and under conditions which guarantee the protection of the environment." Fourth Article 225). is also stipulated that "actions or activities that cause damage to the environment by individuals or legal entities will be subject to punishment through administrative and criminal guarantees" (paragraph 3). In addition, the constitution provides for the civil liability of this group of persons, who are committed to "compensating for the damage to the environment." In addition, the constitution provides for the reconstruction of any damage to the environment regarding the exploitation of mineral resources (the second paragraph of Article 225). Regarding the deployment of nuclear reactors, the last paragraph stipulates that "the location of facilities related to nuclear reactors must be determined following federal law. Otherwise, they should not be implemented" (Suryawan & Ismail, 2020).

Venezuela and the Importance of Natural Resources

Venezuela's environment is famous for its Amazon rainforests, deserts, as well as the world's tallest waterfall, South America's largest lake, and abundant oil, as well as climatic conditions. Venezuela's 1961 constitution emphasizes the need to protect the country's natural resources. Article 106 (Chapter 5, Title 5) of the 1961 Venezuelan Constitution provides for further amendments to the protection of the environment: "The State shall pay attention to the protection and preservation of natural resources in the territory of the territory (Aspan & Yunus, 2019)."

Uruguay and the Common Good of the Nation

Uruguay's 1966 amended constitution recognizes the protection of the environment as a common good. This statement of the Uruguayan constitution expresses the modern content of the right to the environment in this country, which declares the nature of this issue as a common and public interest. Interest cannot be analyzed in the form of the "mutual benefit" of comprehensive members. The constitution further stipulates: "Individuals must refrain from any action that would lead to the destruction, destruction or serious pollution of the environment (Yusa & Hermanto, 2018)."

Suriname, Environmental Protection is the Social Goal of the Government

The 1987 Constitution of Suriname mentions protecting and preserving the environment as one of the social objectives in Article 6, Section 3, Chapter 3. According to this article: "Creating and improving the necessary conditions for the protection of nature and the protection of ecological balance is considered one of the government's social goals (El-Nakhel et al., 2020)."

Peru and the Government's Commitment to Sustainable Operation

The 1993 Peruvian Constitution allows the state to "determine national environmental policy." The law also obliges the government to promote "sustainable exploitation of natural resources" (Article 67). In addition, Article 68 obliges the government to "adequately regulate the protection of biodiversity and natural protected areas and the sustainable development of the Amazon region" (Uribe & Urdinola-Rengifo, 2020).

Paraguay and the Reconciliation of Human Development and Conservation

According to Article 7 of the Paraguayan Constitution, "Everyone has the right to live in a healthy ecologically balanced environment." Therefore, the priority of social interests is "the preservation, restoration and improvement of the environment in such a way as to confuse these objectives with pervasive human development." The constitution imposes restrictions and prohibitions on harmful activities to the environment and legalizes activities that lead to the destruction of the environment and the determination and enforcement of enforcement guarantees for environmental offenses (8). The constitution specifically prohibits the production of toxic waste in the country and, in addition, stipulates that "any damage to the environment shall be subject to an obligation to rehabilitate and pay for the damage" (Valdez-Carrillo et al., 2020).

Panama and the Government's Environmental Commitments

Article 14 of the 1972 Constitution of Panama states: "The government has a fundamental obligation to ensure a healthy, vibrant environment free of air, water, and foodstuffs appropriate to the development of human life." The constitution also stipulates the obligations of the government and of all the inhabitants of the Panamanian territory: "Promoting economic and social development in such a way as to prevent environmental pollution, to preserve the ecological balance and to prevent the destruction of the ecosystem" (115). In addition, Article 116 stipulates that: "The Government shall take the necessary measures to regulate, monitor and implement appropriately to ensure the reasonable use of land, river and marine life resources, forests, and waters, and to prevent improper use." "It ensures their continued protection and recovery." Article 117 also applies to non-renewable natural resources. This article obliges the government to regulate "benefits derived from non-renewable natural resources to prevent the loss of economic, social, and environmental benefits."

Nicaragua; Environmental Rights and Responsibilities of Government and Citizens

Article 60 of the 1986 amended constitution states: "Every Nicaraguan has the right to live in a healthy environment". The constitution also states that "protecting, protecting, and restoring the environment and natural resources is the state's responsibility."

Argentina; Sustainable Development and Its Examples

Article 41 of the 1994 Constitution of Argentina provides that "all inhabitants of the right to a healthy and balanced environment shall be adequately proportioned with human development and beneficial activities which shall satisfy the needs of the present generation without prejudice to the needs of future generations." "They enjoy this right for the rational use of natural resources, for the protection of cultural and natural heritage and biodiversity, and the provision of environmental information and education."

The constitution also makes it the duty of citizens to protect the environment. The constitution states: "As a priority, damage to the environment will be accompanied by an obligation to compensate." Paragraph 4 of Article 41 also prohibits the import of potentially hazardous wastes and radioactive wastes.

Bolivia and Sustainable Exploitation of Natural Resources

The 1967 Bolivian Constitution, with subsequent amendments to Articles 170 and 173, provides for the exploitation of natural resources, which also relates to the field of the environment.

According to Article 170, "the government has to regulate the system of exploitation of renewable natural resources and the regulations related to their protection and expansion." In addition, Article 137 stipulates that "all inhabitants of the national territory are obliged to respect and protect the national heritage."

Colombia and Participation in Environmental Decision Making

Article 79 of the 1991 Constitution of Colombia recognizes the right to the environment in an individual way. According to this article: "Every person has the right to enjoy a healthy environment." Another important issue that is less recognized in the constitution is the issue of participation in environmental decisions. Article 79 of the Constitution stipulates that "the law guarantees participation in decisions that affect the environment." In addition, the constitution provides for the following: From natural resources to ensure sustainable development, protection, restoration, reconstruction, etc. In addition to controlling and being aware of the factors of environmental degradation, imposing legal enforcement guarantees and Outside the territorial territory ... "(Article 80). In addition, while enshrining the government's responsibilities in this regard under Article 95, "the protection of natural and cultural resources and the preservation of a healthy environment is the duty of every Colombian" (Article 95).

Costa Rica and the Right to Environmental Litigation

The Constitution of Costa Rica is "right-oriented" in recognizing the right to a healthy environment, rather than a "founding duty," and another issue evident in this law is the consideration of safeguards and environmental justice. Article 50 of the 1949 Amended Constitution states that "Everyone has the right to a healthy and balanced environment. The government will protect, safeguard and guarantee this right. The government shall also lay down rules and regulations governing responsibilities and safeguards."

Finally, the constitution stipulates that in the case of a claim for damages, "everyone has the right to be blamed for acts that threaten this right (the right to the environment) and also to have the right to sue for damages" (Article 50).

Cuba and Citizen Participation

Article 27 of the 1992 amended Constitution of Cuba deals with the issue of the environment. According to this article, "the government protects the environment and natural resources of the country. There is a close relationship between sustainable economy and social development to ensure the continuity, welfare, and security of the present generation and future generations." Furthermore, like the Iranian constitution, the protection of the environment has been mentioned as a public duty that includes either the government or the citizen (Kial, 2004). "... every citizen must participate in the protection of the climate (atmosphere) and the protection of soil, plants and animals, and the rich potential of nature" (Fernandes, 1992).

Ecuador and the Principle of Precautionary Action

The country's 2008 constitution, Articles 14 and 15, deals with environmental health protection. According to Article 14, the people of Ecuador have the right to live in a healthy, balanced environment that guarantees stability and a good life. In the continuation of this article, some measures in this direction have been identified as the public interest of the people of Ecuador. These include protecting the environment, protecting ecosystems, biodiversity and preventing damage to the environment.

Article 15 also deals with the duties of the government in this regard. The government is committed to promoting and encouraging the public and private sectors to encourage clean technologies. It has also stated that some examples of destructive actions of the environment, including hazardous chemicals and nuclear waste, have been banned from their production, transfer, and import. It should be noted that Article 86 of the 1998 Constitution of Ecuador already recognizes the right to a healthy environment and stipulates that "everyone has the right to live in a healthy and balanced environment and to ensure sustainable development."

Article 90 of the former constitution also expresses precautionary measures and implicitly deals with the principle of precautionary action. According to this article, "in case of doubt about the negative environmental effects and consequences of any act or omission, the government will take precautionary measures, even if there is not enough scientific evidence to cause damage."

El Salvador, Children, and the Environment

Article 34 of the 1983 amended Constitution on the Rights of the Child provides that "every child has the right to a family life and environmental conditions that allow for the full enjoyment of the rights of all persons." Article 69 of the Constitution, regarding the state's duty to protect the environment, enumerates "control over the quality of food products and environmental conditions that affect health and well-being" (Article 69).

Guatemala and the Commitment to Maintaining the Ecological Balance

Guatemala's 1985 amended constitution proclaims the right to health as a fundamental human right that is enforceable without discrimination. Article 97 also stipulates the government's obligations that "the government and its pillars and all the inhabitants of the territory ... shall promote technological, economic and social development, and prevent environmental pollution, and shall maintain its ecological balance." "The government will also adopt the necessary regulations to ensure the exploitation of plants and animals, soil and water in a reasonable manner and to eliminate their depletion."

Guyana and Sustainable Use of Natural Resources

Article 36 of the 1980 Constitution provides by stating the rights of future generations concerning natural resources: "In the interests of present and future generations, the government will support the rational exploitation of soil, mineral and water resources, plants and animals, and will take the necessary measures to ensure, protect and improve the environment." The constitution also makes it the duty of every citizen to participate in activities that lead to the improvement of the environment (Article 25).

Honduras: The Relationship Between the Right to Health and the Environment

Article 145 of the 1982 Constitution of Honduras addresses the issue of the environment. According to this article, the right of every citizen to protect the health of everyone is recognized, and the government is obliged to maintain environmental and satisfactory environmental conditions to protect the health of each individual (López-Feldman et al., 2020).

CONCLUSION

For many years after the orthodox movement, the issue of the environment was not much of a concern to governments. In the framework of the theory of constitutional issues following generational issues, the content of the basic laws is subject to the fundamental issues of political society. Therefore, after the revelation of the environmental crisis in the countries, especially after the '70s, the constitution gradually considered the environment as a fundamental issue. The process can be called environmental grammar. The countries' constitutions have been amended and changed in this regard, and after entering into the norms of reference, they gradually found a special place in the eyes of the judicial authorities and the oversight and guardian institutions of the constitution. This can be described as a transition from formal heterogeneity to substantive heterosexuality. In this article, we tried to analyze some aspects of environmental protection with special reference to the constitutions of Latin American countries. As can be seen, a wide range of environmental issues are addressed in the constitutions of this field; But the common denominator of all of them is the need to protect the order of the environment. This can play a vital role in preserving, improving, and promoting the right to a healthy environment for citizens if this right is truly established.

REFERENCES

Aspan, Z., & Yunus, A. (2019). The right to a good and healthy environment: Revitalizing green constitution. *IOP Conference Series. Earth and Environmental Science*, *343*(1), 012067. doi:10.1088/1755-1315/343/1/012067

Cans, C. (2005). La charte constitutionnelle de l'environnement: Évolution ou révolution du droit français de l'environnement? *Droit de l'environnement*, (131), 194–203.

Carruthers, D. (2001). Environmental politics in Chile: Legacies of dictatorship and democracy. *Third World Quarterly*, *22*(3), 343–358. doi:10.1080/01436590120061642

de Gliniasty, J., & Morand-Deviller, J. (2018). *Les théories jurisprudentielles en droit administratif.* LGDJ, une marque de Lextenso.

Detailed minutes of the deliberations of the Majlis Final Review of the Constitution. (1982). Publications of the General Directorate of Laws of the Islamic Consultative Assembly.

Drummond, J., & Barros-Platiau, A. F. (2006). Brazilian environmental laws and policies, 1934–2002: A critical overview. *Law & Policy*, *28*(1), 83–108. doi:10.1111/j.1467-9930.2005.00218.x

El-Nakhel, C., Pannico, A., Graziani, G., Kyriacou, M. C., Giordano, M., Ritieni, A., De Pascale, S., & Rouphael, Y. (2020). Variation in macronutrient content, phytochemical constitution and in vitro antioxidant capacity of green and red butterhead lettuce dictated by different developmental stages of harvest maturity. *Antioxidants*, *9*(4), 300. doi:10.3390/antiox9040300 PMID:32260224

Fernandes, E. (1992). Law, politics and environmental protection in Brazil. *Journal of Environmental Law*, *4*(1), 41–56. doi:10.1093/jel/4.1.41

Foster, J. B., & Grundmann, R. (2001). Marx's ecology. Materialism & nature. *Canadian Journal of Sociology*, *26*(4), 670. doi:10.2307/3341498

Georgian, A.A. (2009). Duplication and normative diversity in French constitutional law (sources for monitoring parliamentary approvals). *Legal Research Quarterly, 12*(50).

Ghazi Shariat Panahi, S. A. (1995). *Essentials of Constitutional Law* (1st ed.). Yalda Publishing.

Inserguet-Brisset, V. (2005). *Droit de l'environnement*. Presses universitaires de Rennes.

Kial, M., & Kial, M. (2004), Environmental Law in the Perspective of the Constitution. In *Proceedings of the First Iranian Conference on Environmental Law.* Barg Zaytoun Publications.

Laferrière, J. (1947). *Manuel de droit constitutionnel*. Domat-Montchrestien.

López-Feldman, A., Chávez, C., Vélez, M. A., Bejarano, H., Chimeli, A. B., Féres, J., ... Viteri, C. (2020). Environmental impacts and policy responses to Covid-19: A view from Latin America. *Environmental and Resource Economics*, 1–6. PMID:32836838

Mashhadi, A. (2013). *The Right to a Healthy Environment* (1st ed.). Mizan Publications.

Mashhadi, A. (2014). Substantive Constitutionalization of Right to Environment in Iranian & French Law. *Comparative Law Review*, *5*(2), 559–580.

Morand-Deviller, J. (2014). L'environnement dans les constitutions étrangères. *Les nouveaux cahiers du Conseil Constitutionnel*, (2), 83-95.

Ramazani Ghavamabadi, M. H. (2009). L'application du droit à l'environnement en Iran. *Proceeding of International Conference on Human Rights and Environment.*

Randrianandrasana, I. (2016). La protection constitutionnelle de l'environnement à Madagascar. *Revue juridique de lenvironnement, 41*(1), 122-139.

Suryawan, I. G. B., & Ismail, A. (2020). Strengthening environmental law policy and its influence on environmental sustainability performance: Empirical studies of green constitution in adopting countries. *International Journal of Energy Economics and Policy*, *10*(2), 132–138. doi:10.32479/ijeep.8719

Taghizadeh, J. (2007). The Issue of Constitutionalization of the Legal Order. *Journal of Legal Research*, *6*(11), 129–162.

Uribe, N. R., & Urdinola-Rengifo, J. S. (2020). International Environmental Law in Latin America. In Routledge Handbook of International Environmental Law (pp. 263-278). Routledge. doi:10.4324/9781003137825-22

Valdez-Carrillo, M., Abrell, L., Ramírez-Hernández, J., Reyes-López, J. A., & Carreón-Diazconti, C. (2020). Pharmaceuticals as emerging contaminants in the aquatic environment of Latin America: A review. *Environmental Science and Pollution Research International*, *27*(36), 1–29. doi:10.100711356-020-10842-9 PMID:32986197

Yusa, I. G., & Hermanto, B. (2018). Implementasi Green Constitution di Indonesia: Jaminan Hak Konstitusional Pembangunan Lingkungan Hidup Berkelanjutan. *Jurnal Konstitusi*, *15*(2), 306–326. doi:10.31078/jk1524

ADDITIONAL READING

Aspan, Z., & Yunus, A. (2019). The right to a good and healthy environment: Revitalizing green constitution. *IOP Conference Series. Earth and Environmental Science*, *343*(1), 012067. doi:10.1088/1755-1315/343/1/012067

Carruthers, D. (2001). Environmental politics in Chile: Legacies of dictatorship and democracy. *Third World Quarterly*, *22*(3), 343–358. doi:10.1080/01436590120061642

Drummond, J., & Barros-Platiau, A. F. (2006). Brazilian environmental laws and policies, 1934–2002: A critical overview. *Law & Policy*, *28*(1), 83–108. doi:10.1111/j.1467-9930.2005.00218.x

El-Nakhel, C., Pannico, A., Graziani, G., Kyriacou, M. C., Giordano, M., Ritieni, A., De Pascale, S., & Rouphael, Y. (2020). Variation in macronutrient content, phytochemical constitution and in vitro antioxidant capacity of green and red butterhead lettuce dictated by different developmental stages of harvest maturity. *Antioxidants*, *9*(4), 300. doi:10.3390/antiox9040300 PMID:32260224

Fernandes, E. (1992). Law, politics and environmental protection in Brazil. *Journal of Environmental Law*, *4*(1), 41–56. doi:10.1093/jel/4.1.41

Foster, J. B., & Grundmann, R. (2001). Marx's ecology. Materialism & nature. *Canadian Journal of Sociology*, *26*(4), 670. doi:10.2307/3341498

Ghazi Shariat Panahi, S. A. (1995). *Essentials of Constitutional Law* (1st ed.). Yalda Publishing.

Laferrière, J. (1947). *Manuel de droit constitutionnel*. Domat-Montchrestien.

Mashhadi, A. (2013). *The Right to a Healthy Environment* (1st ed.). Mizan Publications.

Mashhadi, A. (2014). Substantive Constitutionalization of Right to Environment in Iranian & French Law. *Comparative Law Review*, *5*(2), 559–580.

Suryawan, I. G. B., & Ismail, A. (2020). Strengthening environmental law policy and its influence on environmental sustainability performance: Empirical studies of green constitution in adopting countries. *International Journal of Energy Economics and Policy*, *10*(2), 132–138. doi:10.32479/ijeep.8719

Taghizadeh, J. (2007). The Issue of Constitutionalization of the Legal Order. *Journal of Legal Research*, *6*(11), 129–162.

Uribe, N. R., & Urdinola-Rengifo, J. S. (2020). International Environmental Law in Latin America. In Routledge Handbook of International Environmental Law (pp. 263-278). Routledge. doi:10.4324/9781003137825-22

Valdez-Carrillo, M., Abrell, L., Ramírez-Hernández, J., Reyes-López, J. A., & Carreón-Diazconti, C. (2020). Pharmaceuticals as emerging contaminants in the aquatic environment of Latin America: A review. *Environmental Science and Pollution Research International*, *27*(36), 1–29. doi:10.100711356-020-10842-9 PMID:32986197

Yusa, I. G., & Hermanto, B. (2018). Implementasi Green Constitution di Indonesia: Jaminan Hak Konstitusional Pembangunan Lingkungan Hidup Berkelanjutan. *Jurnal Konstitusi*, *15*(2), 306–326. doi:10.31078/jk1524

KEY TERMS AND DEFINITIONS

Access Rights: Principle 10 of the Rio Declaration (1992), establishes that the best way to address environmental challenges is with the broad participation of the people involved. To this end, the Principle enshrined three fundamental rights as pillars of sound environmental governance: access to information, access to public participation and access to justice.

Carbon Neutrality: Existence of a balance between the amount of greenhouse gas (GHG) emissions (equivalent carbon dioxide) that is emitted and what is "captured or absorbed" from said emissions. The balance implies that there is "neutrality" of GHG, as its name indicates, but it does not mean that gases are not emitted into the atmosphere. According to Javiera Valencia, geographer of the UACh Austral Patagonia Program, the key point in carbon neutrality is how emissions are captured. In that sense, technology is an option, for example, the application of filters. However, nature plays a crucial role in purifying the air through the process of photosynthesis. Forests, peatlands and oceans are the main carbon sinks.

Climate Action: Any policy, measure or program that aims to reduce greenhouse gases, increase the ability of communities to adapt and build resilience to climate change, or support and finance such plans. Climate action is one of the Sustainable Development Goals (SDGs) that the member states of the United Nations adopted for 2030.

Climate Refuges: Those areas with pristine habitats of high environmental value where ecosystems are healthy and the different species can take refuge from current and future stressors and negative effects of climate change. Ideally, these areas should be protected and no intervention beyond the minimum.

Common Goods: Refers to those goods that are neither public nor private property, but belong to the entire community. They are assets that nature has made common to all people, such as water, air, the sea, riverbanks and ecosystem functions. "Being common prevents them from being sold or appropriated by the State or private, and recognizes everyone's right to access them to satisfy their fundamental rights and to participate in their governance," explains Ezio Costa, lawyer and director FIMA executive.

Ecological Constitution: It is a Constitution that incorporates transversally and as an ordering axis, the protection of the environment, the health of people and the harmony between society and nature. For Ezio Costa, this requires rules in the section on principles, rights, duties and organization of the State. This includes, for example, rules that ensure citizen participation in environmental decisions and autonomy of local and regional governments in the management of the environment and natural resources, among many other things.

Ecosystem Service: Ecosystems provide essential services for the survival and well-being of people. Some examples are: the water cycle, clean air and food. "In order for people to be able to benefit from these services, the State must ensure certain conditions, such as; granting the right to access these services and providing quality education opportunities for the whole society", says María José Brain, in charge of Planning of the Austral Patagonia Program, UACh.

Ecosystem: An ecosystem is a biological system made up of a community of living organisms and the physical environment in which they interact. It is a unit made up of interdependent organisms that share the same habitat (including people). "The Constitution must evolve in its consideration of the territory", says Florencia Ortúzar, "to be considered as much more than the geographical terrain over which Chile has sovereignty". "The territory must be understood in terms of the ecosystems it houses."

Environmental Democracy: Refers to the possibility of exercising the 3 rights of access, for which States must ensure access to information and the ability of people to participate. The Escazú agreement, says Valentina Durán, director of the Environmental Law Center of the University of Chile, seeks to guarantee the exercise of these 3 rights in Latin America and the Caribbean. Durán adds that Constitutions such as the one in France ensure the rights of access to information and environmental participation.

Environmental Justice: The equitable distribution of environmental burdens and benefits among those who inhabit the territory. According to Ezio Costa, this means that there can be no sacrifice zones that take away all the contamination for the benefit of others.

Environmental Law: Environmental law is a collective term encompassing aspects of the law that protect the environment. A related but distinct set of regulatory regimes, now strongly influenced by environmental legal principles, focuses on managing specific natural resources, such as forests, minerals, or fisheries. Other areas, such as environmental impact assessment, may not fit neatly into either category but are nonetheless important components of environmental law. Previous research found that when environmental law reflects moral values for betterment, legal adoption is more likely to be successful, usually in well-developed regions. In less-developed states, changes in moral values are necessary for successful legal implementation when environmental law differs from moral values.

Equity: Defined by UNEP to include intergenerational equity—"the right of future generations to enjoy a fair level of the common patrimony"—and intragenerational equity—"the right of all people within the current generation to fair access to the current generation's entitlement to the Earth's natural resources"—environmental equity considers the present generation under an obligation to account for long-term impacts of activities and to act to sustain the global environment and resource base for future generations. Pollution control and resource management laws may be assessed against this principle.

Healthy Environment: Unlike a pollution-free environment (enshrined in article 19 No. 8 of the current Constitution), this concept refers to an environment whose ecosystem functions fulfill their role, which can provide well-being and health to living beings that they inhabit it A pollution-free environment, on the other hand, has to do with what the State defines as pollution. "With this, there is the absurdity that in the slaughter zones, for example, we can have environments 'free of contamination,' but which, of course, are not healthy," says Florencia Ortúzar, a lawyer at the Inter-American Association for the Defense of the Environment (AIDA).

Intergenerational Justice: Ensure that future generations have the same opportunities as us. "Our use cannot compromise the use of future generations, for example, by destroying ecosystems or cycles of regeneration of resources such as water," says Costa.

Non-Regression Principle: The non-regression principle aims to constantly improve environmental standards by avoiding the rollback of environmental protection or the adoption of retrogressive regulations. "For example, if the values of a norm are revised, the new values must raise standards and never lead to unprotecting the environment", explains Valentina Durán.

Polluter Pays Principle: The polluter pays principle stands for the idea that "the environmental costs of economic activities, including the cost of preventing potential harm, should be internalized rather than imposed upon society at large." All issues related to responsibility for cost for environmental remediation and compliance with pollution control regulations involve this principle.

Precautionary Principle: One of the most commonly encountered and controversial principles of environmental law, the Rio Declaration formulated the precautionary principle as follows, to protect the environment, States shall widely apply the precautionary approach according to their capabilities. Where there are threats of serious or irreversible damage, lack of full scientific certainty shall not be used as a reason for postponing cost-effective measures to prevent environmental degradation. The principle may play a role in any debate over the need for environmental regulation.

Prevention: The concept of prevention etc. can perhaps better be considered an overarching aim that gives rise to a multitude of legal mechanisms, including prior assessment of environmental harm, licensing, or authorization that set out the conditions for operation and the consequences for violation of the conditions, as well as the adoption of strategies and policies. Emission limits and other product or process standards, the use of best available techniques, and similar techniques can all be seen as applications of the concept of prevention.

Public Participation and Transparency: Identified as essential conditions for "accountable governments...industrial concerns," and organizations generally, public participation and transparency are presented by UNEP as requiring "effective protection of the human right to hold and express opinions and to seek, receive and impart ideas, etc. a right of access to appropriate, comprehensible and timely information held by governments and industrial concerns on economic and social policies regarding the sustainable use of natural resources and the protection of the environment, without imposing undue financial burdens upon the applicants and with adequate protection of privacy and business confidentiality," and "effective judicial and administrative proceedings." These principles are present in environmental impact assessment, laws requiring publication and access to relevant environmental data, and administrative procedures.

Right of Access to Environmental Justice: Possibility of people or communities to go before an independent judicial body or court to protect their rights to information and participation and environmental rights in general, through an independent and expeditious judicial process, which contemplates reparation for environmental damage.

Right to a Healthy and Ecologically Balanced Environment: The right to a healthy environment refers to an environment in which life can develop adequately. It is the way in which the right to the environment has been discussed internationally and incorporates notions such as access rights, the right to water and the rights of future generations. In addition, the ecologically balanced environment refers to the intrinsic value of the environment, where it is not only being protected due to its relationship with people, but also because we value it in itself.

Rights of Nature: The discussion about the rights that belong to nature has to do with rethinking the relationship between humanity and the environment that surrounds us, explains Florencia Ortuzar. To grant rights to nature is to recognize it as something valuable per se, beyond its usefulness for humans, as it is traditionally conceived today. By recognizing rights, it ceases to be an object that can be appropriated to be understood as a subject that has the right to develop to its fullest and to be respected and protected. In 2008, Ecuador became the first country in the world to recognize in its Constitution inalienable rights to nature, thus making it a subject of law.

Sustainable Development: Defined by the United Nations Environment Programme as "development that meets the needs of the present without compromising the ability of future generations to meet their own needs," sustainable development may be considered together with the concepts of "integration" (development cannot be considered in isolation from sustainability) and "interdependence" (social and economic development, and environmental protection, are interdependent). Laws mandating environmental impact assessment and requiring or encouraging development to minimize environmental impacts may be assessed against this principle. The modern concept of sustainable development was discussed at the 1972 United Nations Conference on the Human Environment (Stockholm Conference) and the driving force behind the 1983 World Commission on Environment and Development (WCED, or Bruntland Commission). In 1992, the first UN Earth Summit resulted in the Rio Declaration, Principle 3 of which reads: "The right to development must be fulfilled to equitably meet developmental and environmental needs of present and future generations." Sustainable development has been a core concept of international environmental discussion ever since, including at the World Summit on Sustainable Development (Earth Summit 2002) and the United Nations Conference on Sustainable Development (Earth Summit 2012, or Rio+20).

Sustainable or Sustainable Development: According to the UN, development that meets the needs of the present generation without compromising the ability of future generations to meet their own needs.

Transboundary Responsibility: Defined in the international law context as an obligation to protect one's environment. UNEP considers transboundary responsibility at the international level to prevent damage to neighboring environments at the international level as a potential limitation on the rights of the sovereign state. Laws that limit externalities imposed upon human health and the environment may be assessed against this principle.

Chapter 14
Environmental Law and Armed Conflicts

ABSTRACT

The general theory of environmental orthodoxy is based on the idea that by introducing the issue of the environment and the need to protect it in the constitution as a reference norm, we will see the establishment of the legal order in this field. In this chapter, an attempt has been made to analyze the content of the constitutions of Latin America in the context of environmental issues and the conservation of natural resources with a comparative approach and as an example. The basic hypothesis of this chapter is based on the fact that today, paying attention to the issue of the environment at the level of constitutions means following sub-norms and the legal system from the norm of reference and the effects that this right can have in the legal system. This is well reflected in the constitutions of Latin America.

INTRODUCTION

the concern of the international community for the damage to the environment that derives from human activities and the need for an international response to certain environmental problems and the deterioration of the global ecosystem has prompted the emergence and progressive consolidation of a set of international standards, fundamentally of a conventional nature, for the prevention and repair of damage to the environment (Schmitt, 1997).

This concern for the protection of the natural environment against all kinds of human activity also extends to the impact of armed conflicts on that environment, since they can cause some of the most devastating aggressions against the environment. This concern about their preservation in times of war raises many important questions about the application and interpretation of the rules of international environmental law in that case. In particular, one of the most relevant questions, with a view to guaranteeing effective environmental protection, is the question of to what extent the norms of international environmental law remain in force in the event of an armed conflict, since the outbreak of hostilities between States inevitably has repercussions on the applicability of the international norms that bind said subjects. The issue of the applicability of environmental regulations in these circumstances is

DOI: 10.4018/978-1-6684-7188-3.ch014

complex and for many decades we have lacked solid reference criteria. In the words of the United Nations International Law Commission (hereinafter ILC), environmental treaties present a controversial, varied or incipient probability of applicability during an armed conflict, among other reasons because most treaties relating to the environment do not include provisions referring to their applicability in the event of an armed conflict and because these treaties are extremely diverse from the perspective of their objectives, matters and mechanisms. State practice is inconsistent and divided and there is no clear ruling by an international court in this regard; although as far as the doctrine is concerned, the most widespread position is that which maintains that treaties on the environment would be applicable in the event of an international armed conflict. The issue has been the subject of the work of the IDC, favorable to the continuity of environmental treaties in that case, and their proposals have resolved an important gap, providing relative certainty to the issue. And all this, without prejudice to the full validity in time of war of the norms on environmental protection enshrined in International Humanitarian Law, which includes a series of restrictions on environmental destruction, limiting the means and methods of combat and imposing belligerents that take into account the environmental impact of their operations (Bothe et al., 2010).

The purpose of this work is the study of the applicability of international environmental law in times of international armed conflict. The topic is particularly controversial and current. In fact, in terms of the effects of armed conflicts on treaties, environmental agreements constitute the category of treaties that has received the most attention in recent decades and has generated the most debate; and this is so because of the significant increase in the sensitivity of the international community and civil societies in environmental matters and the verification of the devastating damage to the natural environment of certain armed conflicts.

THE EFFECTS OF ARMED CONFLICT ON CONVENTIONAL RULES

Position of the Doctrine

For a long time, it has been difficult to state a general rule on the legal effect of armed conflicts in international conventions. The ILC, at the beginning of its work on the effects of armed conflicts on treaties, found that "[t]he effect of armed conflicts on treaties has been an unresolved and unclear matter in international law for least a century" and that "while state practice and doctrine are abundant, they are inconsistent and evolving. Furthermore, as traditional warfare gives way to modern non-traditional, internal, or non-formal armed conflict, the parameters of the effect of armed conflict on treaties are left in considerable uncertainty.

For its part, the Law of Treaties contains few indications in this regard since the Vienna Convention on the Law of Treaties, of May 23, 1969, is not very helpful in this matter. Indeed, in its article 73, it is declared that: "The provisions of this Convention shall not prejudge any question that in relation to a treaty may arise as a consequence…of the outbreak of hostilities between States."

In the sphere of the applicability of International Environmental Law in the event of war, after the First Persian Gulf War, it was revealed that the issue also required profound clarification, especially on the issue of the effects of the conflict on treaties between States. belligerents, which is considered by the doctrine as an "obscure topic with only the vaguest guiding principles." Given this uncertainty, the positions of the authors were broadly aligned around three theories: namely, the termination theory, the continuation theory and the classification theory (Shelton, 2021).

In the first place, the termination theory had a great prestige and most of the authors of the 19th century ascribed to it. According to this theory, the war was cause for the termination of all treaties in force between the belligerents. The maintenance of treaties between States was incompatible with the state of war and the treaties did not survive the outbreak of hostilities, although the treaties that regulated the relations between belligerent States and neutral States continued in force. At present, in the positions of the International Law Commission, the doctrine, the practice of the States and the jurisprudence, the abandonment of said theory is confirmed. Therefore, the existence of a situation of war between States is not necessarily incompatible with the maintenance in force of treaties between them, although some of them may end and the execution of some treaties may be suspended during times of war.

Second, the continuity theory argues that war suspends treaties, but only under certain circumstances does it become a cause of termination. Supporters of this theory argue that treaties "lose their efficacy in time of war only when their execution is incompatible with the war itself", as would be the case with treaties of alliances or military aid. The Harvard Draft Convention on the Law of Treaties prepared in 1935 by a group of jurists is an example of this theory. Pursuant to article 35 of this draft, treaties that declare that their obligations are enforceable in time of war or that, by their nature or purpose, were manifestly designed by the Parties to be performed in time of war, are not affected by the outbreak of hostilities. The rest of the treaties are suspended until the end of the war, at which time they regain their validity. The resolution of the Institute of International Law of 1986 regarding the effects of armed conflicts on treaties adopts a position close to that previously exposed. The resolution establishes that the outbreak of hostilities does not automatically terminate the treaties between the parties and that after the end of the war they will recover their validity as soon as possible, unless otherwise agreed (Al-Duaij, 2002).

Third, the classification theory posits neither the ipso facto termination nor suspension of treaties, but advocates a classification of treaties, determining the categories of specific treaties that are compatible with a state of war. It defends the use of various criteria as a way to determine the applicability of a specific treaty from the outbreak of hostilities, inter alia, the interpretation of the intention of the Parties, the nature, the object or the scope of the treaty. This war situation can be cause for the termination or suspension of some treaties, but others remain in force. According to Schmitt, many treaties are the expression of common interests that are not related to the causes or effects of wars. Therefore, from the point of view of international law, the objective should be the preservation of those conventions that can survive. The classification theory seems the most appropriate way to do this and to respect international interests in a universal order. Although there is not, at the moment, an international judicial sentence that shows its preference for this current, the truth is that the position of the IDC, part of the practice of the States and the opinions of numerous authors seem to conform to the theory of classification, since it is the theory that best respects the interests of the belligerents and the interests of the international community (Gasser, 1995).

International Law Commission

On May 11, 2011, the International Law Commission approved the title and text of the articles of the project on the effects of armed conflicts on treaties (hereinafter the Project), applicable to the effects of an armed conflict on international relations. between States by virtue of an international treaty. The starting point of the Project is contained in article 3, which indicates that the existence of an armed conflict does not ipso facto lead to the termination of treaties or the suspension of their application between the States that are parties to the treaty. conflict, nor between a State party to the conflict and a State not participating in the conflict. In this way, the IDC clearly opts for the continuity of the validity of international treaties.

The Draft stipulates that if the treaty itself has included provisions on its application in situations of armed conflict, these provisions will apply (art. 4). In the field of treaties on the environment, the closest case to said provision is represented by the International Convention to prevent pollution of sea waters by hydrocarbons, of May 12, 1954, which in its art. XiX. 1) has provided for the total or partial suspension of the same in case of war or hostilities. Aside from this case except national, it can be said that practically all environmental treaties are silent on the issue. In other words, they do not exclude their application in times of armed conflict, although it is true that no agreement has expressly provided for their applicability in times of war. however, it is possible to cite an international agreement that contains provisions indirectly related to this topic. The best example is the Convention on the World Natural and Cultural Heritage, of November 23, 1972, sponsored by the United Nations Organization on Education, Science and Culture (unESCo), which in its article 6.3 indicates that (Bothe, 1991):

Each one of the States Parties to this Convention undertakes not to deliberately take any measure that may cause damage, directly or indirectly, to the cultural and natural heritage referred to in Articles 1 and 2 located in the territory of other States Parties in this Convention.

This agreement does not exclude damage resulting from military activities and could be applicable to hostile actions that damage the natural heritage of other States Parties, excluding from its scope damage caused to States that are not Parties to the Agreement.

It is also necessary to mention in this category certain agreements that are not environmental treaties but include important provisions on its conservation. The Convention on the Law of the Use of Water for Purposes Other than Navigation, of May 21, 1997, dedicates its part iv to environmental preservation and, in general, all its provisions have a certain relevance to the effects of the conservation of watercourses. Its article 29 states: "International watercourses and installations, constructions and other related works shall enjoy the protection conferred by the principles and norms of international law applicable in the event of an international or non-international armed conflict and shall not be used in violation of of those principles and rules (Baker, 1992).

On the other hand, certain treaties on the environment include provisions that are relevant in times of armed conflict, although they do not expressly refer to that state of war, for example, the Protocol on the prevention of marine pollution caused by discharges from ships and aircraft, of February 16, 1976, in its annex i, since it includes a prohibition of the dumping of materials made for bacteriological and chemical warfare, and the Protocol to prevent pollution by dumping in the southern Pacific region, of November 25 of 1986, in its art. 10.1) and 2) and its annex i, which requires a special permit for the dumping of materials produced for bacteriological and chemical warfare. In this category we should also mention the conventions related to marine pollution from ships, whose scope of application excludes state ships or warships, both in time of peace and in time of war, for example, the Convention Convention for the Prevention of Pollution of Sea Waters by Hydrocarbons, of May 12, 1954 or the International Convention for the Prevention of Pollution by Ships, of February 11, 1973.

In addition, there are treaties relating to civil liability for damage, including environmental damage, that are not applicable to damage resulting from wars and armed conflicts, for example, the Convention on Civil Liability in the Field of Nuclear Energy, of July 29, 1960, in its art. 9, or the Convention on civil liability for nuclear damage, of May 29, 1963, in its art. iv.3 a).

It should also be underlined that the Project recognizes that the termination, withdrawal or suspension of a treaty may be partial and certain clauses thereof may remain in force, reinforcing the general principle of the continuity of the validity of treaties in times of armed conflict, provided and when the following circumstances occur (Droege & Tougas, 2013):

a. The treaty contains clauses that are separable from the rest of the treaty with regard to their application;
b. It appears from the treaty or is otherwise established that the acceptance of those clauses has not constituted for the other party or parties to the treaty an essential basis for their consent to be bound by the treaty as a whole; and
c. The continuation of compliance with the rest of the treaty is not unjust.

On the other hand, one of the most outstanding contributions of the IDC Project is the indicative list of pertinent factors that must be taken into account when determining whether a treaty is susceptible to termination, withdrawal or suspension in the event of armed conflict (art. 6). These factors are:

a. The nature of the treaty, and in particular its subject matter, its object and purpose, its content and the number of parties to the treaty; and
b. The characteristics of the armed conflict, such as its territorial extension, its scale and intensity, its duration and, in the case of non-international armed conflicts, the degree of external participation.

One of the factors that is relevant in determining the termination or sus-treaty pension is its subject matter. In this area, article 6 is completed with article 7 and the Draft Annex that indicates a list of treaties that, due to their subject matter, continue to be applied in whole or in part during an armed conflict. The list is based on practice and is indicative only. The treaties that continue to apply in the event of armed conflict are:

a. Treaties on the law of armed conflict, including international humanitarian law treaties;
b. Treaties declaring, creating or regulating a permanent regime or status or related permanent rights, including treaties establishing or modifying land and sea borders;
c. Normative multilateral treaties;
d. International criminal justice treaties;
e. Treaties of friendship, trade and navigation and agreements relating to private rights;
f. Treaties for the international protection of human rights;
g. The treaties related to the international protection of the environment;
h. Treaties relating to international watercourses and related installations and constructions;
i. Treaties relating to aquifers and related installations and constructions;
j. Treaties that are constitutive instruments of international organizations;
k. Treaties relating to the international settlement of disputes by peaceful means, in particular through conciliation, mediation, arbitration and judicial settlement;
l. Treaties relating to diplomatic and consular relations.

Indeed, different groups of treaties remain a priori unaltered by the outbreak of hostilities. It should be noted that the treaties mentioned in the first place, "treaties on the law of armed conflict, including international humanitarian law treaties", were conceived, by their nature and purpose, to be applied in time of war. These treaties are fully applicable in such a situation, as is obvious. As already indicated, International Humanitarian Law incorporates various prohibitions and restrictions directly or indirectly aimed at environmental protection and that are fully in force in times of armed conflict.

On the other hand, and more directly related to the object of this work, it should be emphasized that, in the opinion of the CDi, reflected in the annex to the Project, the treaties related to the protection of the environment continue to be applied in times of armed conflict. It is interesting to note that the Special Rapporteur initially did not find that there were serious reasons for having this category of treaties as a guide to the intention of the parties and instructed the Commission to examine the desirability of including it, emphasizing that "[t]he Most environmental treaties do not contain express provisions on their applicability in the event of armed conflict. The object and modalities of treaties for the protection of the environment are extremely varied. The only general principle is that of the intention of the parties. There were few proposals to delete this category, although the United Kingdom repeatedly found the inclusion of environmental treaties to be problematic and was skeptical about it (Jha, 2014).

The inclusion of environmental treaties in the list of the Draft Annex responded to several considerations. In the first place, the IDC understood that the arguments of the International Court of Justice in its advisory opinion on the Legality of the Threat or Use of Nuclear Weapons are "significant and serve as general and indirect support for the presumption that the treaties relating to to the environment apply in the event of armed conflict'. Although the International Court of Justice has not ruled clearly on this issue, in this advisory opinion, it recognized that the environment is under daily threats and that it is not an abstraction but represents the living space, the quality of life and the health of human beings, including future generations. The Court reiterated principle 24 of the Rio Declaration on Environment and Development and underlined that States must take environmental concerns into account when assessing what is necessary and proportional to meeting legitimate military objectives. In addition, the Court affirmed that the current international law related to the protection and safeguarding of the environment indicates important factors that must be properly taken into consideration in the context of the application of the principles and norms of the Law applicable in armed conflicts. Despite this, the IDC draws attention to the fact that based on the communications of the States, presented in writing in the works related to the advisory opinion, it follows that, among the States, "there was no consensus on the specific legal issue" and there was "no general agreement on the idea that all treaties related to the environment apply both in times of peace and armed conflict, unless expressly provided otherwise".

On the other hand, the inclusion of environmental treaties in the indicative list of the Annex reflects the general opinion among experts, as already indicated. Indeed, in July 1991, the Canadian government organized an international conference of government experts on these issues in Ottawa. The prevailing position among the state delegates was that international law applicable in peacetime remains in force during hostilities. This was the position expressed by the majority at the international non-governmental conference, organized under the auspices of the International Council of Environmental Law and the Environmental Law Commission of the international union for the conservation of nature (IUCN), which took place in December 1991. in Munich. The Conference stressed that international environmental law was applicable between belligerent and non-belligerent States, but that its applicability between belligerents needed to be clarified.

On the other hand, in April 1992 the International Committee of the Red Cross (CiCR) convened a meeting of experts on the matter and the United Nations asked the Committee to submit a report on the issues mentioned. In this document, the CiCR reiterated that international law in force in times of peace is applicable in the event of an armed conflict. The experts of the working group understood that much of international environmental law remained in force in the event of war and that the fundamental treaties had to be analyzed to determine their specific application.

Among the authors, Schmitt indicates that «the approach that best behaves with the reality of armed conflict while fostering world order is one in which a presumption of survivability attaches to peacetime environmental treaties, either absent de facto incompatibility with a state of conflict or express treaty provisions providing for termination». simonDs believes that general regulations on environmental protection are compatible with a state of war, although it seems more appropriate, based on this a priori statement, to prepare a case-by-case study of the compatibility of each treaty with a situation of armed conflict (Hourcle, 2000).

In short, there is a presumption in favor of the continuity of the validity of the treaties related to the international protection of the environment between all the Contracting Parties. Moreover, where this presumption of continuity proves unreasonable or impracticable under given conditions, the resulting position should obviously be a presumption in favor of suspension over termination of the treaty. When belligerent states are unable to fulfill their obligations as a result of the effects of war, some international treaties may be suspended. The suspension will depend on the interpretation of its provisions in the light of articles 6 and 11 of the ILC draft articles. On the other hand, in the opinion of the Special Rapporteur, "the treaty may continue to be applicable in whole or in part, which means that the subsistence of a treaty belonging to a category that appears in the list [of the Annex] may be limited to some of its provisions" and the suspension would affect a part of the provisions. In addition, with the end of hostilities, the belligerent states are expected to resume the obligations suspended during the war. The clearest example of suspended environmental provisions would be those precepts that require notification, consultation or environmental impact assessment before carrying out projects or actions that may damage the environment in the territory of the neighboring State.

THE EFFECTS OF ARMED CONFLICT ON CUSTOMARY RULES

The question of the position and identification of customary norms in international environmental law is a controversial issue. Despite the difficulty in achieving consensus on these issues, there is no doubt that the duty to prevent transboundary harm and the duty to conserve the environment are in the repertoire of customary law.

Duty to Prevent Transboundary Harm

The duty to prevent transboundary damage has a privileged position, both in the field of neighborly relations and cross-border cooperation, where it has its origins, and in international environmental law. In fact, it is a solidly established customary principle in Public International Law since it was enshrined in the arbitration ruling related to the 1941 Trail smelter matter that confronted the United States and Canada. Subsequently, it has been confirmed by different arbitral decisions (Lake Lanós Case) and judicial decisions (Corfu Canal Case). The International Court of Justice itself in 1996 affirmed that the existence of the general obligation to prevent transboundary environmental damage is already part of the corpus of international standards on the environment. In addition, this principle has been included in numerous international instruments, both declaratory and conventional, on environmental matters and has been the subject of constant and general international practice. Principle 2 of the Rio Declaration on Environment and Development states it in the following terms (Bouvier, 1991):

"In accordance with the Charter of the United Nations and the principles of international law, States have the responsibility to ensure that activities carried out within their jurisdiction or under their control do not cause damage to the environment of other States or of areas that are outside the limits of national jurisdiction.

Therefore, it is the expression of a fundamental principle of International Environmental Law whose validity is universally accepted. The question that arises is whether the principle applies in times of international armed conflict. Logically, the answer will depend on the actors involved. Given that the principle refers to the duty of States to "act in such a way that, by them or by persons under their jurisdiction or control, no activities are carried out that cause 'appreciable' damage, environmental or otherwise, beyond the borders to third States, either directly or indirectly in the person or property of their subjects", the principle will not be applicable to relations between belligerent States since the essence of war is the use of one's own territory to defeat the enemy. In relations between belligerents and non-belligerents, nothing prevents, and in fact it is perfectly enforceable, the full applicability of the principle; reinforced, on the other hand, by the inviolability of the territory of neutral States, in accordance with the institution of neutrality. In Falk's opinion, the principle of the prevention of transboundary environmental damage sets forth a customary international norm, which, therefore, is applicable in times of armed conflict to environmental damage that affects non-combatant States. In this sense, in 1983-1984 by mandate of the European Commission, a group of experts worked on the legal aspects of environmental damage derived from military actions in the Iran-Iraqi War, which lasted from 1980 to 1988. Their conclusions state that The general duty to prevent transboundary harm fully applies during an armed conflict to relations between belligerent States and third States (Verwey, 1995).

On the other hand, in reference to this duty, in his dissenting opinion in the ICJ advisory opinion on the legality of the threat and use of nuclear weapons, Judge Weeramantry declared that there are a series of principles on the protection of the environment so strongly rooted in the conscience of humanity that they have become particularly essential norms of general international law; these basic principles that ensure the survival of civilization, and, indeed, of the human species, are already an integral part of such Law. These principles do not depend for their validity on the provisions of a treaty but are part of customary international law and the sine qua non for human survival. The judge concludes by affirming that such principles are not confined to peace or war, but rather cover both situations since they stem from general duties applicable equally in times of peace as in case of war.

Duty to Conserve the Environment

In this line of reflection, based on the study of international documents on the environment, it can be concluded that among these fundamental principles is the one that establishes the duty to protect the environment, not only in relation to other States, but also in the areas subject to its jurisdiction and sovereignty, and even in those areas located beyond any state jurisdiction. This general duty of protection has not always been expressly proclaimed in international texts, but it constitutes an unquestionable presupposition. Based on this premise, the question that arises is whether there is a customary obligation that imposes the obligation to protect the environment in times of international armed conflict. The question, therefore, is not whether there is an obligation to respect the environment of a customary nature, since an opinion in this sense seems well established among the doctrine; the question is whether such an obligation would be fully applicable in times of armed conflict or, more appropriately, whether there is a customary norm that imposes the obligation to respect the environment in times of armed conflict and what degree of protection it affords. It is unquestionable that this affirmation has the support of numerous international instruments, International Humanitarian Law and International Environmental Law; Among them, the Rio Declaration indicates in its principle 24 that (Schafer, 1988):

War is, by definition, the enemy of sustainable development. Consequently, States must respect the provisions of international law that protect the environment in times of armed conflict, and cooperate in its further development, as necessary.

In the World Charter for Nature, it is recognized, in its principle 5, that nature will be protected from the destruction caused by wars or other acts of hostility. Its principle 11.a stipulates that activities that may cause irreversible damage to nature will be avoided. Finally, its principle 20 declares that military activities harmful to nature will be avoided.

For its part, by virtue of resolution 687, the Security Council declared Iraq responsible for the environmental damage and depletion of natural resources suffered by foreign States, citizens and companies. In connection with this decision, Justice Weeramantry, in his dissenting opinion in the case on the legality of the threat or use of nuclear weapons, stated that Iraq's responsibility to which the Security Council alluded in such unequivocal terms was clearly a responsibility derived from customary international law (Mrema et al., 2009).

On the other hand, the Statute of the International Criminal Court includes among the war crimes, described in its article 8, «intentionally launching an attack, knowing that it will cause... extensive, long-lasting and serious damage to the natural environment that would be manifestly excessive in relation to the anticipated concrete and direct military advantage as a whole" (art. 8.2.b.iv) . The protection that the Statute gives to the environment is a unique protection. Indeed, from provision 8.2.b.iv it follows that even in the event that the component of the environment is a military objective and is not definable as a civilian asset or object, if the damage—extensive, lasting, and serious—is manifestly excessive in relation to the anticipated military advantage, then the action is prohibited, despite the fact that said component of the environment represents a military objective. The inclusion of a provision on environmental damage in the statute of the first permanent international criminal court is a significant development. The search for consensus among the States when defining war crimes meant that the Conference included in the statute only those crimes that received almost unanimous support from the delegations. This fact reinforces, to a large extent, not only the position that defends the customary nature of the war crime for serious, long-lasting and extensive damage to the environment, but also the affirmation that there is a customary norm that imposes the preservation of the environment against to extensive, long-lasting and serious damage in times of armed conflict, to all States and in all conflicts regardless of their nature.

But in addition, a large set of international conventions belonging to different branches of Public International Law contribute, from different approaches, to environmental protection through restrictions on military activities, including Additional Protocol I to the Geneva Conventions and the Convention on to the prohibition of using environmental modification techniques for military or other hostile purposes, reinforcing the arguments in favor of the existence of a customary norm that protects the environment in times of international armed conflict against serious, extensive and lasting damage and that proscribes the environmental war, a reflection of that general duty to conserve the environment. In particular, Article 35.3 of Protocol I prohibits the use of methods or means of combat "that have been designed to cause, or are expected to cause extensive, long-lasting and serious damage to the natural environment". For many jurists, this prohibition has not yet reached the status of a customary norm; the opposition of certain States and the CiCR's own opinion are very illustrative of this position. In this sense, in its advisory opinion on the legality of the threat or use of nuclear weapons, the International Court of Justice seems to grant this precept a merely conventional value when it affirms literally, in reference to article 35, that: "All those These measures impose serious limitations on all States that have signed the aforementioned provisions" (Joyner & Kirkhope, 1992).

however, other authors affirm that said environmental provisions "may be declaratory of a rapidly developing customary international law" and, therefore, would bind the States Parties and non-Parties to the Protocol. Also in the report of the Committee of the International Criminal Tribunal for the former Yugoslavia that examined the NATO bombing of the Federal Republic of Yugoslavia in 1999, it is recognized that "Article 55 may, however, reflect current customary law".

CONCLUSION

Although the practice of the States is incoherent and contradictory, and the position of the International Court of Justice is not conclusive, numerous authors have declared themselves in favor of a general presumption that international environmental law applies in times of armed conflict. international. The CiCR also affirmed that international environmental law remains in force during an armed conflict and recommended that "certain provisions of environmental law should not be suspended during hostilities but that the most important 'core norms' must be applied in all circumstances".

The International Law Commission adopted this approach and has enshrined in its Project the continuity of treaties, in general, and a presumption of continuity of treaties on the environment, in particular, but admitting, however, that the applicability can be limited to certain provisions of the treaty and not the totality of its articles.

Despite this, it is also true that certain doubts remain in the air in this matter and there are many difficulties for the application of international environmental law in times of war. In an attempt to clarify the issue, the applicability of international environmental law in times of armed conflict has been compared with the applicability of certain human rights norms in a similar situation. Indeed, in 1992, the CiCR sent a report to the Secretary General of the United Nations in which it compared international environmental law with international human rights law and noted that the core of the latter regulatory body remains in force in case of armed conflict. In this sense, the opinion of the ICJ can be brought up on the way in which international Human Rights Law is applied in times of war. In the Court's advisory opinion on the legality of the threat and use of nuclear weapons, the High Court indicates that:

The protection provided for in the International Covenant on Civil and Political Rights does not cease in time of war, except when article 4 of the Covenant is applied, according to which some provisions may be suspended when there is a situation of national emergency. However, respect for the right to life is not one of those provisions. In principle, the right not to be arbitrarily deprived of life applies also in times of hostilities. Now, the criterion for determining whether the deprivation of life is arbitrary must be [referred] to the applicable lex specialis, namely, the law applicable in the event of an armed conflict, which is intended to govern situations of hostilities. Thus, in a case of loss of life, due to the use of a particular weapon in a war situation, which is considered a case of arbitrary deprivation of life in violation of article 6 of the Covenant, it is something that can only be decide by reference to the law applicable in the event of armed conflict and not by deduction from the provisions of the Covenant.

This interpretation could contribute, in a certain way, to elucidate the way in which international environmental law remains in force in the event of war. In this sense, the duty to protect the environment would also apply after the outbreak of hostilities, but the criterion to determine the legality or not of an action that has damaged the environment would have to be found in the lex specialis, that is, the International human right. This was initially the position of the CDi. Indeed, during the preparation of the Project, the Special Rapporteur included in the proposal article 6 stated:

Applicable law in case of armed conflict. Normative treaties, including those relating to human rights and environmental protection, continue to apply in times of armed conflict, although their application is determined by the applicable lex specialis, i.e., the law applicable in the event of armed conflict.

This proposed article was finally withdrawn by the Special Rapporteur. The doctrine of the lex specialis does not adequately adjust to the fact that the traditional dichotomy between the law of peace and the law of war is being overcome and that there is no general support among States for the theory that defends that the protection of the environment in times of armed conflict is determined exclusively by the law of war.

On the other hand, certain authors have drawn attention to the fact that even if the validity of international environmental law is affirmed, "[t]he task of simply applying the Conventions on environmental protection [in time] of peace to war situations does not seem to be productive, taking into account that the validity of most of the documents examined will be easily and well-foundedly objectionable by the belligerents at their convenience» . In addition, Schmitt stated that, despite the usefulness of certain international provisions on the environment, and even if a consensus were reached on their applicability, the contribution of international environmental law in times of armed conflict to the preservation of the environment would be secondary. Among other reasons, the author argued that International Environmental Law does not intend to respond to attempts at environmental destruction or the instrumentalization of the environment for military or hostile purposes.

Ultimately, the outbreak of hostilities does not automatically terminate or suspend treaties on environmental protection. Certain international norms in this sphere can lose their legal force in the event of armed conflict only if their application is incompatible with war. The applicability will therefore depend on the study of the context of the treaty in question, using factors such as those suggested by the ILC. The norms of International Environmental Law remain in force, regardless of the problems that their specific application may give rise to. Finally, it must be remembered that, in the opinion of the ILC, and following the arguments of the International Court of Justice, "treaties that were compatible with an armed conflict would remain in force, while treaties that were incompatible with the conflict would continue to apply." as an element to determine military necessity and proportionality" since the International Court of Justice affirms that respect for the environment is one of the aspects that must be weighed to conclude whether a military action is in accordance with the principles of necessity and of proportionality.

REFERENCES

Al-Duaij, N. (2002). *Environmental law of armed conflict*. Academic Press.

Baker, B. (1992). Legal protections for the environment in times of armed conflict. *Va. J. Int'l L.*, *33*, 351.

Bothe, M. (1991). The Protection of the Environment in Times of Armed Conflict. *German YB Int'l L.*, *34*, 54.

Bothe, M., Bruch, C., Diamond, J., & Jensen, D. (2010). International law protecting the environment during armed conflict: Gaps and opportunities. *International Review of the Red Cross*, *92*(879), 569–592. doi:10.1017/S1816383110000597

Bouvier, A. (1991). Protection of the natural environment in time of armed conflict. *International Review of the Red Cross (1961-1997)*, *31*(285), 567-578.

Droege, C., & Tougas, M. L. (2013). The Protection of the Natural Environment in Armed Conflict–Existing Rules and need for further legal protection. *Nordic Journal of International Law*, *82*(1), 21–52. doi:10.1163/15718107-08201003

Gasser, H. P. (1995). For better protection of the natural environment in armed conflict: A proposal for action. *The American Journal of International Law*, *89*(3), 637–644. doi:10.2307/2204184

Hourcle, L. R. (2000). Environmental law of war. *Vermont Law Review*, *25*, 653.

Jha, U. C. (2014). *Armed conflict and environmental damage*. Vij Books India Pvt Ltd.

Joyner, C. C., & Kirkhope, J. T. (1992). The Persian Gulf War oil spill: Reassessing the law of environmental protection and the law of armed conflict. *Case W. Res. J. Int'l L.*, *24*, 29.

Mrema, E., Bruch, C., & Diamond, J. (2009). *Protecting the environment during armed conflict: an inventory and analysis of international law*. UNEP/Earthprint.

Schafer, B. K. (1988). The relationship between the international laws of armed conflict and environmental protection: The need to reevaluate what types of conduct are permissible during hostilities. *Cal. W. Int'l LJ*, *19*, 287.

Schmitt, M. N. (1997). Green war: An assessment of the environmental law of international armed conflict. *Yale J. Int'l L.*, *22*, 1.

Shelton, D. (2021). *International environmental law*. Brill.

Verwey, W. D. (1995). Protection of the environment in times of armed conflict: In search of a new legal perspective. *Leiden Journal of International Law*, *8*(1), 7–40. doi:10.1017/S0922156500003083

ADDITIONAL READING

Baker, B. (1992). Legal protections for the environment in times of armed conflict. *Va. J. Int'l L.*, *33*, 351.

Bothe, M. (1991). The Protection of the Environment in Times of Armed Conflict. *German YB Int'l L.*, *34*, 54.

Bothe, M., Bruch, C., Diamond, J., & Jensen, D. (2010). International law protecting the environment during armed conflict: Gaps and opportunities. *International Review of the Red Cross*, *92*(879), 569–592. doi:10.1017/S1816383110000597

Droege, C., & Tougas, M. L. (2013). The Protection of the Natural Environment in Armed Conflict–Existing Rules and need for further legal protection. *Nordic Journal of International Law*, *82*(1), 21–52. doi:10.1163/15718107-08201003

Gasser, H. P. (1995). For better protection of the natural environment in armed conflict: A proposal for action. *The American Journal of International Law*, *89*(3), 637–644. doi:10.2307/2204184

Hourcle, L. R. (2000). Environmental law of war. *Vermont Law Review*, *25*, 653.

Jha, U. C. (2014). *Armed conflict and environmental damage*. Vij Books India Pvt Ltd.

Joyner, C. C., & Kirkhope, J. T. (1992). The Persian Gulf War oil spill: Reassessing the law of environmental protection and the law of armed conflict. *Case W. Res. J. Int'l L.*, *24*, 29.

Mrema, E., Bruch, C., & Diamond, J. (2009). *Protecting the environment during armed conflict: an inventory and analysis of international law*. UNEP/Earthprint.

Schafer, B. K. (1988). The relationship between the international laws of armed conflict and environmental protection: The need to reevaluate what types of conduct are permissible during hostilities. *Cal. W. Int'l LJ*, *19*, 287.

Schmitt, M. N. (1997). Green war: An assessment of the environmental law of international armed conflict. *Yale J. Int'l L.*, *22*, 1.

Shelton, D. (2021). *International environmental law*. Brill.

Verwey, W. D. (1995). Protection of the environment in times of armed conflict: In search of a new legal perspective. *Leiden Journal of International Law*, *8*(1), 7–40. doi:10.1017/S0922156500003083

KEY TERMS AND DEFINITIONS

Environmental Law: Environmental law is a collective term encompassing aspects of the law that protect the environment. A related but distinct set of regulatory regimes, now strongly influenced by environmental legal principles, focuses on managing specific natural resources, such as forests, minerals, or fisheries. Other areas, such as environmental impact assessment, may not fit neatly into either category but are nonetheless important components of environmental law. Previous research found that when environmental law reflects moral values for betterment, legal adoption is more likely to be successful, usually in well-developed regions. In less-developed states, changes in moral values are necessary for successful legal implementation when environmental law differs from moral values.

Equity: Defined by UNEP to include intergenerational equity—"the right of future generations to enjoy a fair level of the common patrimony"—and intragenerational equity—"the right of all people within the current generation to fair access to the current generation's entitlement to the Earth's natural resources"—environmental equity considers the present generation under an obligation to account for long-term impacts of activities and to act to sustain the global environment and resource base for future generations. Pollution control and resource management laws may be assessed against this principle.

Polluter Pays Principle: The polluter pays principle stands for the idea that "the environmental costs of economic activities, including the cost of preventing potential harm, should be internalized rather than imposed upon society at large." All issues related to responsibility for cost for environmental remediation and compliance with pollution control regulations involve this principle.

Precautionary Principle: One of the most commonly encountered and controversial principles of environmental law, the Rio Declaration formulated the precautionary principle as follows, to protect the environment, States shall widely apply the precautionary approach according to their capabilities. Where there are threats of serious or irreversible damage, lack of full scientific certainty shall not be used as a reason for postponing cost-effective measures to prevent environmental degradation. The principle may play a role in any debate over the need for environmental regulation.

Prevention: The concept of prevention etc. can perhaps better be considered an overarching aim that gives rise to a multitude of legal mechanisms, including prior assessment of environmental harm, licensing, or authorization that set out the conditions for operation and the consequences for violation of the conditions, as well as the adoption of strategies and policies. Emission limits and other product or process standards, the use of best available techniques, and similar techniques can all be seen as applications of the concept of prevention.

Public Participation and Transparency: Identified as essential conditions for "accountable governments...industrial concerns," and organizations generally, public participation and transparency are presented by UNEP as requiring "effective protection of the human right to hold and express opinions and to seek, receive and impart ideas, etc. a right of access to appropriate, comprehensible and timely information held by governments and industrial concerns on economic and social policies regarding the sustainable use of natural resources and the protection of the environment, without imposing undue financial burdens upon the applicants and with adequate protection of privacy and business confidentiality," and "effective judicial and administrative proceedings." These principles are present in environmental impact assessment, laws requiring publication and access to relevant environmental data, and administrative procedures.

Sustainable Development: Defined by the United Nations Environment Programme as "development that meets the needs of the present without compromising the ability of future generations to meet their own needs," sustainable development may be considered together with the concepts of "integration" (development cannot be considered in isolation from sustainability) and "interdependence" (social and economic development, and environmental protection, are interdependent). Laws mandating environmental impact assessment and requiring or encouraging development to minimize environmental impacts may be assessed against this principle. The modern concept of sustainable development was discussed at the 1972 United Nations Conference on the Human Environment (Stockholm Conference) and the driving force behind the 1983 World Commission on Environment and Development (WCED, or Bruntland Commission). In 1992, the first UN Earth Summit resulted in the Rio Declaration, Principle 3 of which reads: "The right to development must be fulfilled to equitably meet developmental and environmental needs of present and future generations." Sustainable development has been a core concept of international environmental discussion ever since, including at the World Summit on Sustainable Development (Earth Summit 2002) and the United Nations Conference on Sustainable Development (Earth Summit 2012, or Rio+20).

Transboundary Responsibility: Defined in the international law context as an obligation to protect one's environment. UNEP considers transboundary responsibility at the international level to prevent damage to neighboring environments at the international level as a potential limitation on the rights of the sovereign state. Laws that limit externalities imposed upon human health and the environment may be assessed against this principle.

Chapter 15
Environmental Law and Terrorism

ABSTRACT

Leading research is based on descriptive-analytical principles and the literature review method, which has examined the classical and modern dimensions of terrorism by considering the aspects of this emerging phenomenon. While implementing existing strategies to combat environmental terrorism, this study seeks to provide an effective strategy to combat "green terrorism" or "ecosystem" to provide an effective strategy in the light of new criminal law approaches. The leading research is strengthening practical strategies to combat green terrorism and implementing effective strategies such as environmental litigation, both nationally and internationally, considering legal means. In this regard, green criminalization as a green strategy can be considered an action to combat environmental terrorism.

INTRODUCTION

The environment is increasingly understood and received as a common value and heritage of humanity in general. A value that is the duty of the whole human society to protect and safeguard, but unfortunately today, the increase of environmental damage due to non-observance of correct and clear legal rules has attracted the attention of human societies in the international and global dimensions. At the same time, environmental damage is manifested in environmental crimes. Environmental crimes, which are considered an act of harm to the environment, have a different nature than other crimes and are difficult to recognize due to their complex and multidimensional nature. Therefore, an environmental crime can be referred to as "any current or current type of leave that causes serious damage to the environment and seriously endangers human health (Schofield, 1998)."

Therefore, in general, environmental crimes are divided into two main groups according to their nature:

a. Crimes committed against living organisms in the environment, including humans, plant, and animal organisms.

DOI: 10.4018/978-1-6684-7188-3.ch015

b. Crimes committed against inanimate elements in the environment such as climate, soil, noise, and chemical pollution.

Therefore, the actions violate the rights and security in the field of environment in the mentioned fissures and cause severe damages. Hence, the most important action taken to threaten the environment is "environmental terrorism," which, despite the lack of proliferation and widespread crime worldwide, can significantly impact armed conflict and nuclear and chemical experiments and irreparable damage. Will enter the body of the environment (Schofield, 1998).

Environmental terrorism, however, "refers to any act that uses toxic or hazardous substances to cause harm or threat typically to the environment of humans, animals and the natural environment to severely disrupt public order through intimidation or to cause terror".

Environmental terrorism, like environmental warfare, involves the use of natural elements or forces as weapons. Environmental terrorists, in the name of political or ideological bias, deliberately destroy or change the environment; although such methods have been used for a long time during the war, but have recently been a successful option. They have appeared as terrorist. The end of the Cold War and the change in the face of terrorism are the most important factors contributing to the growth of these groups.

The emergence of this nascent threat has been faster than the law's ability to react. Existing legal mechanisms for countering environmental terrorism rely on existing principles of environmental law and terrorism, which are inadequate to address this particular problem. These legal principles are not adequate to address the threat of environmental terrorism; Because A. The complex set of environmental laws does not act as ordinary criminals effectively within the framework of existing counter-terrorism strategies in countering terrorists; B. These principles do not properly fulfill the function of criminal law. However, it seems that the fight against environmental terrorism requires a new criminal law. The Ecosystem Prohibition Act, or the intentional or inadvertent manipulation or destruction of any aspect of the physical environment, provides a mechanism for punishing environmental terrorists within existing legal structures while at the same time undermining the social ugliness of such an act (Cockayne, 2007).

BACKGROUND AND DEVELOPMENTS

Environmental terrorism, like environmental warfare, involves the use of natural elements for conflict. Environmental terrorism involves targeting the environment itself, such as the deliberate contamination of water or agricultural resources, and using the environment as a conduit for destruction, such as releasing chemical or biological weapons into the atmosphere, the elements of the global ecosystem that have been or can be used for conflict range from complex nuclear factors to simple and efficient water energy. In this regard, the use of such factors has had devastating consequences such as mortality, physical injury, and dangerous ecological hazards (Tyler, 2014).

Emergence of Environmental Terrorism

The environment has been used as a weapon of war for centuries. Throughout history, aggressors and defenders have used the elements of nature against their enemies. For example, in the seventeenth century, the Dutch deliberately submerged all their lowlands to prevent their enemies from advancing. In the Vietnam War, the United States destroyed large areas of southern Vietnam with herbicides to destroy vegetation and forests and deprive its enemy of cover, mobility, and livelihood. Such tactics are now prohibited in international wars by international treaties and international law.10 However, there is no mechanism to deter terrorists from participating in and deliberately destroying the environment. The end of the Cold War and the change in the face of terrorism have made environmental terrorism very likely (Tyler, 2014).

Terrorism After the Cold War

Terrorists, like warring nations before them, want to contain the elements of nature because spending an average amount of time and effort can lead to long-term destructive destruction. In the past, the difficulty of providing the means for this devastation and the geopolitical hardships of the Cold War limited the environmental threat posed by terrorists. However, the end of the Cold War largely eliminated these restrictions. Given the political boundaries, the terrorists and having to operate in the area during the Cold War wanted to limit hasty and reckless attacks. The United States and the Soviet Union have suffered greatly to control terrorism, especially to prevent one-nation states from supporting terrorist activity. Today, religious, ethnic, and nationalist conflicts have been replaced by political boundaries. Terrorist organizations have also resorted to discrediting countries and organized crime to make money. Some countries, which no longer succumb to the demands of the superpowers, immediately support terrorist activities around the world. The end of the Cold War, as a driving force, has also created a favorable environment for the existence of weapons of mass destruction (Tyler, 2014). The threat of nuclear terrorism is not new, but the collapse of the Soviet Union has greatly increased its persistence. In 1994, some argued that "one criminal threat seems to be greater than all other threats, which is the theft or release of radioactive material from Russia and Eastern Europe".

Nuclear proliferation—the sale, theft, or abandonment of nuclear weapons and materials that the Soviet government once centrally held—is a reality. In August 1995, German police discovered 350 grams of nuclear fuel on a commercial flight from Moscow to Munich. The topics of such publication will be discussed in future discussions (Billon & Carter, 2012).

Modern Terrorism

The emerging threat of environmental terrorism arises from the changing nature of terrorism itself. Terrorism has long been a cause of violence or a threat of violence to create an atmosphere of terror. Terrorism generally consists of acts intended to send a message or carry out a particular course of action. According to terrorism expert Brian Jenkins, terrorists have long wanted a spectator population (not a dead population).

But there is growing evidence that many terrorists have changed their target and are now more interested in mass murder. Recent events such as the World Trade Center and the Oklahoma City bombings, and the sarin gas attack on the Tokyo subway underscore this trend. Some experts believe that the old view was that terrorists are concerned about public opinion, but now their preoccupation is with the reward of the Hereafter and not with the reward of this world, and they consider it a religious duty to bring civilization to its knees. Because of this new branch of terrorism, the destructive factors of nature are tempting.

Groups and individuals motivated by promising religious or ideological bias instead of traditional political calculations are more likely to engage in environmental terrorism because they believe there is a moral justification for doing so. The legitimizing force of fanatical ideology empowers these groups to engage in behavior seen by many terrorists as politically immoral or harmful. This does not mean that terrorists with more traditional motives do not value environmental terrorism. However, the broader goals of such organizations often impose restrictions on themselves. Unbridled environmental terrorism can tarnish the image of the group. Endanger group cohesion alienates perceptual components and provokes the intensity of actions that the group can not survive (Coutin, 2005).

There is no need for terrorist acts appropriate for the audience as a limiting factor in religious or ideological terror. Restrictions imposed for political reasons do not apply to those who believe they act for religious or ideological purposes. The number of groups acting for such purposes seems to be increasing. Although the extremism that inspired the World Trade Center bombers is largely known as a bigoted motive, it is by no means the only motive. Millennial sects, that is, groups that believe the new millennium will end civilization, are also prone to mass violence. For example, the Aum Shinrikyo sect, which carried out the 1995 nerve gas attack on the Tokyo subway, uses such beliefs to justify the production and use of chemical weapons. They believed that a chemical attack would reveal secret emotions that they thought were inevitable.

Many internal groups are also motivated by extremist ideologies beyond the traditional terrorist strategy and self-limitation. The right-wing patriotic movement, for example, has been sparked by its paranoid fears of a failed conspiracy involving the federal government, the United Nations, and other sinister elements. This paranoia has fueled the unexpected growth of membership in militant and patriotic organizations in recent years (McCulloch & Pickering, 2009).

This new branch of terrorism is characterized by more fragmented organizational structures in addition to xenophobic motives. Traditional terrorist groups are generally defined as specific organizations with a coherent but simplistic structure in command and control. Based on such a structure, the organization spends time and resources collecting donations, advertising, and planning future actions. The new breed of terrorists has little to do with such organizational considerations, and these groups do not seek legitimacy, recognition, or privilege for their reasons and are themselves reluctant to negotiate.

In April 1995, former CIA Deputy William Stoneman warned of the new terrorist race: "These groups, which are often isolated, are in some ways even more dangerous than the traditional group because they have an official organizational identity." They do not and want to decentralize and segment their activities. Terrorism expert Yuna Alexander described the new method of operation as a "perfectly organized disorder (Boelens et al., 2009)."

TYPES, METHODS, AND METHODS

Nuclear Option

The old scenario of nuclear terrorism involves detonating nuclear devices in an urban environment. Although this is a nuclear option for terrorists, it is by no means the only option. Similar destruction is achieved by sabotaging a commercial nuclear reactor or attacking a carrier of weapons or chemicals.

The act of nuclear terrorism does not require open violence. On Nov. 23, 1995, Chechen guerrilla leader Shamil Basayev told a Russian television station that four boxes of radioactive cesium had been hidden around Moscow. The network later discovered a 32-kilogram box of the substance that emitted more than 300 times the normal amount of radioactivity, which was discovered in Moscow's Isailov

Park. Although the real threat was minimal, it highlighted the danger posed by the terrorists' passive use of nuclear material.

It is easier to make a dirty bomb than that. That is a typical device with a highly radioactive coating. A bomb blast wrapped in radioactive material can be used to contaminate a separate area. However, there is concern that such a device could be used to contaminate buildings or water sources.

On the other hand, terrorists can achieve these goals by using traditional tactics for nuclear purposes. A terrorist group with no desire or ability to build a nuclear weapon can attack a nuclear power plant or uranium processing facility with a portable missile. Although some experts have questioned the effectiveness of such an attack, others have concluded that the threat posed by terrorist attacks on commercial nuclear facilities (Spade, 2015).

Finally, from an environmental point of view, the terrorist nuclear targets do not have much effect. A more relevant issue is the impact that nuclearization has on the environment. At the same time, the myriad variables at play make it virtually impossible to quantify the specific environmental effects of a nuclear terrorism act. In this regard, we can deduce the potential consequences from one of the sad case studies of nuclear history.

In 1986, an experiment at the Chernobyl nuclear power plant in Ukraine went out of control and led to a massive explosion that released several tons of highly radioactive uranium into the atmosphere. At least 70% of the contamination affected Belarus' humans, crops, and animals. Belarus has been a nuclear dumpster since the Chernobyl explosion. Only one percent of Belarus is uninhabited. More than 30,000 hectares of the most fertile agricultural land in Russia, Ukraine, and Belarus have been abandoned, and at least 70,000 square kilometers of farmland have been destroyed.

Scientists at the Belarus National Institute of Agricultural Science and Research in agricultural radiobiology estimate that it will take 600 years to safely re-grow crops in the most polluted areas. Chernobyl precipitation has polluted the entire ecosystem of the region. The Pripyat River, which flows near the power plant and is the main source of drinking and washing water in the region, is now one of the most radioactive rivers in Belarus.

Inside the 30-kilometer no-fly zone around Chernobyl, the coniferous leaves of giant pines grow from mutant pine and birch trees. Fish and livestock that consume irradiated food accumulate radiation in their bodies, entering the food cycle. In essence, the Chernobyl nuclear accident led to ecosystem genocide in the area around the plant.

Although it is unlikely that terrorists will launch a nuclear attack the size of the Chernobyl accident, it is unlikely that a strategic attack could cause similar environmental damage. For example, a nuclear attack in the heart of agriculture in Nebraska or Iowa could infect thousands of acres of the world's most fertile farmland. Similarly, radioactive contamination of the Mississippi or Colorado River sources could have devastating effects on the ecosystems of multiple communities in dozens of states (Edwards et al., 2013).

Chemical and Biological Terrorism

Chemical Factors

Chemical weapons of mass destruction, such as mustard gas, mustard gas, and chlorine, were first used in World War I and the following decades in separate wars. Chemical structures are fast-acting synthetic compounds that were originally designed to poison enemy forces. These weapons range in power from relatively mildly damaged structures such as tear gas to blood and blistering structures such as mustard gas and cyanide. Nervous structures are the most toxic chemical weapons. Nervous structures cause death by attacking the body's central nervous system. Those who inhale the nerve gas or absorb it through the skin begin to tremble uncontrollably and suffocate immediately. Chemical weapons have two types, stable and unstable. A stable structure that lasts for days or even weeks can block large areas of human activity. Unstable structures are quickly blown away or dried by the wind. The components needed to make many of these compounds are marketed. In addition, experts believe that anyone proficient in making pesticides can make chemical weapons. The first major act of chemical terrorism confirms this.

In March 1995, the Shinrikyo sect, an unknown cult of the modern age based in Japan, launched a chemical attack on the Tokyo subway system. Five chemical devices, including lunch boxes and soda boxes, were distributed in many cars inside the underpass at the intersection with Tokyo city center. Many passengers fell to the ground in a matter of minutes, twelve people died in a few days, and about five thousand five hundred people were injured. It took less than a year for members of the Shinerikio Sarin to build an application in that attack. While the attack was opened less than the sect expected, its mere occurrence sent shockwaves worldwide. Shinrikyo set an important tradition by being the first terrorist group to use a weapon of mass destruction. Terrorism experts have long believed that when terrorist operations exceed the threshold, terrorists' use of unconventional weapons will spread rapidly. "The attack on Tokyo had a global theme and content because the massacre showed a ripple of terrorism," said Yona Alexander, a terrorism expert. The FBI said that the number of threats to use chemical weapons had increased exponentially since the incident. Although the attack on Tokyo was a vague distinction from the first major act of chemical terrorism, it was indeed the first strategy to be considered by the terrorist minority over the years. In 1985, for example, federal agents raided the right-wing compound of the Covenant, Sword, and Weapon of God and found, among other things, a barrel containing thirty-five gallons of cyanide intended to poison Washington, DC, and New York City have been (Andreas & Nadelmann, 2008).

Biological Factors

Biological weapons contain pathogenic organisms designed to cause disease and death and do not necessarily start the epidemic process as is often thought. Most biological agents cause death by direct contact. For example, anthrax spores multiply inside the victim's lungs, but the victim does not pass them on to others. Biological weapons can have a devastating effect on the natural environment. The entry of microorganisms alters the ecosystem of an area to such an extent that it may not be habitable for an indefinite period (Farah, 2011). For example, in 1941 and 1942, the British detonated experimental anthrax bombs on the island of Gurnard in Scotland. Millions of anthrax spores were buried in the island's soil. The island is still uninhabitable despite decades of efforts to decontaminate it. Biological weapons, such as anthrax, are easy to produce and inexpensive. The recipe for these agents is available through various sources, and components are typically purchased from commercial vendors or passed from professor to student. Law enforcement officials no longer ask the question, will the act of biological terrorism occur, but when will the act of biological terrorism occur? "The next big thing is that a lot of people come to the hospitals in a very bad condition and die within seventy-two hours," predicts John Sopko, senior legal adviser to the Senate Standing Committee on Minorities. The Office of Technology Assessment (DTA) report on weapons of mass destruction estimates that a warhead of a spotted anthill landing in Washington, DC, will kill between 30,000 and 100,000 people on a day with moderate winds. However, this warhead is not the most probable way that terrorists will use this deadly weapon. A more realistic scenario assessed in the report is a small 220-pound anthropogenic aircraft flying over Washington. Such an aircraft would kill one million people by spraying a cloud of invisible dust in a day with moderate wind. Predictions of biological terrorism that may occur soon are not unfounded. In 1995, four members of the Patriotic Council - an anti-government group - were convicted of plotting to assassinate federal agents by a biological toxin called ricin. This is an extremely powerful agent of ordinary castor seed powder protein extract. If the least amount of ricin is digested, absorbed, or inhaled, it kills the victim by bursting red blood cells. This toxin is six thousand times more toxic than cyanide and has no cure. The patriotic council members planned to mix ricin with a solvent that is easily absorbed through the skin and apply this mixture on the door handle or steering wheel that the victims will be in contact with; for example, 0.7 grams of ricin produced is enough to kill 100 people (White, 2003).

Water-Energy Factors

Water is one of the largest sources of energy stored in the environment. For example, thousands of earthen dams, dams, and dams in the United States have enormous destructive capacity. The static energy of these water bodies, controlled by these barriers, has a destructive force equivalent to a nuclear accident. The Hoover Dam on the Colorado River between Nevada and Arizona prevents thousands of acres of land in both states from being submerged by 9.2 trillion gallons of water. Like other forms of environmental terrorism, the use of water as a weapon has emerged from the battlefield. Throughout history, military commanders have sought to use water energy during the war. The most destructive single action in terms of casualties in war history was the destruction of a dam by the Chinese military in World War II. In June 1938, the Chinese army exploded the Huaiwanko Dam on the Yellow River with dynamite to prevent Japanese invaders (Furman et al., 2012). Although the flood succeeded in stopping the Japanese, the consequences were devastating. The floods destroyed 11 cities and more than 4,000 Chinese villages. Hundreds of thousands of Chinese drowned and millions were left homeless. Millions of hectares of agricultural land were submerged, destroying both agricultural products and non-renewable resources. The Yellow River was not controlled until 1947. During the Korean War, the United States pursued a strategy of destroying the dam against North Korea. The target of this tactic was the environment itself. As an agricultural community, Korea was dependent on land, and the explosion of dams exploited this dependence by wasting valuable agricultural water resources, destroying agricultural land, and cutting off the supply of rice to citizens and soldiers. The attractiveness of the dam is obvious as a target for terrorist activity because at least the effort will lead to massive long-term environmental degradation. In World War II, British commanders described the demolition of the dam as a means to maximize impact with minimal effort. The Federal Agency for Risk Management and Public Safety has determined that an explosive device planted in a large US dam would cause catastrophic destruction downstream. Federal officials have expressed concern that US dams could become targets for terrorist attacks. As history shows, the environmental damage that can result from the destruction of earthen dams, mudslides and dams is enormous. For example, experts predict that if the Glen Canyon Dam collapses, Lake Powell, the second largest reservoir in the United States, will recreate a statue of the Grand Canyon (Aaronson & Shaffer, 2021).

EXISTING DOCTRINE OF DOMESTIC LAW

Counter-Terrorism Rights

The US government operates on the principle that terrorists are ordinary criminals. This strategy of dealing with terrorists as ordinary criminals, someone who participates in a crime for purely personal reasons, is designed to legitimize a terrorist act. Judicial officials ignore the political nature of the crime, suppress the terrorists' strong desire to speak out, and at the same time prevent the mental impact that there is a serious terrorist problem. The focus of counter-terrorism activities is in line with traditional law enforcement, arrest, and punishment strategies. Judicial officials who operate within the scope of this "terrorist means criminal" feature rely on classical criminal law in bringing terrorists to justice. Terrorists are accused of committing any crime that is committed to advance a subsequent terrorist act, not of the very necessity of terrorism. These charges typically focus on transportation, the type of business being attacked, or specific functional weapons such as explosives. The prosecution of customary law crimes such as murder, assassination, and conspiracy are essential elements of the case against terrorists. Terrorists often inadvertently assist judicial authorities in implementing this strategy by engaging in a wide range of criminal behavior leading to a terrorist incident. Terrorism as terrorism did not constitute a federal crime in the United States until the 1996 Counter-Terrorism Act and the Executive Death Penalty were passed. The Counter-Terrorism Act and the death sentence for the Oklahoma City bombing were enacted to provide the federal government with other tools to fight against terrorism (Estevens, 2018).

The law specifically criminalizes the commission of international terrorist activity in the United States. Counter-terrorism law, in addition to criminalizing terrorism, includes provisions that: (a) Prohibits the collection of donations in the United States by foreign groups that, at the discretion of the State Department, are involved in terrorist activities; (b) require plastic explosives to be chemically labeled to make it easier to trace their source to law enforcement; And (c) the use of chemical weapons in the United States or against American citizens abroad. This latest regulation, the criminalization of chemical weapons, reflects the provisions of the 1989 Biological Weapons Act. Congress enacted the Biological Weapons Act to implement parts of the 1972 Convention on Biological Weapons and Biotoxins and the fight against acts of bioterrorism. The key provisions of this law define the design, production, transfer, or informed ownership of any biological agent, toxin, or delivery system as a federal offense. Anti-terrorism legislation and biological weapons are examples of laws designed to combat emerging terrorist threats within the "terrorist means criminal" paradigm. In line with traditional domestic approaches to terrorist activity, these laws focus on arresting and punishing criminals and provide another weapon in the arsenal of lawsuits to prosecutors.

Environmental Protection Rights

The United States is one of the leading countries in environmental protection in terms of legislation and legislation. In practice, however, it has had fundamental shortcomings in many respects. Federal laws governing the environment have been part of American law since the nineteenth century. Environmental protection has expanded dramatically with enacting a set of federal laws in the 1970s, including amendments to the Clean Air Act and the Safe Drinking Water Act. This set of rules creates a comprehensive legal system to control pollution and maintain America's air, water, and soil purity. Those laws that deal with pollution control and environmental protection are enforced by the Environmental Protection

Agency, while the subdivision enforces those that deal more directly with wildlife and other natural resources. Various programs of the Ministry of Interior are implemented. These law enforcement agencies enforce environmental protection laws through various methods, including civil and criminal penalties (Hutchinson & O'malley, 2007).

The most important environmental laws in the United States include criminal penalties for violating these regulations. Without proper documentation, transportation or disposal of hazardous waste is a criminal offense under the Conservation and Recycling of Resources Act. Violations of such criminal regulations are referred to the Ministry of Justice by the relevant organizations for criminal proceedings. Until recently, however, violations of environmental law were rare. In the 1970s, only twenty-five criminal environmental cases were processed by the Ministry of Justice. In 1982, the Environmental Protection Agency referred only twenty environmental criminal cases to the Ministry of Justice. By 1990, however, that number had risen to sixty-five referrals, leading to 134 indictments. This increase in criminal proceedings against corporate polluters can be attributed to several factors, including the failure of civil penalties to deter companies from operating illegally.

In most cases, profitable incentives combined with environmental crime outweigh legal penalties. For companies, these penalties are the cost of trade and business, and in most cases, pass that cost on to the consumer. Criminal penalties are not subject to the same restrictions. Prosecutors believe the defamation, stigma, and waste of good faith combined with criminal proceedings prevent companies and their agents from violating environmental laws. In addition, several environmental laws have allowed prison sentences for corporate officers and employees to the extent of their professional responsibilities. The 1990 Law on Prosecution of Pollution increased the ability of judicial authorities to prosecute individuals violating environmental protection laws. The law quadrupled the number of inspectors assigned to the Environmental Protection Agency, increased technical support for inspectors, and established a national training center to help executives understand the so-called technical set of environmental laws. The primary motivation for environmental law criminalization and an important contributing factor to increased enforcement efforts has been the emergence of environmental self-awareness in the United States (Chiaramonte, 2020). In the last three decades, environmental endowment organizations have created tremendous public awareness of the threatened environment. People now see environmental crimes as more like traditional crimes than mere offenses. A Department of Justice public crime survey ranked illegal environmental activities among violent crimes such as murder and rape and other administrative crimes such as government corruption. Many commentators have argued that post-industrial values that emphasize protecting the environment against the over-exploitation of natural resources are part of a new worldview that commentators have called the New Environmental Index. A brief review of international law seems to confirm the global nature of this process. The international community increasingly recognizes the need to protect the environment and the international law to which it has been developed (Bassiouni, 2008).

Initially, international environmental law took the form of substantive treaties. One of the first treaties to protect the environment was the 1959 Antarctic Treaty. The treaty bans waste disposal and nuclear explosions in Antarctica, making the continent a place for scientific research used only for peaceful purposes. Similarly, the 1972 Convention on the Prevention of Marine Pollution from the Disposal of Waste and Other Substances specifically protects the oceans by controlling the amount and nature of landfills. But the international approach to environmental protection began to change with the United Nations Conference on the Human Environment in Stockholm in 1972. The Stockholm Conference, commonly known as the conference, established the United Nations Environment Program to accelerate environmental action. The conference also issued a statement outlining non-mandatory principles focused on the environment. The first principle of the Declaration states that human beings have the fundamental right to liberty, equality, and adequate living conditions in a quality environment that provides a life of human well-being and dignity. They have a serious responsibility to protect and improve the environment for present and future generations. Are in charge. Some legal commentators believe that the statement and the conduct of countries and international courts suggest that the right to a healthy environment is included in customary international law. Customary international law is defined as a public practice among countries that are practiced so that it provides evidence that this practice is mandatory due to the existence of a legal injunction.

Commentators who have expressed the right to a healthy environment, accepted as a norm of customary international law, have stated that violators of this principle will be punished according to the international crime of ecosystems. The term ecosystems were first coined to classify the great environmental devastation of war, especially the US use of herbicides in Southeast Asia. Ecosystems as a whole are defined as the complete or partial deliberate destruction of any part of the global ecosystem (White, 2012).

Proponents of ecosystems have argued that the fundamental human right to a healthy environment is meaningless unless the law provides criminal redress for violating it. They referred to the substantive jurisprudence relating to such reparations in customary international law. "Convention on the Protection of Fur Floods in the North Pacific (1911)", "International Convention for the Prevention of Oil Pollution (1954)," and "International Convention for the Prevention of Pollution from Ships (1973)" They include criminal regulations. Still, despite the agreement of commentators and the effective support of international authority, ecosystems are not currently recognized as an international crime because many countries are arguing that the right to a healthy environment, which provides the basic legal logic for the crime of ecosystems, has been elevated to the level of customary international law.

INEFFICIENCY OF THE EXISTING LEGAL DOCTRINE

Lack of Applicability of Existing Regulations

The existing set of environmental laws was not intended to control this brutal destruction of the environment committed by terrorists. Extensive legislation created by this set of laws, such as the Comprehensive Environmental Response Act of 1980, Compensation, and Liability, is designed to prevent corporate pollution rather than environmental terrorism. In fact, out of 134 indictments resulting from 65 referrals to the Ministry of Justice in 1990, ninety-eight percent of companies named bosses, owners, deputies, directors, and supervisors as accomplices.

The pursuit of environmental terrorists under the set of environmental protection laws is inconsistent with the index of dealing with terrorists as ordinary criminals. As discussed earlier, terrorist acts have long been prosecuted for crimes committed. The complex legislation, such as those found in the Environmental Code, does not immediately conform to the "terrorist, criminal" index. While equipped with criminal rules, such sets of laws deal with more violations, such as statutory offenses, rather than customary legal offenses. In addition, the criteria of liability, testimony, and evidence that are well established under a set of traditional criminal laws are not very clear under this environmental legislation(Welch, 2003).

A set of environmental laws makes it difficult to prosecute environmental terrorists by forcing judicial officials to monitor unfamiliar territory. Environmental law is technical and complex, and few judicial officials have extensive experience in this field. Without the preamble and the organization's guidelines, the rules of the Environmental Protection Agency reach a total of ten thousand pages in the collection of federal laws and are constantly changing. These bills are so complex that lawmakers, consultants, and environmental lawyers regularly specialize in one or two specific laws.

Judicial officials are likely to avoid the complexity of a set of environmental laws and instead rely solely on a more controllable set of criminal laws to bring environmental terrorists to justice. This way of working cannot reflect social value. Criminal law is not merely a mechanical tool serving the needs of judicial officials. Criminal law reflects social ethics and gives symbolic power and representation to moral values by conveying arguments and knowledge. Failure to prosecute the environmental elements of a terrorist act undermines this expressive function of criminal law (Farah, 2012).

Deficiency in Implementing the Transparent Structure of Criminal law

The purpose of criminal law is to maintain order and security and maintain the moral welfare of the country. Failure to prosecute terrorists for committing crimes against the environment cannot achieve this goal. Criminal law appears for various reasons: protecting the country's interests, deterring, suppressing, and punishing unpleasant activities, establishing order, security, and justice among members of society, and giving symbolic power and representation to the values, beliefs, and prejudices of legislators. Crime is a public act that threatens fundamental social values. Both the law and society represent an element of society's moral disgust against criminals because they have committed an offense. As discussed earlier, people generally consider anti-environmental crimes to be serious criminal offenses (Meron, 1998). For many, environmental protection is as important to our collective well-being as national security, economic well-being, social justice, and even democracy itself. The pursuit of environmental terrorists under the same set of rules specific to corporate polluters does not adequately reflect this position. The Code of Environmental Protection provides for criminal penalties for legal offenses but does not fulfill the expressive function of criminal law. These rules do not give society a proper symbolic image of denouncing society to destroy environmental obligations. Thus, the pursuit of terrorists under this set of laws undermines the strong desire of society to punish environmental crime. Environmental terrorism is condemned by society but not properly punished by law (Cho, 2000).

CONCLUSION

The criminalization of environmental terrorism symbolizes society's denunciation of the deliberate destruction of the environment. Killing ecosystems is treated not as a mere violation of the law but as a traditional crime. This provides practical benefits for judicial officials and further reflects people who now consider committing a crime against the environment a dangerous crime. Although the criminalization of ecosystems is unlikely to deter the actions of environmental terrorists, it does serve both the practical requirements of the terrorism index and the expressive function of criminal law. Therefore, acts of environmental terrorism must be known for what they are (anti-environmental crime). The criminalization of environmental terrorism as an act of ecosystem murder creates social disgust with such behavior while at the same time empowering law enforcement to act within the "terrorist, criminal" index. International law proposes ecosystems as a model for domestic law. International ecosystems are defined based on intentional or inadvertent violation of customary international law. Domestic ecosystems must be based on intentional or inadvertent violation of a specific criminal law prohibiting such behavior. The crime of internal ecosystems should be defined as the intentional or inadvertent manipulation or destruction of any aspect of the physical environment that damages or exploits all or part of any part of the global ecosystem. This definition includes not only destructive acts but also acts that are used as a conduit for destruction. Actions performed with the knowledge of immediate or long-term effects on global ecosystems or with inadvertent disregard for them should be punished as acts of ecosystem killing. The cause is proven through damage to the affected ecosystem or evidence of using the environment as a conduit for terrorist activity. A legal crime of ecosystem empowerment enables law enforcement to combat environmental terrorism within the "terrorist means criminal" index. An environmental terrorist can be arrested and punished as an ordinary criminal. Prosecution due to ecosystems is similar to the prosecution of other ordinary crimes, i.e., it allows the judiciary to avoid the quagmire of a complex set of environmental protection laws. The punishment can be more severe than the one found in the Code of Environmental Protection. Prosecution of ecosystems for crimes committed in one act does not eradicate environmental terrorism but provides the judiciary with another weapon of the indictment, a weapon that strongly denounces the environmental nature of the crime. An application of this rule helps to understand its practical and expressive advantages better. As previously discussed, in 1985, a barrel containing 35 gallons of cyanide was discovered in the compound of the right-wing group Covenant, Sword, and Weapon of God. The cyanide was to poison the water source of Washington, DC, and New York City.

To illustrate, let's assume that the terrorists succeeded in their conspiracy. A cyanide attack on a New York water source has left thousands sick, disrupting the city's water supply for days, and having long-term effects on animal and plant ecosystems. This act of environmental terrorism could be punished as an ecosystem crime, as it involved the deliberate manipulation of the physical environment that has used or damaged parts of the global ecosystem. On the other hand, this action has been done with the knowledge or inadvertent inattention to its immediate or long-term effects on the global ecosystem. The causal link must be established by proving that water poisoning has abused or damaged an aspect of the global ecosystem. Proceedings, in this case, can also convict terrorists of customary law offenses, such as attempted murder or assault, or of violating a set of environmental laws, but the crime of ecosystems is a better option, one that is free. It is associated with a complex set of environmental protection laws and is an option that reflects society's disgust and anger at the environmental aspects of crime.

REFERENCES

Aaronson, E., & Shaffer, G. (2021). Defining crimes in a global age: Criminalization as a transnational legal process. *Law & Social Inquiry*, *46*(2), 455–486. doi:10.1017/lsi.2020.42

Andreas, P., & Nadelmann, E. (2008). *Policing the globe: Criminalization and crime control in international relations*. Oxford University Press.

Bassiouni, M. C. (Ed.). (2008). International criminal law: Vol. 1. *Sources, subjects and contents*. Brill.

Billon, P. L., & Carter, A. (2012). Securing Alberta's tar sands: Resistance and criminalization on a new energy frontier. In *Natural resources and social conflict* (pp. 170–192). Palgrave Macmillan. doi:10.1057/9781137002464_9

Boelens, R., Guevara-Gil, A., & Panfichi, A. (2009). Indigenous water rights in the Andes: Struggles over resources and legitimacy. *Journal of Water Law, 20*(5-6), 268–277.

Chiaramonte, X. (2020). The Struggle for Law: Legal strategies, environmental struggles and climate actions in Italy. *Oñati Socio-Legal Series, 10*(4), 932–954. doi:10.35295/osls.iisl/0000-0000-0000-1153

Cho, B. S. (2000). Emergence of an international environmental criminal law. *UCLA J. Envtl. L. & Pol'y, 19*(1), 11. doi:10.5070/L5191019216

Cockayne, J. (2007). *Transnational organized crime: Multilateral responses to a rising threat*. Academic Press.

Coutin, S. B. (2005). Contesting criminality: Illegal immigration and the spatialization of legality. *Theoretical Criminology, 9*(1), 5–33. doi:10.1177/1362480605046658

Edwards, S. M., Edwards, T. D., & Fields, C. B. (2013). *Environmental crime and criminality: Theoretical and practical issues*. Routledge. doi:10.4324/9780203726754

Estevens, J. (2018). Migration crisis in the EU: Developing a framework for analysis of national security and defence strategies. *Comparative Migration Studies, 6*(1), 1–21. doi:10.118640878-018-0093-3 PMID:30363805

Farah, D. (2011). Terrorist-criminal pipelines and criminalized states: Emerging alliances. *Prism, 2*(3), 15–32.

Farah, D. (2012). *Transnational organized crime, terrorism, and criminalized states in Latin America: An emerging tier-one national security priority*. Strategic Studies Institute, US Army War College.

Furman, R., Ackerman, A. R., Loya, M., Jones, S., & Egi, N. (2012). The criminalization of immigration: Value conflicts for the social work profession. *Journal of Sociology and Social Welfare, 39*(1), 169. PMID:28959084

Hutchinson, S., & O'malley, P. (2007). A crime–terror nexus? Thinking on some of the links between terrorism and criminality. *Studies in Conflict and Terrorism, 30*(12), 1095–1107. doi:10.1080/10576100701670870

McCulloch, J., & Pickering, S. (2009). Pre-Crime and Counter-TerrorismImagining Future Crime in the 'War on Terror'. *British Journal of Criminology, 49*(5), 628–645. doi:10.1093/bjc/azp023

Meron, T. (1998). Is international law moving towards criminalization? *European Journal of International Law, 9*(1), 18–31. doi:10.1093/ejil/9.1.18

Schofield, T. (1998). The environment as an ideological weapon: A proposal to criminalize environmental terrorism. *BC Envtl. Aff. L. Rev., 26*, 619.

Spade, D. (2015). *Normal life*. Duke University Press.

Tyler, T. R. (2014). Reducing corporate criminality: The role of values. *The American Criminal Law Review, 51*, 267.

Welch, M. (2003). Ironies of social control and the criminalization of immigrants. *Crime, Law, and Social Change, 39*(4), 319–337. doi:10.1023/A:1024068321783

White, R. (2003). Environmental issues and the criminological imagination. *Theoretical Criminology*, *7*(4), 483–506. doi:10.1177/13624806030074005

White, R. (2012). Environmental activism and resistance to state–corporate crime. In *State crime and resistance* (pp. 141–153). Routledge.

ADDITIONAL READING

Aaronson, E., & Shaffer, G. (2021). Defining crimes in a global age: Criminalization as a transnational legal process. *Law & Social Inquiry*, *46*(2), 455–486. doi:10.1017/lsi.2020.42

Andreas, P., & Nadelmann, E. (2008). *Policing the globe: Criminalization and crime control in international relations*. Oxford University Press.

Bassiouni, M. C. (Ed.). (2008). International criminal law: Vol. 1. *Sources, subjects and contents*. Brill.

Billon, P. L., & Carter, A. (2012). Securing Alberta's tar sands: Resistance and criminalization on a new energy frontier. In *Natural resources and social conflict* (pp. 170–192). Palgrave Macmillan. doi:10.1057/9781137002464_9

Boelens, R., Guevara-Gil, A., & Panfichi, A. (2009). Indigenous water rights in the Andes: Struggles over resources and legitimacy. *Journal of Water Law*, *20*(5-6), 268–277.

Chiaramonte, X. (2020). The Struggle for Law: Legal strategies, environmental struggles and climate actions in Italy. *Oñati Socio-Legal Series*, *10*(4), 932–954. doi:10.35295/osls.iisl/0000-0000-0000-1153

Cho, B. S. (2000). Emergence of an international environmental criminal law. *UCLA J. Envtl. L. & Pol'y*, *19*(1), 11. doi:10.5070/L5191019216

Coutin, S. B. (2005). Contesting criminality: Illegal immigration and the spatialization of legality. *Theoretical Criminology*, *9*(1), 5–33. doi:10.1177/1362480605046658

Edwards, S. M., Edwards, T. D., & Fields, C. B. (2013). *Environmental crime and criminality: Theoretical and practical issues*. Routledge. doi:10.4324/9780203726754

Estevens, J. (2018). Migration crisis in the EU: Developing a framework for analysis of national security and defence strategies. *Comparative Migration Studies*, *6*(1), 1–21. doi:10.118640878-018-0093-3 PMID:30363805

KEY TERMS AND DEFINITIONS

Environmental Law: Environmental law is a collective term encompassing aspects of the law that protect the environment. A related but distinct set of regulatory regimes, now strongly influenced by environmental legal principles, focuses on managing specific natural resources, such as forests, minerals, or fisheries. Other areas, such as environmental impact assessment, may not fit neatly into either category but are nonetheless important components of environmental law. Previous research found that when environmental law reflects moral values for betterment, legal adoption is more likely to be successful, usually in well-developed regions. In less-developed states, changes in moral values are necessary for successful legal implementation when environmental law differs from moral values.

Equity: Defined by UNEP to include intergenerational equity—"the right of future generations to enjoy a fair level of the common patrimony"—and intragenerational equity—"the right of all people within the current generation to fair access to the current generation's entitlement to the Earth's natural resources"—environmental equity considers the present generation under an obligation to account for long-term impacts of activities and to act to sustain the global environment and resource base for future generations. Pollution control and resource management laws may be assessed against this principle.

Polluter Pays Principle: The polluter pays principle stands for the idea that "the environmental costs of economic activities, including the cost of preventing potential harm, should be internalized rather than imposed upon society at large." All issues related to responsibility for cost for environmental remediation and compliance with pollution control regulations involve this principle.

Precautionary Principle: One of the most commonly encountered and controversial principles of environmental law, the Rio Declaration formulated the precautionary principle as follows, to protect the environment, States shall widely apply the precautionary approach according to their capabilities. Where there are threats of serious or irreversible damage, lack of full scientific certainty shall not be used as a reason for postponing cost-effective measures to prevent environmental degradation. The principle may play a role in any debate over the need for environmental regulation.

Prevention: The concept of prevention etc. can perhaps better be considered an overarching aim that gives rise to a multitude of legal mechanisms, including prior assessment of environmental harm, licensing, or authorization that set out the conditions for operation and the consequences for violation of the conditions, as well as the adoption of strategies and policies. Emission limits and other product or process standards, the use of best available techniques, and similar techniques can all be seen as applications of the concept of prevention.

Public Participation and Transparency: Identified as essential conditions for "accountable governments...industrial concerns," and organizations generally, public participation and transparency are presented by UNEP as requiring "effective protection of the human right to hold and express opinions and to seek, receive and impart ideas, etc. a right of access to appropriate, comprehensible and timely information held by governments and industrial concerns on economic and social policies regarding the sustainable use of natural resources and the protection of the environment, without imposing undue financial burdens upon the applicants and with adequate protection of privacy and business confidentiality," and "effective judicial and administrative proceedings." These principles are present in environmental impact assessment, laws requiring publication and access to relevant environmental data, and administrative procedures.

Sustainable Development: Defined by the United Nations Environment Programme as "development that meets the needs of the present without compromising the ability of future generations to meet their own needs," sustainable development may be considered together with the concepts of "integration" (development cannot be considered in isolation from sustainability) and "interdependence" (social and economic development, and environmental protection, are interdependent). Laws mandating environmental impact assessment and requiring or encouraging development to minimize environmental impacts may be assessed against this principle. The modern concept of sustainable development was discussed at the 1972 United Nations Conference on the Human Environment (Stockholm Conference) and the driving force behind the 1983 World Commission on Environment and Development (WCED, or Bruntland

Commission). In 1992, the first UN Earth Summit resulted in the Rio Declaration, Principle 3 of which reads: "The right to development must be fulfilled to equitably meet developmental and environmental needs of present and future generations." Sustainable development has been a core concept of international environmental discussion ever since, including at the World Summit on Sustainable Development (Earth Summit 2002) and the United Nations Conference on Sustainable Development (Earth Summit 2012, or Rio+20).

Transboundary Responsibility: Defined in the international law context as an obligation to protect one's environment. UNEP considers transboundary responsibility at the international level to prevent damage to neighboring environments at the international level as a potential limitation on the rights of the sovereign state. Laws that limit externalities imposed upon human health and the environment may be assessed against this principle.

Conclusion

Environmental law is a continuing source of controversy. Debates over the necessity, fairness, and cost of environmental regulation are ongoing, as well as regarding the appropriateness of regulations vs. market solutions to achieve even agreed-upon ends.

Allegations of scientific uncertainty fuel the ongoing debate over greenhouse gas regulation, and are a major factor in debates over whether to ban particular pesticides. In cases where the science is well-settled, it is not unusual to find that corporations intentionally hide or distort the facts, or sow confusion.

It is very common for regulated industry to argue against environmental regulation on the basis of cost. Difficulties arise in performing cost-benefit analysis of environmental issues. It is difficult to quantify the value of an environmental value such as a healthy ecosystem, clean air, or species diversity. Many environmentalists' response to pitting economy vs. ecology is summed up by former Senator and founder of Earth Day Gaylord Nelson, "The economy is a wholly owned subsidiary of the environment, not the other way around." Furthermore, environmental issues are seen by many as having an ethical or moral dimension, which would transcend financial cost. Even so, there are some efforts underway to systemically recognize environmental costs and assets, and account for them properly in economic terms.

While affected industries spark controversy in fighting regulation, there are also many environmentalists and public interest groups who believe that current regulations are inadequate, and advocate for stronger protection. Environmental law conferences—such as the annual Public Interest Environmental Law Conference in Eugene, Oregon—typically have this focus, also connecting environmental law with class, race, and other issues.

An additional debate is to what extent environmental laws are fair to all regulated parties. For instance, researchers Preston Teeter and Jorgen Sandberg highlight how smaller organizations can often incur disproportionately larger costs as a result of environmental regulations, which can ultimately create an additional barrier to entry for new firms, thus stifling competition and innovation.

Early examples of legal enactments designed to consciously preserve the environment, for its own sake or human enjoyment, are found throughout history. In the common law, the primary protection was found in the law of nuisance, but this only allowed for private actions for damages or injunctions if there was harm to land. Thus, smells emanating from pigsties, strict liability against dumping rubbish, or damage from exploding dams. Private enforcement, however, was limited and found to be woefully inadequate to deal with major environmental threats, particularly threats to common resources. During the "Great Stink" of 1858, the dumping of sewerage into the River Thames began to smell so ghastly in the summer heat that Parliament had to be evacuated. Ironically, the Metropolitan Commission of Sewers Act 1848 had allowed the Metropolitan Commission for Sewers to close cesspits around the city in an attempt to "clean up" but this simply led people to pollute the river. In 19 days, Parliament passed a further Act to build the London sewerage system. London also suffered from terrible air pollution, and this culminated in the "Great Smog" of 1952, which in turn triggered its own legislative response: the Clean Air Act 1956. The basic regulatory structure was to set limits on emissions for households and businesses (particularly burning of coal) while an inspectorate would enforce compliance.

Global and regional environmental issues are increasingly the subject of international law. Debates over environmental concerns implicate core principles of international law and have been the subject of numerous international agreements and declarations.

Customary international law is an important source of international environmental law. These are the norms and rules that countries follow as a matter of custom and they are so prevalent that they bind all states in the world. When a principle becomes customary law is not clear cut and many arguments are put forward by states not wishing to be bound. Examples of customary international law relevant to the environment include the duty to warn other states promptly about icons of an environmental nature and environmental damages to which another state or states may be exposed, and Principle 21 of the Stockholm Declaration ('good neighbourliness' or sic utere).

Given that customary international law is not static but ever evolving and the continued increase of air pollution (carbon dioxide) causing climate changes, has led to discussions on whether basic customary principles of international law, such as the jus cogens (peremptory norms) and erga omnes principles could be applicable for enforcing international environmental law.

Numerous legally binding international agreements encompass a wide variety of issue-areas, from terrestrial, marine and atmospheric pollution through to wildlife and biodiversity protection. International environmental agreements are generally multilateral (or sometimes bilateral) treaties (a.k.a. convention, agreement, protocol, etc.). Protocols are subsidiary agreements built from a primary treaty. They exist in many areas of international law but are especially useful in the environmental field, where they may be used to regularly incorporate recent scientific knowledge. They also permit countries to reach an agreement on a framework that would be contentious if every detail were to be agreed upon in advance. The most widely known protocol in international environmental law is the Kyoto Protocol, which followed from the United Nations Framework Convention on Climate Change.

While the bodies that proposed, argued, agreed upon, and ultimately adopted existing international agreements vary according to each agreement, certain conferences, including 1972's United Nations Conference on the Human Environment, 1983's World Commission on Environment and Development, 1992's United Nations Conference on Environment and Development, and 2002's World Summit on Sustainable Development have been particularly important. Multilateral environmental agreements sometimes create an International Organization, Institution or Body responsible for implementing the agreement. Major examples are the Convention on International Trade in Endangered Species of Wild Fauna and Flora (CITES) and the International Union for Conservation of Nature (IUCN).

International environmental law also includes the opinions of international courts and tribunals. While there are few and they have limited authority, the decisions carry much weight with legal commentators and are quite influential on the development of international environmental law. One of the biggest challenges in international decisions is to determine an adequate compensation for environmental damages. The courts include the International Court of Justice (ICJ), the International Tribunal for the Law of the Sea (ITLOS), the European Court of Justice, European Court of Human Rights and other regional treaty tribunals

Although numerous international environmental treaties have been concluded, effective agreements remain difficult to achieve for a variety of reasons. Because environmental problems ignore political boundaries, they can be adequately addressed only with the cooperation of numerous governments, among which there may be serious disagreements on important points of environmental policy. Furthermore, because the measures necessary to address environmental problems typically result in social and economic hardships in the countries that adopt them, many countries, particularly in the developing world, have been reluctant to enter into environmental treaties. Since the 1970s a growing number of environmental treaties have incorporated provisions designed to encourage their adoption by developing countries. Such measures include financial cooperation, technology transfer, and differential implementation schedules and obligations.

The greatest challenge to the effectiveness of environmental treaties is compliance. Although treaties can attempt to enforce compliance through mechanisms such as sanctions, such measures usually are of limited usefulness, in part because countries in compliance with a treaty may be unwilling or unable to impose the sanctions called for by the treaty. In general, the threat of sanctions is less important to most countries than the possibility that by violating their international obligations they risk losing their good standing in the international community. Enforcement mechanisms other than sanctions have been difficult to establish, usually because they would require countries to cede significant aspects of their national sovereignty to foreign or international organizations. In most agreements, therefore, enforcement is treated as a domestic issue, an approach that effectively allows each country to define compliance in whatever way best serves its national interest. Despite this difficulty, international environmental treaties and agreements are likely to grow in importance as international environmental problems become more acute.

Many areas of international environmental law remain underdeveloped. Although international agreements have helped to make the laws and regulations applicable to some types of environmentally harmful activity more or less consistent in different countries, those applicable to other such activities can differ in dramatic ways. Because in most cases the damage caused by environmentally harmful activities cannot be contained within national boundaries, the lack of consistency in the law has led to situations in which activities that are legal in some countries result in illegal or otherwise unacceptable levels of environmental damage in neighbouring countries.

This problem became particularly acute with the adoption of free trade agreements beginning in the early 1990s. The North American Free Trade Agreement (NAFTA), for example, resulted in the creation of large numbers of maquiladoras—factories jointly owned by U.S. and Mexican corporations and operated in Mexico—inside a 60-mile- (100-km) wide free trade zone along the U.S.-Mexican border. Because Mexico's government lacked both the resources and the political will to enforce the country's environmental laws, the maquiladoras were able to pollute surrounding areas with relative impunity, often dumping hazardous wastes on the ground or directly into waterways, where they were carried into U.S. territory. Prior to NAFTA's adoption in 1992, the prospect of problems such as these led negotiators to append a so-called "side agreement" to the treaty, which pledged environmental cooperation between the signatory states. Meanwhile, in Europe concerns about the apparent connection between free trade agreements and environmental degradation fueled opposition to the Maastricht Treaty, which created the EU and expanded its jurisdiction.

The design and application of modern environmental law have been shaped by a set of principles and concepts outlined in publications such as Our Common Future (1987), published by the World Commission on Environment and Development, and the Earth Summit's Rio Declaration (1992).

The Precautionary Principle: As discussed above, environmental law regularly operates in areas complicated by high levels of scientific uncertainty. In the case of many activities that entail some change to the environment, it is impossible to determine precisely what effects the activity will have on the quality of the environment or on human health. It is generally impossible to know, for example, whether a certain level of air pollution will result in an increase in mortality from respiratory disease, whether a certain level of water pollution will reduce a healthy fish population, or whether oil development in an environmentally sensitive area will significantly disturb the native wildlife. The precautionary principle requires that, if there is a strong suspicion that a certain activity may have environmentally harmful consequences, it is better to control that activity now rather than to wait for incontrovertible scientific evidence. This principle is expressed in the Rio Declaration, which stipulates that, where there are "threats of serious or irreversible damage, lack of full scientific certainty shall not be used as a reason for postponing cost-effective measures to prevent environmental degradation." In the United States the precautionary principle was incorporated into the design of habitat-conservation plans required under the aegis of the Endangered Species Act. In 1989 the EC invoked the precautionary principle when it banned the importation of U.S. hormone-fed beef, and in 2000 the organization adopted the principle as a "full-fledged and general principle of international law." In 1999 Australia and New Zealand invoked the precautionary principle in their suit against Japan for its alleged overfishing of southern bluefin tuna.

The Prevention Principle: Although much environmental legislation is drafted in response to catastrophes, preventing environmental harm is cheaper, easier, and less environmentally dangerous than reacting to environmental harm that already has taken place. The prevention principle is the fundamental notion behind laws regulating the generation, transportation, treatment, storage, and disposal of hazardous waste and laws regulating the use of pesticides. The principle was the foundation of the Basel Convention on the Control of Transboundary Movements of Hazardous Wastes and their Disposal (1989), which sought to minimize the production of hazardous waste and to combat illegal dumping. The prevention principle also was an important element of the EC's Third Environmental Action Programme, which was adopted in 1983.

The "Polluter Pays" Principle: Since the early 1970s the "polluter pays" principle has been a dominant concept in environmental law. Many economists claim that much environmental harm is caused by producers who "externalize" the costs of their activities. For example, factories that emit unfiltered exhaust into the atmosphere or discharge untreated chemicals into a river pay little to dispose of their waste. Instead, the cost of waste disposal in the form of pollution is borne by the entire community. Similarly, the driver of an automobile bears the costs of fuel and maintenance but externalizes the costs associated with the gases emitted from the tailpipe. Accordingly, the purpose of many environmental regulations is to force polluters to bear the real costs of their pollution, though such costs often are difficult to calculate precisely. In theory, such measures encourage producers of pollution to make cleaner products or to use cleaner technologies. The "polluter pays" principle underlies U.S. laws requiring the cleanup of releases of hazardous substances, including oil. One such law, the Oil Pollution Act (1990), was passed in reaction to the spillage of some 11 million gallons (41 million litres) of oil into Prince William Sound in Alaska in 1989. The "polluter pays" principle also guides the policies of the EU and other governments throughout the world. A 1991 ordinance in Germany, for example, held businesses responsible for the costs of recycling or disposing of their products' packaging, up to the end of the product's life cycle; however, the German Federal Constitutional Court struck down the regulation as unconstitutional. Such policies also have been adopted at the regional or state level; in 1996 the U.S. state of Florida, in order to protect its environmentally sensitive Everglades region, incorporated a limited "polluter pays" provision into its constitution.

The Integration Principle: Environmental protection requires that due consideration be given to the potential consequences of environmentally fateful decisions. Various jurisdictions (e.g., the United States and the EU) and business organizations (e.g., the U.S. Chamber of Commerce) have integrated environmental considerations into their decision-making processes through environmental-impact-assessment mandates and other provisions.

The Public Participation Principle: Decisions about environmental protection often formally integrate the views of the public. Generally, government decisions to set environmental standards for specific types of pollution, to permit significant environmentally damaging activities, or to preserve significant resources are made only after the impending decision has been formally and publicly announced and the public has been given the opportunity to influence the decision through written comments or hearings. In many countries citizens may challenge in court or before administrative bodies government decisions affecting the environment. These citizen lawsuits have become an important component of environmental decision making at both the national and the international level.

Public participation in environmental decision making has been facilitated in Europe and North America by laws that mandate extensive public access to government information on the environment. Similar measures at the international level include the Rio Declaration and the 1998 Århus Convention, which committed the 40 European signatory states to increase the environmental information available to the public and to enhance the public's ability to participate in government decisions that affect the environment. During the 1990s the Internet became a primary vehicle for disseminating environmental information to the public.

Sustainable Development: Sustainable development is an approach to economic planning that attempts to foster economic growth while preserving the quality of the environment for future generations. Despite its enormous popularity in the last two decades of the 20th century, the concept of sustainable development proved difficult to apply in many cases, primarily because the results of long-term sustainability analyses depend on the particular resources focused upon. For example, a forest that will provide a sustained yield of timber in perpetuity may not support native bird populations, and a mineral deposit that will eventually be exhausted may nevertheless support more or less sustainable communities. Sustainability was the focus of the 1992 Earth Summit and later was central to a multitude of environmental studies.

One of the most important areas of the law of sustainable development is ecotourism. Although tourism poses the threat of environmental harm from pollution and the overuse of natural resources, it also can create economic incentives for the preservation of the environment in developing countries and increase awareness of unique and fragile ecosystems throughout the world. In 1995 the World Conference on Sustainable Tourism, held on the island of Lanzarote in the Canary Islands, adopted a charter that encouraged the development of laws that would promote the dual goals of economic development through tourism and protection of the environment. Two years later, in the Malé Declaration on Sustainable Tourism, 27 Asian-Pacific countries pledged themselves to a set of principles that included fostering awareness of environmental ethics in tourism, reducing waste, promoting natural and cultural diversity, and supporting local economies and local community involvement. Highlighting the growing importance of sustainable tourism, the World Tourism Organization declared 2002 the International Year of Ecotourism.

Compilation of References

Aaronson, E., & Shaffer, G. (2021). Defining crimes in a global age: Criminalization as a transnational legal process. *Law & Social Inquiry*, *46*(2), 455–486. doi:10.1017/lsi.2020.42

Abate, C., Patel, L., Rauscher, F. J. III, & Curran, T. (1990). Redox regulation of fos and jun DNA-binding activity in vitro. *Science*, *249*(4973), 1157–1161. doi:10.1126cience.2118682 PMID:2118682

Abels, G. (2011). Gender equality policy. In Policies within the EU Multi-Level System (pp. 325-348). Nomos Verlagsgesellschaft mbH & Co. KG. doi:10.5771/9783845228266-325

Adib, R., Murdock, H. E., Appavou, F., Brown, A., Epp, B., Leidreiter, A. & Farrell, T. C. (2015). *Renewables 2015 global status report*. REN21 Secretariat Press.

Adler, R. W. (2010). Climate change and the hegemony of state water law. *Stan. Envtl. LJ*, *29*, 1.

Agyeman, J., Schlosberg, D., Craven, L., & Matthews, C. (2016). Trends and directions in environmental justice: From inequity to everyday life, community, and just sustainabilities. *Annual Review of Environment and Resources*, *41*(1), 321–340. doi:10.1146/annurev-environ-110615-090052

Aichele, R., & Felbermayr, G. (2013). The Effect of the Kyoto Protocol on Carbon Emissions. *Journal of Policy Analysis and Management*, *32*(4), 731–757. doi:10.1002/pam.21720

Alan, B. (2017). *Human rights and the environment: where next?* Routledge.

Alan, B. (2017). *Human rights and the environment: Where next?* Routledge.

Al-Duaij, N. (2002). *Environmental law of armed conflict*. Academic Press.

Alston, P. (2020). The Committee on Economic, Social and Cultural Rights. *NYU Law and Economics Research Paper*, *15*, 20-24.

Andreas, P., & Nadelmann, E. (2008). *Policing the globe: Criminalization and crime control in international relations*. Oxford University Press.

Antkowiak, T. M. (2007). Remedial approaches to human rights violations: The Inter-american Court of human rights and beyond. *Colum. J. Transnat'l L.*, *46*, 351.

Argumedo, A., Swiderska, K., Pimbert, M., Song, Y., & Pant, R. (2011). *Implementing Farmers Rights under the FAO International Treaty on PGRFA: The Need for a Broad Approach Based on Biocultural Heritage. International Institute for Environment and Development*. IIED.

Arora-Jonsson, S. (2014, November). Forty years of gender research and environmental policy: Where do we stand? In *Women's Studies International Forum* (Vol. 47, pp. 295-308). Pergamon.

Asadi, R. (2021). Analysis of Challenges of Environmental Non-Governmental Organizations in Iran with Riggs Prismatic Society Theory. *Research Political Geography Quarterly, 6*(2).

Asano, T. (Ed.). (1998). *Wastewater reclamation and reuse: water quality management library* (Vol. 10). CRC Press.

Aspan, Z., & Yunus, A. (2019). The right to a good and healthy environment: Revitalizing green constitution. *IOP Conference Series. Earth and Environmental Science, 343*(1), 012067. doi:10.1088/1755-1315/343/1/012067

Aukusti Lehtinen, A. (2006). 'Green waves' and globalization: A Nordic view on environmental justice. *Norsk Geografisk Tidsskrift-Norwegian Journal of Geography, 60*(1), 46–56. doi:10.1080/00291950600548881

Badwaza, Y. M. (2005). *Public interest litigation as practised by South African human rights NGOs: any lessons for Ethiopia?* [Doctoral dissertation]. University of Pretoria.

Baker, B. (1992). Legal protections for the environment in times of armed conflict. *Va. J. Int'l L., 33*, 351.

Baker, L. A. (Ed.). (2009). *The water environment of cities*. Springer. doi:10.1007/978-0-387-84891-4

Ball, S., & Bell, S. (1994). *Environmental Law*. The Law and Policy Relating to the Protection of the Environment.

Barboza, J. (2010). *The Environment, Risk and Liability in International law*. Martinus Nijhoff Publishers.

Bardgett, R. D., & Gibson, D. J. (2017). Plant ecological solutions to global food security. *Journal of Ecology, 105*(4), 859–864. doi:10.1111/1365-2745.12812

Barnett, J. (2003). Security and climate change. *Global Environmental Change, 13*(1), 7–17. doi:10.1016/S0959-3780(02)00080-8

Bassiouni, M. C. (Ed.). (2008). International criminal law: Vol. 1. *Sources, subjects and contents*. Brill.

Bekker, G. (2007). The African court on human and peoples' rights: Safeguarding the interests of African states. *Journal of African Law, 51*(1), 151–172. doi:10.1017/S0021855306000210

Benjamin, A. H., Marques, C. L., & Tinker, C. (2004). The water giant awakes: An overview of water law in Brazil. Tex L. *Rev., 83*, 2185.

Bentata, P. (2014). Liability as a complement to environmental regulation: An empirical study of the French legal system. *Environmental Economics and Policy Studies, 16*(3), 201–228. doi:10.100710018-013-0073-7

Betlem, G. (1995). Dutch Soil Protection Act 1994. *The. Eur. Envtl. L. Rev., 4*, 232.

Betlem, G. (1995). Standing for Ecosystems—Going Dutch. *The Cambridge Law Journal, 54*(1), 153–170. doi:10.1017/S0008197300083197

Bhagwati, P. N. (1984). Judicial activism and public interest litigation. *Colum. J. Transnat'l L., 23*, 561.

Billon, P. L., & Carter, A. (2012). Securing Alberta's tar sands: Resistance and criminalization on a new energy frontier. In *Natural resources and social conflict* (pp. 170–192). Palgrave Macmillan. doi:10.1057/9781137002464_9

Birnie, P. W., & Boyle, A. E. (2002). *International Law and the Environment*. Oxford University Press.

Biswas, M. R. (1981). The United Nations Conference on New and Renewable Sources of Energy: A review. *Mazingira, 5*, 11–16.

Björnsdóttir, A. L. (2013). *The UN Security Council and Climate Change: Rising Seas Levels, Shrinking Resources, and the 'Green Helmets'* [Master's thesis].

Blumm, M. C. (1988). Public property and the democratization of western water law: A modern view of the public trust doctrine. *Envtl. L.*, *19*, 573.

Boelens, R., Guevara-Gil, A., & Panfichi, A. (2009). Indigenous water rights in the Andes: Struggles over resources and legitimacy. *Journal of Water Law*, *20*(5-6), 268–277.

Borowy, I. (2014). *Defining sustainable development for our common future. A history of the world commission on environment and development.* Routledge Press.

Bothe, M. (1991). The Protection of the Environment in Times of Armed Conflict. *German YB Int'l L.*, *34*, 54.

Bothe, M., Bruch, C., Diamond, J., & Jensen, D. (2010). International law protecting the environment during armed conflict: Gaps and opportunities. *International Review of the Red Cross*, *92*(879), 569–592. doi:10.1017/S1816383110000597

Boutillon, S. (2001). The precautionary principle: Development of an international standard. *Mich. J. Int'l L.*, *23*, 429.

Bouvier, A. (1991). Protection of the natural environment in time of armed conflict. *International Review of the Red Cross (1961-1997)*, *31*(285), 567-578.

Boyle, A., & Redgwell, C. (2021). *Birnie, Boyle, and Redgwell's International Law and the Environment.* Oxford University Press. doi:10.1093/he/9780199594016.001.0001

Bradbrook, A. (2008). *The development of renewable energy technologies and energy efficiency measures through public international law.* Oxford University Press. doi:10.1093/acprof:oso/9780199532698.003.0006

Bradbrook, A. J. (1996). Energy law as an academic discipline. *Journal of Energy & Natural Resources Law*, *14*(2), 193–217. doi:10.1080/02646811.1996.11433062

Bradbrook, A. J. (2013). *International Law and Renewable Energy: Filling the Void.* Duncker & Humblot Press.

British Petroleum. (2001). *Statistical Review of World Energy.* British Petroleum press.

Bromideh, A. A. (2011). The widespread challenges of NGOs in developing countries: Case studies from Iran. *International NGO Journal*, *6*(9), 197–202.

Brownlie, I. (2008). *Principles of Public International Law.* Oxford University Press.

Brown, O., Hammill, A., & McLeman, R. (2007). Climate change as the 'new'security threat: Implications for Africa. *International Affairs*, *83*(6), 1141–1154. doi:10.1111/j.1468-2346.2007.00678.x

Brown, O., & McLeman, R. (2009). A recurring anarchy? The emergence of climate change as a threat to international peace and security: Analysis. *Conflict Security and Development*, *9*(3), 289–305. doi:10.1080/14678800903142680

Bruce, S. (2013). International law and renewable energy: Facilitating sustainable energy for all. *Melbourne Journal of International Law*, *14*, 18.

Brulle, R. J., & Pellow, D. N. (2006). Environmental justice. *Annual Review of Public Health*, *27*, 103–124. doi:10.1146/annurev.publhealth.27.021405.102124 PMID:16533111

Brzoska, M. (2009). The securitization of climate change and the power of conceptions of security. *S&F Sicherheit und Frieden*, *27*(3), 137–145. doi:10.5771/0175-274x-2009-3-137

Buckingham, S., & Le Masson, V. (Eds.). (2017). *Understanding climate change through gender relations.* Taylor & Francis. doi:10.4324/9781315661605

Buergenthal, T. (1982). The Inter-American Court of Human Rights. *The American Journal of International Law*, *76*(2), 231–245. doi:10.2307/2201452

Buhaug, H., Gleditsch, N. P., & Theisen, O. M. (2008). *Implications of climate change for armed conflict*. World Bank.

Bullock, J. M., Dhanjal-Adams, K. L., Milne, A., Oliver, T. H., Todman, L. C., Whitmore, A. P., & Pywell, R. F. (2017). Resilience and food security: Rethinking an ecological concept. *Journal of Ecology*, *105*(4), 880–884. doi:10.1111/1365-2745.12791

Cameron, J., & Abouchar, J. (1991). The precautionary principle: A fundamental principle of law and policy for the protection of the global environment. *BC Int'l & Comp. L. Rev.*, *14*, 1.

Cans, C. (2005). La charte constitutionnelle de l'environnement: Évolution ou révolution du droit français de l'environnement? *Droit de l'environnement*, (131), 194–203.

Carolan, M. S. (2013). *Reclaiming food security*. Routledge. doi:10.4324/9780203387931

Carruthers, D. (2001). Environmental politics in Chile: Legacies of dictatorship and democracy. *Third World Quarterly*, *22*(3), 343–358. doi:10.1080/01436590120061642

Carruthers, D. V. (2008). The globalization of environmental justice: Lessons from the US-Mexico border. *Society & Natural Resources*, *21*(7), 556–568. doi:10.1080/08941920701648812

Cherp, A., Jewell, J., & Goldthau, A. (2011). Governing global energy: Systems, transitions, complexity. *Global Policy*, *2*(1), 75–88. doi:10.1111/j.1758-5899.2010.00059.x

Chiaramonte, X. (2020). The Struggle for Law: Legal strategies, environmental struggles and climate actions in Italy. *Oñati Socio-Legal Series*, *10*(4), 932–954. doi:10.35295/osls.iisl/0000-0000-0000-1153

Chikozho, C., Saruchera, D., Danga, L., & da Silva, C. (2018). *A Compendium of the South African water law review post-1994*. Water Research Commission.

Cho, B. S. (2000). Emergence of an international environmental criminal law. *UCLA J. Envtl. L. & Pol'y*, *19*(1), 11. doi:10.5070/L5191019216

Chong, J. (2014). Climate-readiness, competition and sustainability: An analysis of the legal and regulatory frameworks for providing water services in Sydney. *Water Policy*, *16*(1), 1–18. doi:10.2166/wp.2013.058

Christoffersen, J., & Madsen, M. R. (Eds.). (2011). *The European court of human rights between law and politics*. Oxford University Press. doi:10.1093/acprof:oso/9780199694495.001.0001

Chu, J., Chen, J., Wang, C., & Fu, P. (2004). Wastewater reuse potential analysis: Implications for China's water resources management. *Water Research*, *38*(11), 2746–2756. doi:10.1016/j.watres.2004.04.002 PMID:15207605

Cockayne, J. (2007). *Transnational organized crime: Multilateral responses to a rising threat*. Academic Press.

Conca, K., Thwaites, J., & Lee, G. (2016). Bully Pulpit or Bull in a China Shop? Climate change and the UN Security Council. *Annu. Meet. Acad. Counc. United Nations Syst, 1*.

Conca, K., Thwaites, J., & Lee, G. (2017). Climate change and the UN Security Council: Bully pulpit or bull in a china shop? *Global Environmental Politics*, *17*(2), 1–20. doi:10.1162/GLEP_a_00398

Conway, D. (2010). The United Nations Security Council and climate change: Challenges and opportunities. *Climate Law*, *1*(3), 375–407. doi:10.1163/CL-2010-018

Copeland, C. (2008). *Cruise ship pollution: Background, laws and regulations, and key issues.* Congressional Research Service.

Cordonier Segger, M. C., & Khalfan, A. (2004). *Sustainable development law: principles, practices and prospects.* Oxford university press.

Cottier, T., Malumfashi, G., Matteotti-Berkutova, S., Nartova, O., De Sepibus, J., & Bigdeli, S. Z. (2011). *Energy in WTO law and policy: The prospects of international trade regulation from fragmentation to coherence.* WTO Press. doi:10.1017/CBO9780511792496

Cousins, S. (2013). UN Security Council: Playing a role in the international climate change regime? *Global Change, Peace & Security, 25*(2), 191–210. doi:10.1080/14781158.2013.787058

Coutin, S. B. (2005). Contesting criminality: Illegal immigration and the spatialization of legality. *Theoretical Criminology, 9*(1), 5–33. doi:10.1177/1362480605046658

Dankelman, I. (Ed.). (2010). *Gender and climate change: An introduction.* Routledge.

Dasgupta, C. (2012). Present at the creation: the making of the UN Framework Convention on Climate Change. In *Handbook of Climate Change and India* (pp. 113–122). Routledge.

Davies, K., & Riddell, T. (2017). The Warming War: How Climate Change is Creating Threats to International Peace and Security. *Geo. Envtl. L. Rev., 30*, 47.

de Chazournes, L. B. (1998). *Kyoto protocol to the united nations framework convention on climate change.* UN's Audiovisual Library of International Law. http://untreaty. un.org/cod/avl/ha/kpccc/kpccc.html

de Gliniasty, J., & Morand-Deviller, J. (2018). *Les théories jurisprudentielles en droit administratif.* LGDJ, une marque de Lextenso.

Detailed minutes of the deliberations of the Majlis Final Review of the Constitution. (1982). Publications of the General Directorate of Laws of the Islamic Consultative Assembly.

Detraz, N. (2017). *Gender and the Environment.* John Wiley & Sons.

Ditzen, U. (1971). Environmental Protection in West Germany. *Business Lawyer, 27*, 833.

Dodds, F., & Sherman, R. (2009). *Climate Change and Energy Insecurity:" The Challenge for Peace, Security and Development.* Routledge. doi:10.4324/9781849774406

Doh, J. P., & Guay, T. R. (2004). Globalization and corporate social responsibility: How non-governmental organizations influence labor and environmental codes of conduct. In *Management and international review* (pp. 7–29). Gabler Verlag. doi:10.1007/978-3-322-90997-8_2

Drake, L. (2016). International law and the renewable energy sector. The Oxford Handbook of International Climate Change Law, 1, 357.

Droege, C., & Tougas, M. L. (2013). The Protection of the Natural Environment in Armed Conflict–Existing Rules and need for further legal protection. *Nordic Journal of International Law, 82*(1), 21–52. doi:10.1163/15718107-08201003

Drummond, J., & Barros-Platiau, A. F. (2006). Brazilian environmental laws and policies, 1934–2002: A critical overview. *Law & Policy, 28*(1), 83–108. doi:10.1111/j.1467-9930.2005.00218.x

Eckstein, G. E. (2009). Water scarcity, conflict, and security in a climate change world: Challenges and opportunities for international law and policy. *Wis. Int'l LJ, 27*, 409.

Edele, A. (2005). *Non-governmental organizations in China*. Geneva, Switzerland: The Programme on NGOs and Civil Society, Centre for Applied Studies in International Negotiation, CASIN.

Edenhofer, O., Pichs-Madruga, R., Sokona, Y., Seyboth, K., Kadner, S., Zwickel, T., & Matschoss, P. (Eds.). (2011). *Renewable energy sources and climate change mitigation: Special report of the intergovernmental panel on climate change*. Cambridge University Press. doi:10.1017/CBO9781139151153

Edwards, S. M., Edwards, T. D., & Fields, C. B. (2013). *Environmental crime and criminality: Theoretical and practical issues*. Routledge. doi:10.4324/9780203726754

El-Nakhel, C., Pannico, A., Graziani, G., Kyriacou, M. C., Giordano, M., Ritieni, A., De Pascale, S., & Rouphael, Y. (2020). Variation in macronutrient content, phytochemical constitution and in vitro antioxidant capacity of green and red butterhead lettuce dictated by different developmental stages of harvest maturity. *Antioxidants*, *9*(4), 300. doi:10.3390/antiox9040300 PMID:32260224

Elver, H. (2008). International environmental law, water and the future. In *International Law and the Third World* (pp. 191–208). Routledge-Cavendish.

Eno, R. W. (2002). The Jurisdiction of the African Court on Human and Peoples'. *Rights. Afr. Hum. Rts. LJ*, *2*, 223.

Estevens, J. (2018). Migration crisis in the EU: Developing a framework for analysis of national security and defence strategies. *Comparative Migration Studies*, *6*(1), 1–21. doi:10.118640878-018-0093-3 PMID:30363805

Esty, D. C. (1998). Non-governmental organizations at the World Trade Organization: Cooperation, competition, or exclusion. *J. Int'l Econ. L.*, *1*(1), 123–148. doi:10.1093/jiel/1.1.123

Evans, M. D. (Ed.). (2014). *International law*. Oxford University Press. doi:10.1093/he/9780199654673.001.0001

Faber, D. (2005). Building a transnational environmental justice movement: Obstacles and opportunities in the age of globalization. *Coalitions across borders: Transnational protest and the neoliberal order*, 43-68.

Faber, D. R., & McCarthy, D. (2012). Neo-liberalism, globalization and the struggle for ecological democracy: linking sustainability and environmental justice. In *Just sustainabilities* (pp. 55–80). Routledge.

Farah, D. (2011). Terrorist-criminal pipelines and criminalized states: Emerging alliances. *Prism*, *2*(3), 15–32.

Farah, D. (2012). *Transnational organized crime, terrorism, and criminalized states in Latin America: An emerging tier-one national security priority*. Strategic Studies Institute, US Army War College.

Fernandes, E. (1992). Law, politics and environmental protection in Brazil. *Journal of Environmental Law*, *4*(1), 41–56. doi:10.1093/jel/4.1.41

Fitzgerald, E. A. (1996). The Constitutionality of Toxic Substances Regulation Under the Canadian Environmental Protection Act. *U. Brit. Colum. L. Rev.*, *30*, 55.

Foster, J. B., & Grundmann, R. (2001). Marx's ecology. Materialism & nature. *Canadian Journal of Sociology*, *26*(4), 670. doi:10.2307/3341498

Freckmann, A., & Wegerich, T. (2001). *The German Legal System*. International Business Lawyer.

Furman, R., Ackerman, A. R., Loya, M., Jones, S., & Egi, N. (2012). The criminalization of immigration: Value conflicts for the social work profession. *Journal of Sociology and Social Welfare*, *39*(1), 169. PMID:28959084

Galizzi, P. (2012). *Missing in action: Gender in international environmental law*. Fordham Law Legal Studies Research Paper, (2779320).

Gallegos, T. J., Varela, B. A., Haines, S. S., & Engle, M. A. (2015). Hydraulic fracturing water use variability in the U nited S tates and potential environmental implications. *Water Resources Research, 51*(7), 5839–5845. doi:10.1002/2015WR017278 PMID:26937056

Garcia, D. (2018). Lethal artificial intelligence and change: The future of international peace and security. *International Studies Review, 20*(2), 334–341. doi:10.1093/isr/viy029

Gasser, H. P. (1995). For better protection of the natural environment in armed conflict: A proposal for action. *The American Journal of International Law, 89*(3), 637–644. doi:10.2307/2204184

Georgian, A.A. (2009). Duplication and normative diversity in French constitutional law (sources for monitoring parliamentary approvals). *Legal Research Quarterly, 12*(50).

Ghazi Shariat Panahi, S. A. (1995). *Essentials of Constitutional Law* (1st ed.). Yalda Publishing.

Gielen, D., Boshell, F., Saygin, D., Bazilian, M. D., Wagner, N., & Gorini, R. (2019). The role of renewable energy in the global energy transformation. *Energy Strategy Reviews, 24*, 38–50. doi:10.1016/j.esr.2019.01.006

Gleick, P. H. (1998). The human right to water. *Water Policy, 1*(5), 487–503. doi:10.1016/S1366-7017(99)00008-2

Godden, L. (2005). Water law reform in Australia and South Africa: Sustainability, efficiency and social justice. *Journal of Environmental Law, 17*(2), 181–205. doi:10.1093/envlaw/eqi016

Goldfarb, W. (2020). *Water law*. CRC Press. doi:10.1201/9781003069829

Goldston, J. A. (2006). Public interest litigation in Central and Eastern Europe: Roots, prospects, and challenges. *Hum. Rts. Q., 28*(2), 492–527. doi:10.1353/hrq.2006.0018

Gonzalez, C. G. (2011). Climate change, food security, and agrobiodiversity: Toward a just, resilient, and sustainable food system. *Fordham Environmental Law Review*, 493-522.

Gottlieb, R., & Fisher, A. (1996). Community food security and environmental justice: Searching for a common discourse. *Agriculture and Human Values, 13*(3), 23–32. doi:10.1007/BF01538224

Gould, C. C. (2010). Moral issues in globalization. The Oxford handbook of business ethics, 305-334.

Greer, S., & Wildhaber, L. (2012). Revisiting the debate about 'constitutionalising'the European Court of Human Rights. *Human Rights Law Review, 12*(4), 655–687. doi:10.1093/hrlr/ngs034

Großmann, K., Padmanabhan, M., & Afiff, S. (2017). Gender, ethnicity, and environmental transformations in Indonesia and beyond. *ASEAS-Austrian Journal of South-East Asian Studies, 10*(1), 1–10.

Gunningham, N. (2012). Confronting the challenge of energy governance. *Transnational Environmental Law, 1*(1), 119–135. doi:10.1017/S2047102511000124

Guo, J., Bao, Y., & Wang, M. (2018). Steel slag in China: Treatment, recycling, and management. *Waste Management, 78*, 318-330.

Gupta, A. (2000). Governing trade in genetically modified organisms: The Cartagena Protocol on Biosafety. *Environment, 42*(4), 22–33. doi:10.1080/00139150009604881

Gupta, S. (2009). Environmental law and policy: Climate change as a threat to international peace and security. *Perspectives on Global Issues, 4*(1), 7–17.

Hagen, J. J. (2016). Queering women, peace and security. *International Affairs, 92*(2), 313–332. doi:10.1111/1468-2346.12551

Hashemi, F., Sadighi, H., Chizari, M., & Abbasi, E. (2017). Influencing Factors on Emerging Capabilities of Environmental Non-Governmental Organizations (ENGOs): Using Grounded Theory. *Journal of Applied Environmental Biological Sciences*, *7*(3), 173–184.

Hashemi, F., Sadighi, H., Chizari, M., & Abbasi, E. (2019). Influencing Factors on Emerging Capabilities of Environmental Non-Governmental Organizations (ENGOs): Using Grounded Theory. *OIDA International Journal of Sustainable Development*, *12*(03), 39–54.

Hashemi, F., Sadighi, H., Chizari, M., & Abbasi, E. (2019). The relationship between ENGOs and Government in Iran. *Heliyon*, *5*(12), e02844. doi:10.1016/j.heliyon.2019.e02844 PMID:31890931

Hassan, P., & Azfar, A. (2003). Securing environmental rights through public interest litigation in South Asia. *Va. Envtl. LJ*, *22*, 215.

Heffron, R. J., & Talus, K. (2016). The evolution of energy law and energy jurisprudence: Insights for energy analysts and researchers. *Energy Research & Social Science*, *19*, 1–10. doi:10.1016/j.erss.2016.05.004

Heidari, F., Dabiri, F., & Heidari, M. (2017). Legal system governing on water pollution in Iran. *Journal of Geoscience and Environment Protection*, *5*(09), 36–59. doi:10.4236/gep.2017.59004

Hendry, J. R. (2006). Taking aim at business: What factors lead environmental non-governmental organizations to target particular firms? *Business & Society*, *45*(1), 47–86. doi:10.1177/0007650305281849

Holden, E., Linnerud, K., & Banister, D. (2014). Sustainable development: Our common future revisited. *Global Environmental Change*, *26*, 130–139. doi:10.1016/j.gloenvcha.2014.04.006

Hourcle, L. R. (2000). Environmental law of war. *Vermont Law Review*, *25*, 653.

Hutchinson, S., & O'malley, P. (2007). A crime–terror nexus? Thinking on some of the links between terrorism and criminality. *Studies in Conflict and Terrorism*, *30*(12), 1095–1107. doi:10.1080/10576100701670870

Ide, T., & Scheffran, J. (2014). On climate, conflict and cumulation: Suggestions for integrative cumulation of knowledge in the research on climate change and violent conflict. *Global Change, Peace & Security*, *26*(3), 263–279. doi:10.1080/14781158.2014.924917

Inserguet-Brisset, V. (2005). *Droit de l'environnement*. Presses universitaires de Rennes.

Intergovernmental Panel on Climate Change. (2007). *Climate change: The physical science basis: Summary for policymakers*. IPCC Press.

International Energy Agency & Birol, F. (2013). World energy outlook 2013. International Energy Agency Press.

International Energy Agency. (2014). *World energy outlook 2014: Executive summary*. International Energy Agency Press.

International Energy Agency. (2015). *Energy and climate change: World energy outlook special report*. International Energy Agency Press.

International Renewable Energy Agency. (2017). *Renewable Energy Highlights*. International Renewable Energy Agency Press.

Jacquemont, F., & Caparrós, A. (2002). The convention on biological diversity and the climate change convention 10 years after Rio: Towards a synergy of the two regimes. *Rev. Eur. Comp. & Int'l Envtl. L.*, *11*(2), 169–180. doi:10.1111/1467-9388.00315

Jarass, H. D., & DiMento, J. (1993). Through Comparative Lawyers' Goggles: A Primer on German Environmental Law. *Geo. Int'l Envtl. L. Rev.*, *6*, 47.

Jha, U. C. (2014). *Armed conflict and environmental damage*. Vij Books India Pvt Ltd.

Johnson, H., & Walters, R. (2014). Food security. In *The Handbook of Security* (pp. 404–426). Palgrave Macmillan. doi:10.1007/978-1-349-67284-4_19

Joyner, C. C., & Kirkhope, J. T. (1992). The Persian Gulf War oil spill: Reassessing the law of environmental protection and the law of armed conflict. *Case W. Res. J. Int'l L.*, *24*, 29.

Joyner, C. C., & Little, G. E. (1996). It's Not Nice to Fool Mother Nature-The Mystique of Feminist Approaches to International Environmental Law. *BU Int'l LJ*, *14*, 223.

Kattelus, M., Rahaman, M. M., & Varis, O. (2014, May). M yanmar under reform: Emerging pressures on water, energy and food security. *Natural Resources Forum*, *38*(2), 85–98. doi:10.1111/1477-8947.12032

Kelsen, H. (2000). *The law of the United Nations: a critical analysis of its fundamental problems: with supplement* (Vol. 11). The Lawbook Exchange, Ltd.

Kendall, R. (2012). Climate change as a security threat to the Pacific Islands. *New Zealand Journal of Environmental Law*, *16*, 83–116.

Khan, M., Chaudhry, M. N., Ahmad, S. R., & Saif, S. (2020). The role of and challenges facing non-governmental organizations in the environmental impact assessment process in Punjab, Pakistan. *Impact Assessment and Project Appraisal*, *38*(1), 57–70. doi:10.1080/14615517.2019.1684096

Kial, M., & Kial, M. (2004), Environmental Law in the Perspective of the Constitution. In *Proceedings of the First Iranian Conference on Environmental Law*. Barg Zaytoun Publications.

Klein, C. A., Angelo, M. J., & Hamann, R. (2009). Modernizing water law: The example of Florida. *Florida Law Review*, *61*, 403.

Koskenniemi, M. (2007). The fate of public international law: Between technique and politics. *The Modern Law Review*, *70*(1), 1–30. doi:10.1111/j.1468-2230.2006.00624.x

Kostka, G. (2016). Command without control: The case of C hina's environmental target system. *Regulation & Governance*, *10*(1), 58–74. doi:10.1111/rego.12082

Kostruba, A. (2019). The rule of law and its impact on socio-economic, environmental, gender and cultural issues. Kostruba AV, The Rule of Law and its Impact on Socio-Economic, Environmental, Gender and Cultural Issues. *Space and Culture, India*, *7*(2), 1–2.

Koutalakis, C., Buzogany, A., & Börzel, T. A. (2010). When soft regulation is not enough: The integrated pollution prevention and control directive of the European Union. *Regulation & Governance*, *4*(3), 329–344. doi:10.1111/j.1748-5991.2010.01084.x

Krasnova, I. O. (2014). Environmental security as a legal category. *Lex Russica*, (5), 543–555.

Krenkel, P. (2012). *Water quality management*. Elsevier.

Kriebel, D., Tickner, J., Epstein, P., Lemons, J., Levins, R., Loechler, E. L., Quinn, M., Rudel, R., Schettler, T., & Stoto, M. (2001). The precautionary principle in environmental science. *Environmental Health Perspectives*, *109*(9), 871–876. doi:10.1289/ehp.01109871 PMID:11673114

Kurian, P. A. (2018). *Engendering the environment? Gender in the World Bank's environmental policies*. Routledge. doi:10.4324/9781315185101

Kurtz, H. E. (2007). Gender and environmental justice in Louisiana: Blurring the boundaries of public and private spheres. *Gender, Place and Culture*, *14*(4), 409–426. doi:10.1080/09663690701439710

Laferrière, J. (1947). *Manuel de droit constitutionnel*. Domat-Montchrestien.

Lawrence, G. (2013). *Food security, nutrition and sustainability*. Earthscan. doi:10.4324/9781849774499

Lazarova, V., & Bahri, A. (Eds.). (2004). *Water reuse for irrigation: agriculture, landscapes, and turf grass*. CRC Press. doi:10.1201/9780203499405

Leach, P. (2011). *Taking a case to the European Court of Human Rights*. Oxford University Press.

Leal-Arcas, R., & Minas, S. (2016). The micro level: Insights from specific policy areas: Mapping the international and European governance of renewable energy. *Yearbook of European Law*, *35*(1), 621–666. doi:10.1093/yel/yew022

Lepard, B. D. (2010). *Customary International Law: A New Theory with Practical Applications*. Cambridge University Press. doi:10.1017/CBO9780511804717

Lewis, M. K. (2020). Why China Should Unsign the International Covenant on Civil and Political Rights. *Vanderbilt Journal of Transnational Law*, *53*, 131.

Liberatore, A. (2013). Climate change, security and peace: The role of the European Union. *Review of European Studies*, *5*(3), 83. doi:10.5539/res.v5n3p83

Li, G., He, Q., Shao, S., & Cao, J. (2018). Environmental non-governmental organizations and urban environmental governance: Evidence from China. *Journal of Environmental Management*, *206*, 1296–1307. doi:10.1016/j.jenvman.2017.09.076 PMID:28993017

Lindblom, A. K. (2005). *Non-governmental organisations in international law*. Cambridge University Press.

Lindquist, C. H., & Lindquist, C. A. (1997). Gender differences in distress: Mental health consequences of environmental stress among jail inmates. *Behavioral Sciences & the Law*, *15*(4), 503–523. doi:10.1002/(SICI)1099-0798(199723/09)15:4<503::AID-BSL281>3.0.CO;2-H PMID:9433751

Liu, C. (2018). Are women greener? Corporate gender diversity and environmental violations. *Journal of Corporate Finance*, *52*, 118–142. doi:10.1016/j.jcorpfin.2018.08.004

Lixinski, L. (2010). Treaty interpretation by the inter-American court of human rights: Expansionism at the service of the unity of international law. *European Journal of International Law*, *21*(3), 585–604. doi:10.1093/ejil/chq047

López-Feldman, A., Chávez, C., Vélez, M. A., Bejarano, H., Chimeli, A. B., Féres, J., ... Viteri, C. (2020). Environmental impacts and policy responses to Covid-19: A view from Latin America. *Environmental and Resource Economics*, 1–6. PMID:32836838

Lund, H. (2007). Renewable energy strategies for sustainable development. *Energy*, *32*(6), 912–919. doi:10.1016/j.energy.2006.10.017

Maas, A., & Scheffran, J. (2012). Climate conflicts 2.0? Climate engineering as a challenge for international peace and security. *Sicherheit und Frieden (S+ F)/Security and Peace*, 193-200.

MacGregor, S. (Ed.). (2017). *Routledge handbook of gender and environment*. Taylor & Francis. doi:10.4324/9781315886572

Maguire, R., & Jessup, B. (2021). Gender, race and environmental law: A feminist critique. In *International Women's Rights Law and Gender Equality* (pp. 107–127). Routledge.

Malanczuk, P. (2002). *Akehurst's modern introduction to international law*. Routledge. doi:10.4324/9780203427712

Mallory, C. L. (1999). *Toward an ecofeminist environmental jurisprudence: Nature, law, and gender*. University of North Texas.

Mashhadi, A. (2013). *The Right to a Healthy Environment* (1st ed.). Mizan Publications.

Mashhadi, A. (2014). Substantive Constitutionalization of Right to Environment in Iranian & French Law. *Comparative Law Review*, *5*(2), 559–580.

Matlock, M. M., Henke, K. R., & Atwood, D. A. (2002). Effectiveness of commercial reagents for heavy metal removal from water with new insights for future chelate designs. *Journal of Hazardous Materials*, *92*(2), 129–142. doi:10.1016/S0304-3894(01)00389-2 PMID:11992699

McBeath, J. H., McBeath, J., & McBeath, J. (2010). *Environmental change and food security in China*. Springer. doi:10.1007/978-1-4020-9180-3

McClanahan, B., & Brisman, A. (2015). Climate change and peacemaking criminology: Ecophilosophy, peace and security in the "war on climate change". *Critical Criminology*, *23*(4), 417–431. doi:10.100710612-015-9291-6

McCulloch, J., & Pickering, S. (2009). Pre-Crime and Counter-TerrorismImagining Future Crime in the 'War on Terror'. *British Journal of Criminology*, *49*(5), 628–645. doi:10.1093/bjc/azp023

McDonald, M. (2013). Discourses of climate security. *Political Geography*, *33*, 42–51. doi:10.1016/j.polgeo.2013.01.002

McGlade, C., & Ekins, P. (2015). The geographical distribution of fossil fuels unused when limiting global warming to 2 C. *Nature*, *517*(7533), 187–190. doi:10.1038/nature14016 PMID:25567285

McInerney, T. (2007). Putting regulation before responsibility: Towards binding norms of corporate social responsibility. *Cornell Int'l LJ*, *40*, 171.

McIntyre, O. (2010). The proceduralisation and growing maturity of international water law: Case concerning pulp mills on the river Uruguay (Argentina v Uruguay), International Court of Justice, 20 April 2010. *Journal of Environmental Law*, *22*(3), 475–497. doi:10.1093/jel/eqq019

McIntyre, O. (2016). *Environmental protection of international watercourses under international law*. Routledge. doi:10.4324/9781315580043

Meron, T. (1998). Is international law moving towards criminalization? *European Journal of International Law*, *9*(1), 18–31. doi:10.1093/ejil/9.1.18

Metson, G. S., Iwaniec, D. M., Baker, L. A., Bennett, E. M., Childers, D. L., Cordell, D., Grimm, N. B., Grove, J. M., Nidzgorski, D. A., & White, S. (2015). Urban phosphorus sustainability: Systemically incorporating social, ecological, and technological factors into phosphorus flow analysis. *Environmental Science & Policy*, *47*, 1–11. doi:10.1016/j.envsci.2014.10.005

Michalena, E., & Hills, J. M. (2013). *Introduction: Renewable Energy Governance: Is it Blocking the Technically Feasible in Renewable Energy Governance*. Springer Press.

Moghaddam, M. R. A., Maknoun, R., & Tahershamsi, A. (2008). Environmental engineering education in Iran: Needs, problems and solutions. *Environmental Engineering and Management Journal*, *7*(6), 775–779. doi:10.30638/eemj.2008.103

Moghimi, S. M. (2007). The relationship between environmental factors and organizational entrepreneurship in non-governmental organizations (NGOs) in Iran. *Iranian Journal of Management Studies*, *1*(1), 39–55.

Mohai, P., Pellow, D., & Roberts, J. T. (2009). Environmental justice. *Annual Review of Environment and Resources*, *34*(1), 405–430. doi:10.1146/annurev-environ-082508-094348

Mohseni, R., & Shokri, M. (2013). Study of carbon dioxide emissions in Iran with a fuzzy approach. *Iranian Journal of Energy*, *16*(1), 1–16.

Mombo, F., Speelman, S., Huylenbroeck, G. V., Hella, J., & Pantaleo, M. (2011). *Ratification of the Ramsar convention and sustainable wetlands management: Situation analysis of the Kilombero Valley wetlands in Tanzania*. Academic Press.

Moradi, A., & Aminian, M. (2009). Iran's greenhouse gas emissions in 2009. *Nesha Alam Magazine*, *13*, 55–59.

Morand-Deviller, J. (2014). L'environnement dans les constitutions étrangères. *Les nouveaux cahiers du Conseil Constitutionnel*, (2), 83-95.

Morrow, K. (2017). Integrating gender issues into the global climate change regime. *Understanding climate change through gender relations*, 31-44.

Mowbray, A. (2005). The Creativity of the European Court of Human Rights. *Human Rights Law Review*, *5*(1), 57–79. doi:10.1093/hrlrev/ngi003

Mrema, E., Bruch, C., & Diamond, J. (2009). *Protecting the environment during armed conflict: an inventory and analysis of international law*. UNEP/Earthprint.

Mulyana, I. (2016). The Development of International Law in the Field of Renewable Energy. *Hasanuddin Law Review*, *1*(1), 38–60. doi:10.20956/halrev.v1i1.213

Nadeem, O., & Hameed, R. (2008). Evaluation of environmental impact assessment system in Pakistan. *Environmental Impact Assessment Review*, *28*(8), 562–571. doi:10.1016/j.eiar.2008.02.003

Naff, T., & Dellapenna, J. (2017). Can there be confluence? A comparative consideration of Western and Islamic fresh water law. In *International Law and Islamic Law* (pp. 281–305). Routledge. doi:10.4324/9781315092515-15

Najafifard, M., & Mashhadi, A. (2014). *Green Economy Based on Sustainable Development in the Light of the Rio+20 Declaration. International Conference and Online Green Economy*, Tehran, Iran.

Nevitt, M. (2020). Is Climate Change a Threat to International Peace and Security? *Mich. J. Int'l L.*, *42*, 527. doi:10.2139srn.3689320

Nevitt, M. P. (2020). On Environmental Law, Climate Change, & National Security Law. *Harv. Envtl. L. Rev.*, *44*, 321.

Newman, M. K., Lucas, A., LaDuke, W., Berila, B., Di Chiro, G., Gaard, G., & Sturgeon, N. (2004). *New perspectives on environmental justice: Gender, sexuality, and activism*. Rutgers University Press.

Ng, T. (2010). Safeguarding peace and security in our warming world: A role for the Security Council. *Journal of Conflict and Security Law*, *15*(2), 275–300. doi:10.1093/jcsl/krq010

Nhamo, G., Nhemachena, C., Nhamo, S., Mjimba, V., & Savić, I. (2020). *SDG7-Ensure Access to Affordable, Reliable, Sustainable, and Modern Energy*. Emerald Group Publishing. doi:10.1108/9781789737998

Nnoko-Mewanu, J., Téllez-Chávez, L., & Rall, K. (2021). Protect rights and advance gender equality to mitigate climate change. *Nature Climate Change*, *11*(5), 368–370. doi:10.103841558-021-01043-4

Nomura, K. (2007). Democratisation and environmental non-governmental organisations in Indonesia. *Journal of Contemporary Asia, 37*(4), 495–517. doi:10.1080/00472330701546566

Norgaard, K., & York, R. (2005). Gender equality and state environmentalism. *Gender & Society, 19*(4), 506–522. doi:10.1177/0891243204273612

Norouzi, N., & Movahedian, H. (2021). Right to Education in Mother Language: In the Light of Judicial and Legal Structures. In Handbook of Research on Novel Practices and Current Successes in Achieving the Sustainable Development Goals (pp. 223-241). IGI Global.

Norouzi, N. (2021). Post-COVID-19 and globalization of oil and natural gas trade: Challenges, opportunities, lessons, regulations, and strategies. *International Journal of Energy Research, 45*(10), 14338–14356. doi:10.1002/er.6762 PMID:34219899

Norouzi, N., & Ataei, E. (2021). Covid-19 Crisis and Environmental law: Opportunities and challenges. *Hasanuddin Law Review, 7*(1), 46–60. doi:10.20956/halrev.v7i1.2772

Norouzi, N., Khanmohammadi, H. U., & Ataei, E. (2021). The Law in the Face of the COVID-19 Pandemic: Early Lessons from Uruguay. *Hasanuddin Law Review, 7*(2), 75–88. doi:10.20956/halrev.v7i2.2827

Olsson, O., Weichgrebe, D., & Rosenwinkel, K. H. (2013). Hydraulic fracturing wastewater in Germany: Composition, treatment, concerns. *Environmental Earth Sciences, 70*(8), 3895–3906. doi:10.100712665-013-2535-4

Omorogbe, Y. O. (2008). *Promoting sustainable development through the use of renewable energy: The role of the law.* Oxford University Press. doi:10.1093/acprof:oso/9780199532698.003.0003

Orakhelashvili, A. (2018). *Akehurst's Modern Introduction to International Law*. Routledge. doi:10.4324/9780429439391

Ormazdi, M., Pourfikouhi, A., & Amar, T. (2013). The necessities of verifying the policies of Non Governmental Organizations development in planning and management of rural tourism of Iran. *Life Science Journal, 10*(3s).

Ostling, M. (1994). Decision-Making in Dutch and Swedish Environmental Law. *Tilburg Foreign L. Rev., 4*(3), 209–234. doi:10.1163/221125995X00013

Ottinger, R. L., Robinson, N., & Tafur, V. (2005). *Compendium of sustainable energy laws*. Cambridge University Press. doi:10.1017/CBO9780511664885

Pachauri, R. K., Allen, M. R., Barros, V. R., Broome, J., Cramer, W., Christ, R., & van Ypserle, J. P. (2014). *Climate change 2014: synthesis report. Contribution of Working Groups I, II and III to the fifth assessment report of the Intergovernmental Panel on Climate Change*. IPCC.

Pasqualucci, J. M. (2012). *The practice and procedure of the Inter-American Court of Human Rights*. Cambridge University Press. doi:10.1017/CBO9780511843884

Pellow, D. N., & Brulle, R. J. (2005). *Power, justice, and the environment: toward critical environmental justice studies*. Academic Press.

Penny, C. K. (2007). Greening the security council: Climate change as an emerging "threat to international peace and security". *International Environmental Agreement: Politics, Law and Economics, 7*(1), 35–71. doi:10.100710784-006-9029-8

Penny, C. K. (2018). Climate change as a 'threat to international peace and security. In *Climate change and the UN Security Council*. Edward Elgar Publishing. doi:10.4337/9781785364648.00009

Pourbafrani, H., & Hemati, M. (2016). *A Critique on Iranian Criminal Policy towards Environmental Crimes*. Academic Press.

Pourhoseyni, B., Masoud, G., & Shekarchizadeh, M. (2019). The Role of Non-Governmental Organizations in Iran's Criminal Policy towards Environmental Protection. *Iranian Journal of Medical Law*, *13*, 337–351.

Prasad, P. M. (2004). Environmental protection: The role of liability system in India. *Economic and Political Weekly*, 257–269.

Prasad, P. M. (2006). Environment protection: Role of regulatory system in India. *Economic and Political Weekly*, 1278–1288.

Rabani, H., Jalalian, A., & Pournouri, M. (2020). Typology of Environmental Crimes in Iran (Case Study: Crimes Related to Environmental Pollution). *Anthropogenic Pollution*, *4*(2), 78–83.

Rahimi, N., Kargari, N., & Khodi, M. (2004). A Study of the PAC Development Mechanism in the Kyoto Protocol and the Financing of Projects. *Iranian Journal of Energy*, *21*, 57–71.

Ramazani Ghavamabadi, M. H. (2009). L'application du droit à l'environnement en Iran. *Proceeding of International Conference on Human Rights and Environment.*

Randrianandrasana, I. (2016). La protection constitutionnelle de l'environnement à Madagascar. *Revue juridique de lenvironnement, 41*(1), 122-139.

Read, R., & O'Riordan, T. (2017). The precautionary principle under fire. *Environment*, *59*(5), 4–15. doi:10.1080/00139157.2017.1350005

Rhyner, C. R., Schwartz, L. J., Wenger, R. B., & Kohrell, M. G. (2017). *Waste management and resource recovery*. CRC Press. doi:10.1201/9780203734278

Richter, B. (2010). *Beyond Smoke and Mirrors: Climate Change and Energy in the 21st century*. Cambridge University Press. doi:10.1017/CBO9780511802638

Ringler, C., Biswas, A. K., & Cline, S. A. (Eds.). (2010). *Global change: Impacts on water and food security*. Springer. doi:10.1007/978-3-642-04615-5

Rosiek, K. (2020). Directions and challenges in the management of municipal sewage sludge in Poland in the context of the circular economy. *Sustainability*, *12*(9), 3686. doi:10.3390u12093686

Rourke, J. T., & Boyer, M. A. (2008). *International politics on the world stage*. McGraw-Hill.

Rowe, D. R., & Abdel-Magid, I. M. (2020). *Handbook of wastewater reclamation and reuse*. CRC press. doi:10.1201/9780138752514

Russell, D. L. (2019). *Practical wastewater treatment*. John Wiley & Sons.

Ruta, M., & Venables, A. J. (2012). International trade in natural resources: Practice and policy. *Annual Review of Resource Economics*, *4*(1), 331–352. doi:10.1146/annurev-resource-110811-114526

Ryan, S. E. (2014). Rethinking gender and identity in energy studies. *Energy Research & Social Science*, *1*, 96–105. doi:10.1016/j.erss.2014.02.008

Salgot, M., & Folch, M. (2018). Wastewater treatment and water reuse. *Current Opinion in Environmental Science & Health*, *2*, 64–74. doi:10.1016/j.coesh.2018.03.005

Sankar, U. (1998). *Laws and institutions relating to environmental protection in India. The role of law and legal institutions in Asian economic development*. Erasmus University.

Santilli, J. (2012). *Agrobiodiversity and the Law: Regulating genetic resources, food security and cultural diversity*. Routledge. doi:10.4324/9780203155257

Saul, B. (2009). Climate Change, Conflict and Security: International Law Challenges. *NZ Armed FL Rev.*, *9*, 1.

Schachter, O. (1951). *The Law of the United Nations*. Academic Press.

Schafer, B. K. (1988). The relationship between the international laws of armed conflict and environmental protection: The need to reevaluate what types of conduct are permissible during hostilities. *Cal. W. Int'l LJ*, *19*, 287.

Schall, C. (2008). Public interest litigation concerning environmental matters before human rights courts: A promising future concept? *Journal of Environmental Law*, *20*(3), 417–453. doi:10.1093/jel/eqn025

Schlosberg, D., & Carruthers, D. (2010). Indigenous struggles, environmental justice, and community capabilities. *Global Environmental Politics*, *10*(4), 12–35. doi:10.1162/GLEP_a_00029

Schmitt, M. N. (1997). Green war: An assessment of the environmental law of international armed conflict. *Yale J. Int'l L.*, *22*, 1.

Schofield, T. (1998). The environment as an ideological weapon: A proposal to criminalize environmental terrorism. *BC Envtl. Aff. L. Rev.*, *26*, 619.

Schrijver, N. (1997). *Sovereignty Over Natural Resources: Balancing Rights and Duties*. Cambridge University Press. doi:10.1017/CBO9780511560118

Scott, S. V. (2008). Climate change and peak oil as threats to international peace and security: Is it time for the security council to legislate? *Melbourne Journal of International Law*, *9*(2), 495–514.

Shamsaii, M. (2006). International Law and Sustainable Development. *Law and Politics Research*, *19*, 7–24.

Shelton, D. (2021). *International environmental law*. Brill.

Sherk, G. W. (1990). Eastern water law: trends in state legislation. *Virginia Environmental Law Journal*, 287-321.

Shohani, A., Ataei, E., & Norouzi, N. (2021). Prevention and Suppression of Environmental Crimes in the Light of the Actions of Non-Governmental Organizations in the Iranian Legal System. *Research Journal of Ecology and Environmental Sciences*, *1*(1), 57–70.

Shupe, S. J. (1982). Waste in Western Water Law: A Blueprint for Change. Or. *Law Review*, *61*, 483.

Sial, S. A., Zaidi, S. M. A., & Taimour, S. (2018). *Review of existing environmental laws and regulations in Pakistan*. WWF-Pakistan.

Sicurelli, D. (2016). *The European Union's Africa policies: norms, interests and impact*. Routledge. doi:10.4324/9781315239828

Sindico, F. (2007). Climate change: A security (council) issue. *Carbon & Climate L. Rev.,* 29.

Smith, Z.A. & Taylor, K.D. (2008). *Renewable and Alternative Energy Resources: A Reference Handbook*. ABC-CLIO Press.

Smith, D., Hinz, H., Mulema, J., Weyl, P., & Ryan, M. J. (2018). Biological control and the Nagoya Protocol on access and benefit sharing–a case of effective due diligence. *Biocontrol Science and Technology*, *28*(10), 914–926. doi:10.1080/09583157.2018.1460317

Smith, S., Nickson, T. E., & Challender, M. (2021). Germplasm exchange is critical to conservation of biodiversity and global food security. *Agronomy Journal*, *113*(4), 2969–2979. doi:10.1002/agj2.20761

Song, J., Sun, Y., & Jin, L. (2017). PESTEL analysis of the development of the waste-to-energy incineration industry in China. *Renewable & Sustainable Energy Reviews*, *80*, 276–289. doi:10.1016/j.rser.2017.05.066

Spade, D. (2015). *Normal life*. Duke University Press.

Spataru, C. (2017). *Whole energy system dynamics: Theory, modelling and policy*. Taylor & Francis Press. doi:10.4324/9781315755809

Spellman, F. R. (2008). *Handbook of water and wastewater treatment plant operations*. CRC press. doi:10.1201/9781420075311

Spiro, P. J. (2007). Non-governmental organizations and civil society. The Oxford Handbook of International Environmental Law, 770-790.

Subramaniam, Y., Masron, T. A., & Azman, N. H. N. (2020). Biofuels, environmental sustainability, and food security: A review of 51 countries. *Energy Research & Social Science*, *68*, 101549. doi:10.1016/j.erss.2020.101549

Sun, Y., Chen, Z., Wu, G., Wu, Q., Zhang, F., Niu, Z., & Hu, H. Y. (2016). Characteristics of water quality of municipal wastewater treatment plants in China: Implications for resources utilization and management. *Journal of Cleaner Production*, *131*, 1–9. doi:10.1016/j.jclepro.2016.05.068

Suryawan, I. G. B., & Ismail, A. (2020). Strengthening environmental law policy and its influence on environmental sustainability performance: Empirical studies of green constitution in adopting countries. *International Journal of Energy Economics and Policy*, *10*(2), 132–138. doi:10.32479/ijeep.8719

Swanston, S. F. (1993). Race, gender, age, and disproportionate impact: What can we do about the failure to protect the most vulnerable. *The Fordham Urban Law Journal*, *21*, 577.

Sze, J. (2006). *Noxious New York: The racial politics of urban health and environmental justice*. MIT Press. doi:10.7551/mitpress/5055.001.0001

Taghizadeh, J. (2007). The Issue of Constitutionalization of the Legal Order. *Journal of Legal Research*, *6*(11), 129–162.

Tahbaz, M. (2016). Environmental challenges in today's Iran. *Iranian Studies*, *49*(6), 943–961. doi:10.1080/00210862.2016.1241624

Tan, P., & Jackson, S. (2013). Impossible dreaming: Does Australia's water law and policy fulfil Indigenous aspirations. *Environment and Planning Law Journal*, *30*(2), 132–149.

Tarlock, A. D. (1975). Recent developments in the recognition of instream uses in western water law. *Utah L. Rev.,* 871.

Tarlock, A. D. (1990). Western Water Law, Global Warming, and Growth Limitations. *Loy. LAL Rev.*, *24*, 979.

Tarlock, A. D. (1992). The role of non-governmental organizations in the development of international environmental law. *Chi.- Kent L. Rev.*, *68*, 61.

Tayebi, S., & Zarabi, M. (2018). Environmental Diplomacy and Climate Change; Constructive strategic approach to reducer. *Human & Environment*, *16*(4), 159–170.

Telesetsky, A. (1999). Kyoto Protocol. *Ecology Law Quarterly*, *26*, 797–813.

Temper, L., Del Bene, D., & Martinez-Alier, J. (2015). Mapping the frontiers and front lines of global environmental justice: The EJAtlas. *Journal of Political Ecology*, *22*(1), 255–278. doi:10.2458/v22i1.21108

Tignino, M. (2010). Water, international peace, and security. *International Review of the Red Cross*, *92*(879), 647–674. doi:10.1017/S181638311000055X

Trianni, A., Negri, M., & Cagno, E. (2021). What factors affect the selection of industrial wastewater treatment configuration? *Journal of Environmental Management*, *285*, 112099. doi:10.1016/j.jenvman.2021.112099 PMID:33588160

Tyler, T. R. (2014). Reducing corporate criminality: The role of values. *The American Criminal Law Review*, *51*, 267.

Udombana, N. J. (2000). Toward the African Court on Human and Peoples'. *Rights: Better Late Than Never. Yale Hum. Rts. & Dev. LJ*, *3*, 45.

United Nations (2010). *Delivering on Energy: An Overview of Activities of UN Energy and Its Members*. United Nations Press.

United Nations. (1972). *United Nations Conference on the Human Environment*. United Nations Press.

United nations. (1988). *Protection of global climate for present and future generations of mankind*. General Assembly Press.

United Nations. (2010). *Delivering on Energy: An Overview of Activities of UN Energy and Its Members*. United Nations Press.

United Nations. (2012). *United Nations General Assembly Declares 2014–2024 decade of Sustainable Energy for All*. United Nations Press.

United Nations. (2015a). *Transforming our world: The 2023 agenda for sustainable development*. United nation press.

United Nations. (2015b). *Transforming our world: The 2030 agenda for sustainable development*. Seventieth General Assembly Press.

United Nations. (2016). *Sustainable development goals report 2016*. United nation press.

Uribe, N. R., & Urdinola-Rengifo, J. S. (2020). International Environmental Law in Latin America. In Routledge Handbook of International Environmental Law (pp. 263-278). Routledge. doi:10.4324/9781003137825-22

Valdez-Carrillo, M., Abrell, L., Ramírez-Hernández, J., Reyes-López, J. A., & Carreón-Diazconti, C. (2020). Pharmaceuticals as emerging contaminants in the aquatic environment of Latin America: A review. *Environmental Science and Pollution Research International*, *27*(36), 1–29. doi:10.100711356-020-10842-9 PMID:32986197

van Rijswick, M. H. (2010). Interaction between European and Dutch water law. In *Water Policy in the Netherlands* (pp. 218–238). Routledge.

Venancio, M. D., Pope, K., & Sieber, S. (2018). Brazil's new government threatens food security and biodiversity. *Nature*, *564*(7734), 39–40. doi:10.1038/d41586-018-07611-7 PMID:30518897

Verschuuren, J. (2010). The Dutch "Crisis and Recovery Act": Economic recovery and legal crisis? *Potchefstroom Electronic Law Journal/Potchefstroomse Elektroniese Regsblad, 13*(5).

Verwey, W. D. (1995). Protection of the environment in times of armed conflict: In search of a new legal perspective. *Leiden Journal of International Law*, *8*(1), 7–40. doi:10.1017/S0922156500003083

Vidar, M. (2022). Soil and agriculture governance and food security. *Soil Security*, *6*, 100027. doi:10.1016/j.soisec.2021.100027

Villarín, M. C., & Merel, S. (2020). Paradigm shifts and current challenges in wastewater management. *Journal of Hazardous Materials*, *390*, 122139. doi:10.1016/j.jhazmat.2020.122139 PMID:32007860

Vogel, D. (1983). Cooperative regulation: Environmental protection in Great Britain. *The Public Interest*, *72*, 88.

Wachira, G. M. (2008). *African Court on Human and Peoples' Rights: Ten years on and still no justice*. Minority Rights Group International.

Wagner, S. (1999). Forest Legislation in a Constitutional State-The example of the Federal Republic of Germany. *Forstwissenschaftliche Beiträge der Professur Forstökonomie und Forstpolitik der ETH Zürich*, *21*, 41–48.

Walker, G. (2009). Globalizing environmental justice: The geography and politics of frame contextualization and evolution. *Global Social Policy*, *9*(3), 355–382. doi:10.1177/1468018109343640

Wanjiru, L. (2012). Gender, climate change and sustainable development. *Fordham Environmental Law Review*, 1-6.

Warren, P. D. (2015). *Climate Change and International Peace and Security: Possible Roles for the UN Security Council in Addressing Climate Change*. Academic Press.

Warschauer, M., & Liaw, M. L. (2010). *Emerging Technologies in Adult Literacy and Language Education*. National Institute for Literacy Press. doi:10.1037/e529982011-001

Watkins, K. (2007). *Human Development Report 2007/2008 - Fighting Climate Change: Human solidarity in a divided world*. United Nations Press.

Weissbrodt, D., & Kruger, M. (2017). Norms on the responsibilities of transnational corporations and other business enterprises with regard to human rights. In *Globaization and International Investment* (pp. 199–220). Routledge.

Welch, M. (2003). Ironies of social control and the criminalization of immigrants. *Crime, Law, and Social Change*, *39*(4), 319–337. doi:10.1023/A:1024068321783

White, R. (2003). Environmental issues and the criminological imagination. *Theoretical Criminology*, *7*(4), 483–506. doi:10.1177/13624806030074005

White, R. (2012). Environmental activism and resistance to state–corporate crime. In *State crime and resistance* (pp. 141–153). Routledge.

Whiteside, K. H. (2006). *Precautionary politics: principle and practice in confronting environmental risk*. Mit Press.

Wilkins, G. (2010). *Technology transfer for renewable energy*. Taylor & Francis. doi:10.4324/9781849776288

Wilkinson, D. (2005). *Environment and law*. Routledge. doi:10.4324/9780203994443

Wingfield, S., Martínez-Moscoso, A., Quiroga, D., & Ochoa-Herrera, V. (2021). Challenges to water management in Ecuador: Legal authorization, quality parameters, and socio-political responses. *Water (Basel)*, *13*(8), 1017. doi:10.3390/w13081017

Wu, J., Chang, I. S., Yilihamu, Q., & Zhou, Y. (2017). Study on the practice of public participation in environmental impact assessment by environmental non-governmental organizations in China. *Renewable & Sustainable Energy Reviews*, *74*, 186–200. doi:10.1016/j.rser.2017.01.178

Yanfang, L., & Wei, C. (2011). Framework of Laws and Policies on Renewable Energy and Relevant Systems in China under the Background of Climate Change. *Vermont Journal of Environmental Law*, *13*(4), 823–865. doi:10.2307/vermjenvilaw.13.4.823

Yong, Y. S., Lim, Y. A., & Ilankoon, I. M. S. K. (2019). An analysis of electronic waste management strategies and recycling operations in Malaysia: Challenges and future prospects. *Journal of Cleaner Production*, *224*, 151–166. doi:10.1016/j.jclepro.2019.03.205

Yusa, I. G., & Hermanto, B. (2018). Implementasi Green Constitution di Indonesia: Jaminan Hak Konstitusional Pembangunan Lingkungan Hidup Berkelanjutan. *Jurnal Konstitusi*, *15*(2), 306–326. doi:10.31078/jk1524

About the Authors

Nima Norouzi graduated his primary, high school, and college education from the NODET (National organization for Intellectual people) primary Education system. he took the national university entrance exam and, with rank 365 (among 164000), entered the Amirkabir University of Technology and started his education in system management and engineering in the B.Sc degree which he continued this filed in his M.Sc degree in the same university. After his graduation with his B.Sc and M.Sc degree, and because of his various educational and research field successes, he has been appreciated many times by the university. While studying his M.Sc Degree, Nima started a L.L.B course in Islamic Azad university. Then Nima got his S.J.D in Private and Islamic law from university of Tehran. Nima currently does reading and research in energy and environmental law. He is currently a researcher in the environmental field at Islamic Azad University and University of Tehran.(c5699294-9d45-49e1-bbba-5338453c258c)

Hussein Movahedian graduated with a Juris Doctor and Associate Master in Private Law in Imam Sadegh University. He currently is affiliated as a senior researcher of Islamic Azad University and is well-established in environmental law. Several well-respected papers of his are published in this field. His main research field is the environmental responsibility of governments and corporations in contracts. (a4fd05b6-a6b2-4a7a-b314-a5a4706b14e0)

Index

S

T

Recommended Reference Books

IGI Global's reference books are available in three unique pricing formats:
Print Only, E-Book Only, or Print + E-Book.

Order direct through IGI Global's Online Bookstore at
www.igi-global.com or through your preferred provider.

Smart Cities, Citizen Welfare, and the Implementation of Sustainable Development Goals

ISBN: 9781799877851
EISBN: 9781799877875
© 2022; 402 pp.
List Price: US$ **215**

Achieving Sustainability Using Creativity, Innovation, and Education: A Multidisciplinary Approach

ISBN: 9781799879633
EISBN: 9781799879657
© 2022; 260 pp.
List Price: US$ **240**

Trends and Innovations in Urban E-Planning

ISBN: 9781799890904
EISBN: 9781799890928
© 2022; 307 pp.
List Price: US$ **195**

Research Anthology on Citizen Engagement and Activism for Social Change

ISBN: 9781668437063
EISBN: 9781668437070
© 2022; 1,611 pp.
List Price: US$ **1,040**

Innovative Strategic Planning and International Collaboration for the Mitigation of Global Crises

ISBN: 9781799883395
EISBN: 9781799883418
© 2022; 327 pp.
List Price: US$ **215**

Faith-Based Influences on Legislative Decision Making: Emerging Research and Opportunities

ISBN: 9781799868071
EISBN: 9781799868095
© 2022; 353 pp.
List Price: US$ **215**

Do you want to stay current on the latest research trends, product announcements, news, and special offers?
Join IGI Global's mailing list to receive customized recommendations, exclusive discounts, and more.
Sign up at: **www.igi-global.com/newsletters**.

Publisher of Timely, Peer-Reviewed Inclusive Research Since 1988

IGI Global
PUBLISHER of TIMELY KNOWLEDGE

www.igi-global.com   Sign up at www.igi-global.com/newsletters   facebook.com/igiglobal   twitter.com/igiglobal   linkedin.com/igiglobal

Ingram Content Group UK Ltd.
Milton Keynes UK
UKHW051859190423
420461UK00008B/226

9 781668 471883